MYOCARDIAL ENERGY METABOLISM

MYOCARDIAL ENERGY METABOLISM

edited by

JAN WILLEM DE JONG

Cardiochemical Laboratory,
Thorax Center, Erasmus University Rotterdam
Rotterdam, The Netherlands

1988 **MARTINUS NIJHOFF PUBLISHERS**
a member of the KLUWER ACADEMIC PUBLISHERS GROUP
DORDRECHT / BOSTON / LANCASTER

Library of Congress Cataloging in Publication Data

Myocardial energy metabolism.

   (Developments in cardiovascular medicine)
   Includes index.
   1. Heart--Muscle--Metabolism. 2. Energy metabolism.
I. Jong, Jan Willem de, 1942-   . II. Series.
[DNLM: 1. Energy Metabolism. 2. Myocardium--metabolism.
W1 DE997VME / WG 280 M9962]
QP114.M48M96 1988   612'.17   88-19564

ISBN-13: 978-94-010-7087-4   e-ISBN-13: 978-94-009-1319-6
DOI: 10.1007/978-94-009-1319-6

Published by Martinus Nijhoff Publishers,
P.O. Box 17, 3300 AA Dordrecht, The Netherlands.

Sold and distributed in the U.S.A. and Canada
by Kluwer Academic Publishers,
101 Philip Drive, Norwell, MA 02061, U.S.A.

In all other countries, sold and distributed
by Kluwer Academic Publishers Group,
P.O. Box 322, 3300 AH Dordrecht, The Netherlands.

For Lieneke

MYOCARDIAL ENERGY METABOLISM offers a scientific survey of the biochemistry of the heart, with an emphasis on the energy aspects. This approach may be looked upon as complementary to the classical haemodynamic view of the heart. Sections on *Basic mechanisms*, *Models and techniques*, and on *Clinical implications* make this book a worthwhile acquisition for cardiologists, cardiac surgeons, basic scientists in cardiovascular physiology, as well as for students with a more than average 'heart' for the organ.

Not all basic questions with regard to formation, breakdown and usage of energy (ATP) in the heart have been asked or answered, but invited experts from the fields of cardiac biochemistry, (electro)physiology, pharmacology, anesthesiology and clinical cardiological research have extensively covered the subject from their points of view. Furthermore, MYOCARDIAL ENERGY METABOLISM shows how basic scientific knowledge may be integrated into cardiological practice. Especially the potential that new techniques like NMR and PET-scanning offer to the cardiologist of the near future becomes clear when the heart is studied with regard to its energy metabolism.

I wish to thank the authors for their effort. I am especially grateful to Peter Achterberg for helpful discussion and contribution.

Rotterdam, April 1988   Jan Willem de Jong

Table of Contents

List of Contributors

Achterberg, Peter W.,
Cardiochemical Laboratory, Thoraxcenter, Erasmus University Rotterdam, P.O. Box 1738, 3000 DR Rotterdam, The Netherlands

Belardinelli, Luiz,
Departments of Medicine (Division of Cardiology) and Physiology, Medical Center, University of Virginia, P.O. Box 456, Charlottesville, Virginia 22908, U.S.A.
Presently at: Department of Medicine, University of Florida, P.O. Box J-277, JHMHC, Gainesville, Florida 36210, U.S.A.

Bucx, Jeroen J.J.,
Department of Medicine, Health Sciences Centre, University of Calgary School of Medicine, 3330 Hospital Drive N.W., Calgary, Alberta T2N 4N1, Canada
Presently at: Thoraxcenter, Erasmus University Rotterdam, P.O. Box 1738, 3000 DR Rotterdam, The Netherlands

Bünger, Rolf,
Department of Physiology, School of Medicine, Uniformed Services University of the Health Sciences, 4301 Jones Bridge Road, Bethesda, Maryland 20814-4799, U.S.A.

Czarnecki, Wlodzimierz,
II Department of Cardiology, Medical Centre of Postgraduate Education, Grochowski Hospital, Grenadierow 51/59, 04-073 Warsaw, Poland
Presently at: Department of Medicine, Health Sciences Centre, University of Calgary School of Medicine, 3330 Hospital Drive N.W., Calgary, Alberta T2N 4N1, Canada

De Jong, Jan Willem,
Cardiochemical Laboratory, Thoraxcenter, Erasmus University Rotterdam, P.O. Box 1738, 3000 DR Rotterdam, The Netherlands

De Tombe, Peter P.,
Department of Medicine, Health Sciences Centre, University of Calgary School of Medicine, 3330 Hospital Drive N.W., Calgary, Alberta T2N 4N1, Canada

Ferrari, Roberto,
Chair of Cardiology, University of Brescia, 25100 Brescia, Italy

Harmsen, Eef,
Department of Biochemistry, University of Oxford, South Parks Road, Oxford OX1 3QU, U.K.
Presently at: Department of Medicine, Health Sciences Centre, University of Calgary School of Medicine, 3330 Hospital Drive N.W., Calgary, Alberta T2N 4N1, Canada

Hartman, David A.,
Department of Physiology, School of Medicine, Uniformed Services University of the Health Sciences, 4301 Jones Bridge Road, Bethesda, Maryland 20814-4799, U.S.A.

Huizer, Tom,
Cardiochemical Laboratory, Thoraxcenter, Erasmus University Rotterdam, P.O. Box 1738, 3000 DR Rotterdam, The Netherlands

Ito, Bruce R.,
Department of Experimental Cardiology, Max-Planck-Institute, Benekestrasse 2, D-6350 Bad Nauheim, F.R.G.
Presently at: University of California San Diego, Department of Cardiovascular Medicine, La Jolla, California 92093, U.S.A.

Kirkels, Johan H.,
Interuniversity Cardiology Institute of the Netherlands, c/o University Hospital, Catharijnesingel 101, 3511 GV Utrecht, The Netherlands

Lamers, Jos M.J.,
Department of Biochemistry I, Erasmus University Rotterdam, P.O. Box 1738, 3000 DR Rotterdam, The Netherlands

Leijendekker, Willem J.,
Department of Medicine, Health Sciences Centre, University of Calgary School of Medicine, 3330 Hospital Drive N.W., Calgary, Alberta T2N 4N1, Canada

Mallet, Robert T.,
Department of Physiology, School of Medicine, Uniformed Services University of the Health Sciences, 4301 Jones Bridge Road, Bethesda, Maryland 20814-4799, U.S.A.

Melin, Jaques A.,
Positron Emission Tomography Laboratory, University of Louvain Medical School, Avenue Hippocrate 55, P.O. Box 5550, B-1200 Brussels, Belgium

Merin, Robert G.,
Department of Anesthesiology, University of Texas Medical School at Houston, 6431 Fannin, Rm 5.020, Houston, Texas 77030, U.S.A.

Olsson, Ray A.,
Suncoast Cardiovascular Research Laboratory, Department of Internal Medicine and Biochemistry, University of South Florida College of Medicine, 12901 Bruce B. Dows Boulevard., Tampa, Florida 33612, U.S.A.

Pinson, Arié,
Laboratory for Myocardial Research, The Hebrew University - Hadassah Medical School, P.O. Box 1172, 91010 Jerusalem, Israel

Piper, H. Michael,
Physiologisches Institut I, Universität Düsseldorf, Moorenstrasse 5, D-4000 Düsseldorf 1, F.R.G.

Ruigrok, Tom J.C.,
Department of Cardiology, University Hospital, Catharijnesingel 101, 3511 GV Utrecht, The Netherlands

Schaper, Wolfgang,
Department of Experimental Cardiology, Max-Planck-Institute, Benekestrasse 2, D-6350 Bad Nauheim, F.R.G.

Seymour, Anne-Marie L.,
Department of Biochemistry, University of Oxford, South Parks Road, Oxford OX1 3QU, U.K.

Siegmund, Berthold,
Physiologisches Institut I, Universität Düsseldorf, Moorenstrasse 5, D-4000 Düsseldorf 1, F.R.G.

Skladanowski, Andrzej C.,
Department of Biochemistry, Medical School Gdansk, Debinki 1, 80-211 Gdansk, Poland
Presently at: Department of Cardiology, University of Wales College of Medicine, Heath Park, Cardiff CF4 4XN, U.K.

Spahr, Rolf,
Physiologisches Institut I, Universität Düsseldorf, Moorenstrasse 5, D-4000 Düsseldorf 1, F.R.G.

Sperling, Oded,
Department of Clinical Biochemistry, Beilinson Medical Center, 49100 Petah-Tikva, Israel; and Department of Chemical Pathology, Sackler School of Medicine, Tel-Aviv University, Ramat Aviv, Israel

Taegtmeyer, Heinrich,
Division of Cardiology, Department of Internal Medicine, University of Texas Medical School at Houston, 6431 Fannin, MSB 1.246, Houston, Texas 77030, U.S.A.

Ter Keurs, Henk E.D.J.,
Department of Medicine, Health Sciences Centre, University of Calgary School of Medicine, 3330 Hospital Drive N.W., Calgary, Alberta T2N 4N1, Canada

Van der Giessen, Willem J.,
Laboratory for Experimental Cardiology, Thoraxcenter, Erasmus University Rotterdam, P.O. Box 1738, 3000 DR Rotterdam, The Netherlands

Van der Meer, Peter,
Cardiochemical Laboratory, Thoraxcenter, Erasmus University Rotterdam, P.O. Box 1738, 3000 DR Rotterdam, The Netherlands

Van der Veen, Frits H.,
Department of Physiology, University of Limburg, P.O. Box 616, 6200 MD Maastricht, The Netherlands

Van der Vusse, Ger J.,
Department of Physiology, University of Limburg, P.O. Box 616, 6200 MD Maastricht, The Netherlands

Van Echteld, Cees J.A.,
Interuniversity Cardiology Institute of the Netherlands and Department of Cardiology, University Hospital, Catherijnesingel 101, 3511 GV Utrecht, The Netherlands

Verdouw, Pieter D.,
Laboratory for Experimental Cardiology, Thoraxcenter, Erasmus University Rotterdam, P.O. Box 1738, 3000 DR Rotterdam, The Netherlands

West, G. Alexander,
Departments of Medicine (Division of Cardiology) and Physiology, Medical Center, University of Virginia, P.O. Box 456, Charlottesville, Virginia 22908, U.S.A.

Wyns, William,
Positron Emission Tomography Laboratory, University of Louvain Medical School,
Avenue Hippocrate 55, P.O. Box 5550, B-1200 Brussels, Belgium

Zimmer, Heinz-Gerd,
Department of Physiology, University of Munich, Pettenkoferstrasse 12, D-8000
Munich 2, F.R.G.

Basic Mechanisms

Chapter 1

ATP-Metabolism in Normoxic and Ischemic Heart

J.W. de Jong and P.W. Achterberg, Cardiochemical Laboratory,
Thoraxcenter, Erasmus University, Rotterdam, The Netherlands

ATP is the immediate fuel for the heart pump. The balance between ATP-synthesis and -utilization is delicate. Under normal circumstances the energy for ATP-synthesis derives from oxidative phosphorylation. Anaerobic glycolysis can only partly substitute for this process. During insufficient perfusion of cardiac tissue, ATP breaks down via ADP and AMP to adenosine, inosine, hypoxanthine, xanthine and urate. The nucleosides and purine bases leave the heart. They are useful for the diagnosis of ischemic heart disease. Various drugs, such as beta blockers and calcium entry blockers, can reduce ATP-catabolism. After restoration of blood supply to the ischemic heart, ATP-content remains low for a long time. During ischemia the decrease in ATP correlates with the decrease in heart function. This relationship is not so obvious in the reperfusion phase. Theoretically, a decreased myocardial ATP-content can be replenished by a) (ribo)phosphorylation of adenosine and adenine, b) salvage of inosine and hypoxanthine, and c) de novo synthesis from, a.o., amino acids. For therapeutic purposes inosine is a good candidate.

Energy Sources

The energy, needed for the continuous contraction of the heart, derives from the breakdown of ATP to ADP. Under aerobic conditions oxygen-dependent catabolism of nutrients produces most ATP in the numerous mitochondria in heart muscle (see Fig. 1). This oxidative phosphorylation is under triple control, i.e., by ADP, oxygen and substrate (4). Mitochondrial ATP moves via the adenine nucleotide translocator and creatine kinases to the cytosol, for use by the contractile proteins.

Heart supplies the other tissues with the prerequisites of normal cellular function. It pumps oxygen-rich and nutrient-containing blood through the body. The heart muscle itself receives oxygen and nutrients via the coronary circulation. Myocardial oxygen use (for contraction), oxygen supply and coronary flow are in delicate equilibrium. A shift occurs if, for instance, the amount of cardiac work varies. Catabolism of ATP to ADP during myocardial contraction happens at a very fast rate. About 4-8% of the total ATP (20-25 umol/g dry wt.) breaks down to ADP during a single contraction (8). The ADP formed can be reconverted back to ATP through the action of creatine kinase with breakdown of creatine phosphate to creatine. In this way creatine phosphate provides a temporary buffer against rapid ATP-depletion (Fig. 1). A more important function of creatine kinase is ATP transport from mitochondria to actomyosine (Chapter 7). Elevated ADP levels can also be reconverted to ATP by the enzyme adenylate kinase with simultaneous formation of AMP. This AMP can be deaminated to IMP [a major pathway in skeletal muscle (see Chapter 6)], or dephosphorylated to adenosine. The heart retains the major part of the nucleotides, in contrast to adenosine and other non-phosphorylated purines.

ATP catabolism

The heart releases adenosine and other purines as soon as ATP-breakdown exceeds ATP-production. Their appearance in the coronary circulation correlates with increased coronary flow (Chapter 8) and thereby increased oxygen supply to the heart. This enhances ATP-formation, and decreases both adenosine release and coronary flow. Adenosine breaks down quickly to inosine and further to hypoxanthine, xanthine and urate (Fig. 2). These compounds also appear in the coronary effluent Xanthine and urate can only be formed if the enzyme xanthine oxidase/dehydrogenase is present in heart.

As mentioned before, AMP breakdown to inosine occurs either via adenosine or IMP. 5'-nucleotidases, present in the cell membrane and the cytosol, catalyze the former pathway. A soluble AMP deaminase catalyzes the latter pathway. Regulation of these enzymes takes place by ATP and the adenylate energy charge (11,18). Adenosine forms also from S-adenosylhomocysteine (SAH). The enzyme involved, SAH-hydrolase, is rather inactive, but it is important for adenosine production during normoxia (1). Inosine derives from either adenosine (catalyzed by an active adenosine deaminase), or from IMP [catalyzed by 5'-nucleotidase (2,18)].

In turn inosine breaks down to hypoxanthine, a reaction catalyzed by the reversible nucleoside phosphorylase reaction (2,10,12). The next step is hypoxanthine catabolism to xanthine and further to urate. Xanthine reductase catalyzes both reactions (16). There is considerable species variation in activity of the cardiac enzyme; it is virtually absent from rabbit heart (15). During ischemia conversion of the reductase to the oxidase form can take place; the oxidase generates superoxide radicals during urate formation (14). Clinical trials with cardiac xanthine oxidase inhibitors should wait until one has properly demonstrated the enzyme in human heart. Relatively little information is available on cardiac XMP and GMP catabolism (19).

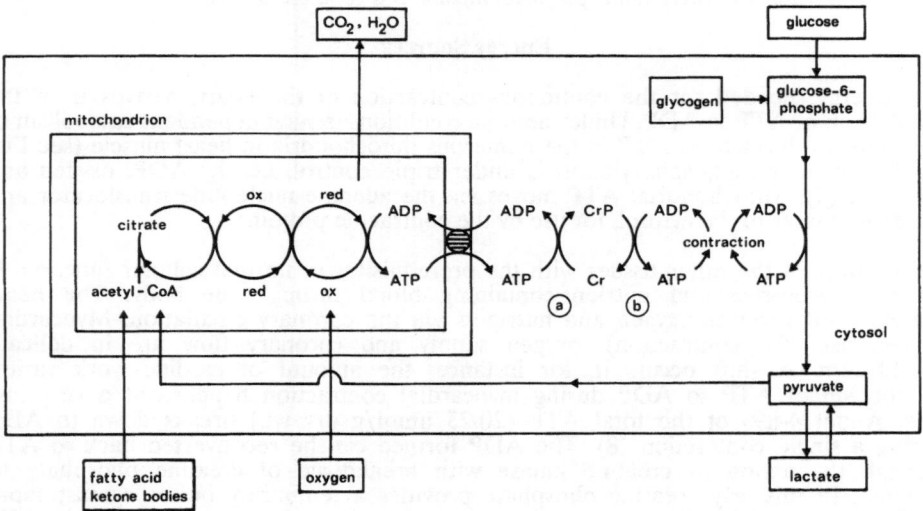

Fig. 1. Major pathways of ATP-formation for myocardial contraction. Mitochondria produce ATP by oxidative phosphorylation; the so-called adenine nucleotide translocator transports it to the cytosol. Immediate transfer of the "high-energy" phosphate group of ATP to creatine phosphate (CrP) occurs via the mitochondrial creatine kinase (CK), bound to the mitochondria (a). Cytosolic creatine kinase (b) reconverts this creatine phosphate to ATP. Myocardial contraction consumes ATP. Fatty acids, ketone bodies, glucose and lactate break down to carbon dioxide and water in the presence of sufficient amounts of oxygen (see Chapters 3 and 4). During hypoxia or ischemia, lactate leaves the heart. ATP-production via anaerobic glycolysis is insufficient. Ox, red = oxidized and reduced components of the respiratory chain, respectively.

The release of adenosine and catabolites (i.e., inosine, hypoxanthine, xanthine and urate) is useful to get an indication of the energy status of the heart. Similarly, with cardiac ATP and creatine phosphate determination, for example in perioperative biopsies, one can assess the "energetic quality" of heart tissue (Chapters 11 and 22). Both in experimental and clinical studies, the increased release of purines is a good indicator of myocardial hypoxia or ischemia (Chapter 21).

Drugs, found to protect the ischemic heart, decrease ischemia-induced purine release. Calcium entry blockers reduce net ATP-breakdown and purine release (5,6) by either increasing ATP-formation through coronary vasodilation, by decreasing ATP-breakdown through overload reduction, or through negative inotropy (Chapter 24). Beta-blockers decrease purine release (20) induced by ischemia. Cardiac purine release is a valuable tool in pharmacological studies.

ATP Anabolism

Several of the purines mentioned above can be incorporated into the myocardial ATP-pool, which diminishes severely during prolonged ischemia. In the heart only a small amount of adenosine forms constantly from AMP, and rephosphorylates immediately to AMP (3). An active adenosine kinase in the cytosol catalyzes the phosphorylation of adenosine (7). The enzyme has a low K_m (0.5 uM). This pathway is useful to raise a decreased myocardial adenine nucleotide pool (13), but hypotension and renal vasoconstriction are potential problems. In addition, adenosine elicits electrophysiological effects (Chapter 9).

Salvage of hypoxanthine, or its precursor inosine, is feasible through its reaction with 5-phosphoribosyl-1-pyrophosphate (PRPP), catalyzed by a reversible hypoxanthine phosphoribosyltransferase. The product, IMP, converts to AMP via adenylosuccinate (enzymes involved: adenylosuccinate synthase and lyase). The PRPP concentration is low (Chapter 20), and adenylosuccinate metabolism is slow. Therefore, the overall conversion of inosine and hypoxanthine to AMP is not a rapid process, but it is promising for cardioprotection (see Chapter 26). Almost nothing is known about xanthine and guanine salvage in heart tissue (19). Hypoxanthine accomplishes active regeneration of GTP in post-ischemic hearts (10).

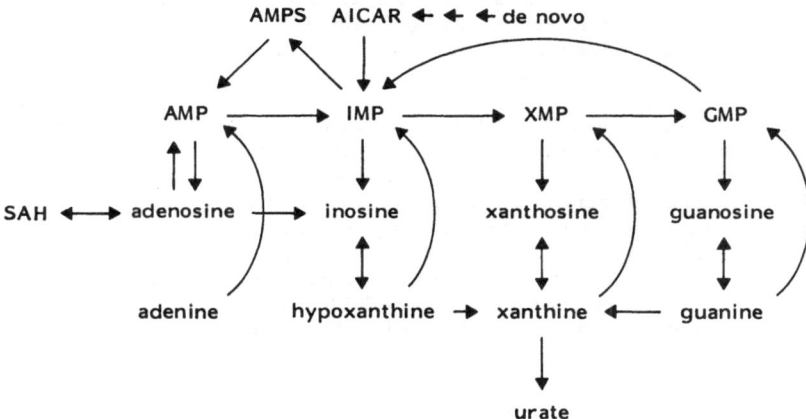

Fig. 2. Scheme depicting AMP catabolism and anabolism. AICAR = 5-amino-4-imidazolecarboxamide ribotide, AMPS = adenylosuccinate, SAH = S-adenosylhomocysteine.

In principle, ribophosphorylation of adenine [catalyzed by adenine phosphoribosyl-transferase (17)], could also increase the cardiac AMP level. However, adenine feeding is toxic (21).

Another pathway leading to AMP - de novo synthesis of the purine skeleton from small molecules, such as glycine (22) - is very slow. It is a multistep synthesis, involving many ATP's. PRPP is rate-limiting. A trial to short-cut this pathway with 5-amino-4-imidazolecarboxamide riboside (AICAriboside), which can be phosphory-lated to the intermediate AICAR (see Fig. 2), has met with little success (19). Ribose speeds up salvage and de novo synthesis (Chapter 10).

ATP and Function

The ATP-level after hypoxia and reperfusion of the heart correlate well with myocardial function (see Chapter 5). Stimulation of myocardial ATP-pool resynthesis by giving purines could promote the velocity and quality of post-ischemic cardiac recovery (Chapters 10 and 23). It could also provide additional protection against a lethal second ischemic attack. Moreover, myocardial protection or preservation during open-heart surgery or heart transplantation could also gain from well-defined purine administration, or prevention of purine release (Chapter 26).

References

1. Achterberg PW, De Tombe PP, Harmsen E, De Jong JW: Myocardial S-adenosylhomocysteine hydrolase is important for adenosine production during normoxia. Biochim Biophys Acta 840:393-400, 1985
2. Achterberg PW, Harmsen E, de Jong JW: Adenosine deaminase inhibition and myocardial purine releaseduring normoxia and ischaemia. Cardiovasc Res 19:593-598, 1985
3. Achterberg PW, Stroeve RJ, De Jong JW: Myocardial adenosine cycling rates during normoxia and under conditions of stimulated purine release. Biochem J 235:13-17, 1986
4. Chance B, Leigh JS Jr, Kent J, McCully K: Metabolic control principles and ^{31}P NMR. Fed Proc 45:2915-2920, 1986
5. De Jong JW, Harmsen E, De Tombe PP: Diltiazem administered before or during myocardial ischemia decreases adenine nucleotide catabolism, J Mol Cell Cardiol 16:363-370, 1984
6. De Jong JW, Huizer T, Tijssen JGP: Energy conservation by nisoldipine in ischaemic heart. Br J Pharmacol 83:943-949, 1984
7. De Jong JW, Kalkman C: Myocardial adenosine kinase: Activity and localization determined with rapid radiometric assay. Biochim Biophys Acta 320:388-396, 1973
8. Fossel ET, Morgan HE, Ingwall JS: Measurement of changes in high-energy phosphates in the cardiac cycle by using gated ^{31}P nuclear magnetic resonance. Proc Natl Acad Sci USA 77:3654-3658, 1980
9. Harmsen E, De Tombe PP, De Jong JW: Synergistic effect of nifedipine and propranolol on adenosine (catabolite) release from ischemic rat heart. Eur J Pharmacol 90:401-409, 1983
10. Harmsen E, De Tombe PP, De Jong JW, Achterberg PW: Enhanced ATP and GTP synthesis from hypoxanthine or inosine after myocardial ischemia. Am J Physiol 246:H37-H43, 1984
11. Harmsen E, Verwoerd TC, Achterberg PW, De Jong JW: Regulation of porcine heart and skeletal muscle AMP-deaminase by adenylate energy charge. Comp Biochem Physiol 75B:1-3, 1983
12. Maguire MH, Lukas MC, Rettie JF: Adenine nucleotide salvage synthesis in the rat heart; pathways of adenosine salvage. Biochim Biophys Acta 262:108-115, 1972
13. Mauser M, Hoffmeister HM, Nienaber C, Schaper W: Influence of ribose, adenosine, and "AICAR" on the rate of myocardial adenosine triphosphate synthesis during reperfusion after coronary artery occlusion in the dog. Circ Res 56:220-230, 1985
14. McCord JM: Oxygen-derived free radicals in post-ischemic tissue injury. N Engl J Med 312:159-163, 1985
15. Schoutsen B, De Jong JW: Age-dependent increase in xanthine oxidoreductase differs in various heart cell types. Circ Res 61:604-607, 1987
16. Schoutsen B, De Jong JW, Harmsen E, De Tombe PP, Achterberg PW: Myocardial xanthine oxidase/dehydrogenase. Biochim Biophys Acta 762:519-524, 1983

17. Schrader J, Gerlach E: Compartmentation of cardiac adenine nucleotides and formation of adenosine. Pflügers Arch 367:129-135, 1976
18. Schütz W, Schrader J, Gerlach E: Different sites of adenosine formation in the heart. Am J Physiol 240:H963-H970, 1981
19. Swain JL, Holmes EW: Nucleotide metabolism in cardiac muscle. In: Fozzard HA, Haber E, Jennings RB, Katz AM, Morgan HE, eds: The heart and cardiovascular system. New York: Raven Press, 1986:911-929
20. Wangler RD, DeWitt DF, Sparks HV: Effect of beta-adrenergic blockade on nucleoside release from the hypoperfused isolated heart. Am J Physiol 247:H330-H336, 1984
21. Yokazawa T, Zheng PD, Oura H, Koizumi F: Animal model of adenine-induced chronic renal failure in rats. Nephron 44:P230-P234, 1986
22. Zimmer H-G: Normalization of depressed heart function in rats by ribose. Science 220:81-82, 1983

Chapter 2

Development of Myocardial Energy Metabolism

P.W. Achterberg, Cardiochemical Laboratory, Thoraxcenter,
Erasmus University Rotterdam, The Netherlands

Age-dependent changes in metabolism, structure and function of the heart may be responsible for differences in tolerance to hypoxia or ischemia. Presumably, the newborn heart has a relatively higher capacity for anaerobic glycolysis and a lower maximal capacity for aerobic ATP-formation than the adult heart. Work performance of the in vivo heart under physiological conditions shows an age-dependent increase. Hypoxic ATP-breakdown rate seems lower in newborn than adult hearts under conditions where anaerobic glycolysis is not higher in newborn than adult hearts. Pathological conditions may influence normal cardiac development and energy metabolism, leading to alterations in the supposedly superior tolerance of the newborn heart to ischemia. During hypoxia and ischemia, the balance between ATP-formation and ATP-breakdown determines the eventual outcome. Developmental differences exist in ATP- and adenosine catabolism. These may contribute to age-dependent differences in cardiac hemodynamics and coronary flow regulation. Further investigations should show whether and how both formation and breakdown of ATP are altered in newborn versus adult hearts under clinically relevant conditions.

Metabolic and Functional Development of the Heart

Birth and the perinatal changes in cardiopulmonary circulation submit the newborn heart to increased work load. Adaptations such as increased cell number follow, as well as changes in intracellular structures and activities of various enzymes. The number and size of mitochondria in cardiac myocytes increase (11). A parallel increase is found in the activity of mitochondrial proteins, related to the oxidative formation of ATP (19). The ratio of mitochondrial over total cardiac protein increases from 0.24 at birth to 0.47 in the adult rat heart. Foetal and neonatal heart has been repeatedly found to have a higher glycogen content and a higher capacity for anaerobic ATP-formation. The rate of anaerobic glycolysis decreases markedly in rats by the day of birth (21,22) with a further steady decline during the first postnatal week. Postnatal lactate dehydrogenase (LDH) and hexokinase activities of the guinea-pig heart decrease in a similar way (4). Increases in LDH-activity in rat heart between 1 and 6 weeks of age have also been reported (3), however. Anaerobic glycolysis is important for cardiac energy production during the hypoxic stress around birth. The increase in fetal rabbit cardiac glycogen content by glucose infusion to the mother, six hours before delivery, causes a better ability of newborn hearts to sustain hypoxic beating (17). In fetal lambs a linear relationship between ATP and glycogen contents exists during hypoxia in utero (23), as well as a correlation between the available glycogen and fetal cardiovascular functioning. Similarly, alterations in fetal heart rate and ECG-patterns, signifying ischemia, are related to ATP and glycogen contents (35).

The above-mentioned postnatal decline in cardiac anaerobic capacity and rise in mitochondrial mass is linked with alterations in fatty acid metabolizing capacity. A decreased carnitine-dependent fatty acid oxidation is found in calf-heart mitochondria as compared to adult bovine mitochondria. Following the suggestion that palmitoylcarnitine synthesis is rate-limiting in newborn hearts (43), a steady, postnatal increase in palmitoyltransferase activity has been observed in rat- (42) and guinea-pig (4) heart. The general inference from the above (and many other) studies is that the fetal and neonatal heart can presumably synthetize more ATP by anaerobic glycolysis than the adult heart, e.g., during hypoxia (26,27). One needs to bear in mind, however, that anaerobic ATP-formation from glucose would need to have an at least fifteen-fold higher glucose-utilization than oxidative breakdown of glucose to generate a similar amount of ATP. The crucial question, therefore, is how much ATP needs to be generated in hypoxic or ischemic newborn and adult myocardium. The answer is, of course, that the amount of ATP synthetized should preferably equal the amount broken down. The myofibrillar contractile ATP-ases and sodium- and calcium-pumping ATP-ases are the main contributors to ATP-breakdown. Their activity directly relates to the work performance of the heart. A doubling of rat cardiac Ca^{2+}-stimulated myofibrillar ATP-ase activity takes place between one and six weeks of age (3) and comparable changes also occur in rabbit heart (32). Moreover, rat cardiac actin, determined as protein-bound ADP, increases with age too (41). Fetal ventricular myocardium contains a higher proportion of non-contractile over contractile mass. The newborn sheep heart presumably contains fewer contractile units or sarcomeres (16). This would predispose to the observed smaller maximal velocity of shortening in isolated muscle of the fetal lamb heart (16) and the altered preload-developed pressure relation in isolated newborn rat hearts (25). Developmental changes in the anatomy and function of sarcoplasmic reticulum and calcium homeostasis are implicated as the direct cause of differences in newborn and adult heart function (5,15,32,38), which in addition can cause differences in ATP-turnover (46).

Does the newborn heart need to synthetize similar amounts of ATP as the adult heart does? It is generally assumed that heart rate is a major determinant of work performance in the developing heart. In rats, the spontaneous heart rate increases during the first ten postnatal days from around 300 to 425 beats/min. Thereafter it gradually decreases in the next two months to around 350. Simultaneously, the resting systolic blood pressure increases over the first 60 postnatal days from around 40 to 110 mmHg, with parallel changes in diastolic blood pressure (from 20 to 80 mmHg, ref. 31). A rough indicator of cardiac work performance (i.e., heart rate times blood pressure) therefore increases nearly three-fold over the first 20 to 30 postnatal days. It appears therefore that the newborn heart has a lower need for ATP-formation than adult heart under similar physiological conditions.

In the next paragraph we will discuss the potential differences in tolerance to hypoxia and ischemia of the newborn and adult heart as reflected in cardiac ATP-contents. From the above, however, it is clear that increased capacity for anaerobic glycolysis will only be the assumed (26,27) relevant factor in tolerance to hypoxia, if the rates of ATP-breakdown are comparable in newborn and adult heart. This is most probably not found under physiological or clinical conditions. In addition it remains to be demonstrated, whether the observed developmental differences in hearts of experimental animals are relevant to human pediatric cardiology. And, if so, on what developmental time scale.

Developmental Differences in Cardiac Tolerance to Hypoxia and Ischemia

Supposedly the newborn heart has a greater tolerance to ischemia or hypoxia than the adult heart (26,27) as reflected by better function and ATP-content. Better hypoxic ATP-preservation might result from increased ATP-formation by anaerobic glycolysis as suggested by higher lactate release from hypoxic newborn versus adult heart preparations (26,27). However, the newborn heart seems less tolerant to ischemia as judged from ischemic lactate accumulation (10,45) and/or the time of onset of ischemic contracture (10). Lactate accumulation has been indicated as a

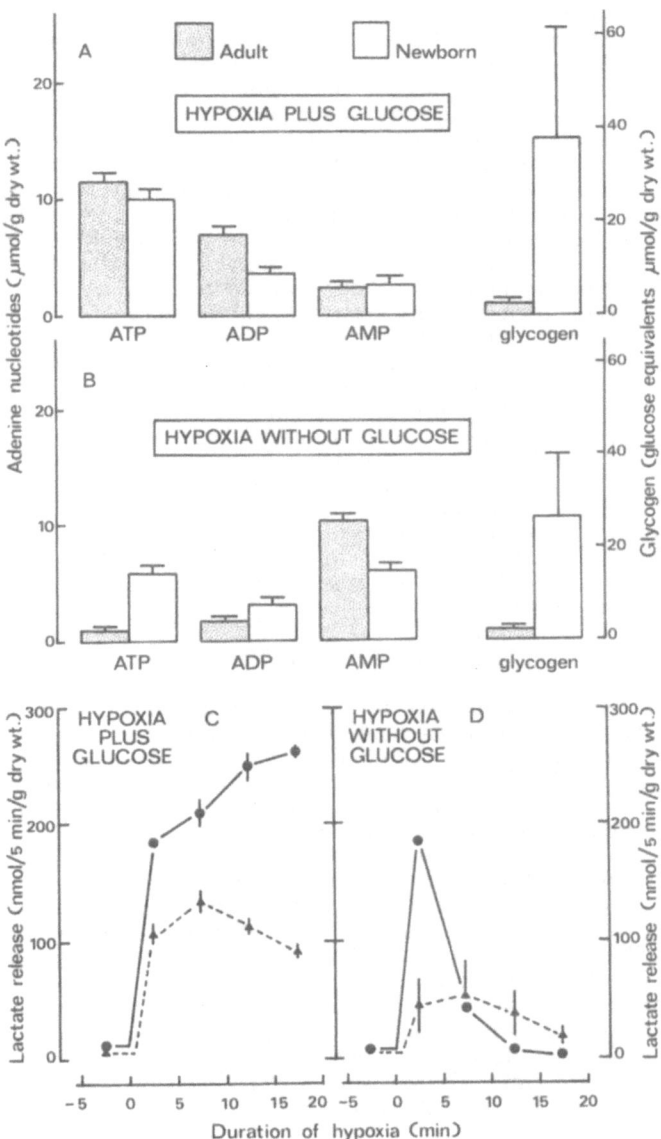

Fig. 1. Energy metabolism in adult and newborn hearts during hypoxia in the presence or absence of glucose. (A,B): Adenine nucleotides and glycogen after 20 min hypoxia plus (A) or without (B) glucose (10 mM). (C,D): Lactate release from adult (circles) and newborn (triangles) hearts during hypoxia plus (C) or without (D) glucose. Newborn hearts have higher glycogen contents (A,B) and lower lactate release (C,D) than adult hearts. Moreover, they have similar (A) or higher (B) ATP-contents. This means that newborn hearts have lower rates of ATP-breakdown during hypoxia than adult hearts. Means ± S.E.M. (n = 4).

major factor in determining cardiac function after ischemia and reperfusion (33). Because of the relatively large cytosolic volume in newborn than adult hearts, a larger amount of lactate in newborn heart may reflect a similar increase in the actual concentration of lactate, however. Other factors, which can influence the outcome of ischemia, apart from the degree of coronary flow reduction and species used, are the occurrence of acidosis and calcium-accumulation, the temperature, work-performance, pre-ischemic quality of the heart and probably age. Our studies on isolated, perfused adult and newborn (10-days old) rat hearts show that 15 min of global ischemia at a fixed heart rate leads to a similar decrease in adult and newborn ATP-content (12). Hypothermia during ischemia has a similar protective effect on ATP-loss in ischemic adult and newborn hearts. A second series of experiments (Fig. 1) demonstrates that after 20 min of hypoxia (300 beats/min) in the presence of glucose, adult and newborn hearts display a similar decrease in ATP-content. Adult hearts, however, produce far more lactate and brake down their glycogen content faster than newborn hearts. In the absence of substrate, however, newborn hearts preserve higher ATP-contents than adult hearts, without faster glycogen breakdown. We have to conclude therefore that newborn hearts have a lower rate of hypoxic ATP-breakdown, because we find no indications of faster anaerobic ATP-synthesis. Lower myocardial energy demand of hypoxic newborn than adult heart has also been inferred (46) from studies on rabbit heart ventricular septal preparations. During ischemia in adult heart, glycolysis will eventually be strongly inhibited (34), whereby the onset of inhibition is related to the extent of flow reduction and the duration of ischemia. No comparable studies have been performed in immature hearts yet. How different the conclusions from experiments on age-dependent tolerance to ischemia sometimes may be, the difference between synthesis and breakdown of ATP is a major factor determining the outcome of ischemia in both adult and newborn hearts.

Development of Coronary Vasoactivity

Contrasting reports have appeared on the development of adenosine-mediated coronary vasodilation. In newborn lamb heart, adenosine is not involved in the metabolic regulation of coronary flow (14). Other authors reach the opposite conclusion for this species (30) and for one-week-old guinea-pig hearts (40). Finally, in rabbit heart the response to infused adenosine increases with age (8). In experiments on isolated ten-day-old and adult rat hearts, we observe large, age-dependent differences in the composition of purines released, both during post-ischemic reperfusion (12) and during hypoxia. At similar rates of total purine release, newborn hearts release far less adenosine than adult hearts do (Fig. 2), but similar amounts of inosine. It is not yet clear, whether the cause is more active deamination and/or phosphorylation of adenosine, or a shift from breakdown of AMP to breakdown of IMP in newborn hearts. The catabolism of infused adenosine is very rapid, even in adult rat heart (1). This may greatly influence the adenosine concentration at the site of its vasodilatory receptor. The differences in myocardial adenosine production described here, may influence the sensitivity of the newborn coronary system towards infused adenosine. It may even be the cause of different autoregulatory properties of the newborn heart.

Myocardial Reperfusion Damage

Oxygen-radical mediated reperfusion damage may contribute to post-ischemic cardiac dysfunction (20). The enzyme xanthine oxidase is repeatedly implicated as a source of cardiac oxygen-radicals during reperfusion (9). We find the enzyme to be present in adult rat heart (37). However, from both direct measurements of xanthine oxidase activity in homogenates and from the determination of the enzyme's products xanthine and urate in perfusates, we must conclude that xanthine oxidase/-dehydrogenase is virtually absent from newborn hearts (2,36). Other sources of oxygen-radicals may well be present in both adult and newborn hearts, however. We find that during reperfusion after ischemia, newborn hearts release significantly less ATP-catabolites than adult hearts do (2). This strongly suggests a better tolerance of the newborn heart to reperfusion damage. However, young patients, undergoing open-heart surgery for correction of Tetralogy of Fallot, show increased myocardial lipid peroxidation (13) after reperfusion.

Pathological Development of the Heart

Myocardial ischemia in infancy can be subdivided into several groups (18). First, there is ischemia associated with birth asphyxia (hypoxia), cerebrovascular accidents and respiratory distress of the newborn. Secondly, myocardial ischemia can be associated with congenital or acquired abnormalities of coronary arteries. Thirdly, of course there is the group of congenital heart diseases, involving left or right ventricular outflow obstruction. Especially in the latter group, persistent alterations in volume or pressure load cause hypo- or hypertrophic abnormalities. These may alter the sensitivity to ischemic stress. In addition, pediatric cardiac pathology has been observed in a variety of congenital disorders of metabolism (29). Each of these could specifically influence cardiac tolerance to ischemia. Because a large group of pediatric patients with congenital anatomical malformations needs cardiac surgery, insight into the factors, which determine the quality of cardioplegic arrest, is of utmost importance. The potential complications of cardioplegia in infants and young children extend beyond developmental differences in structure, metabolism and function of the heart (7). The following aspects can contribute to different outcome of newborn and adult cardioplegia: The size of the patients (relevant to continuation of hypothermia); the size and anatomy of the heart (collaterals can cause rewarming); the underlying disease (hypertrophy); the properties of the newborn circulation (increased chance of oedema in newborns).

The newborn heart can react to hypoxia and hyperoxia by cell-proliferation and cell-differentiation, respectively. The adult heart appears unable to increase its cell-number, but will react to hypoxia by cellular hypertrophy (24). Hypoxia, moreover,

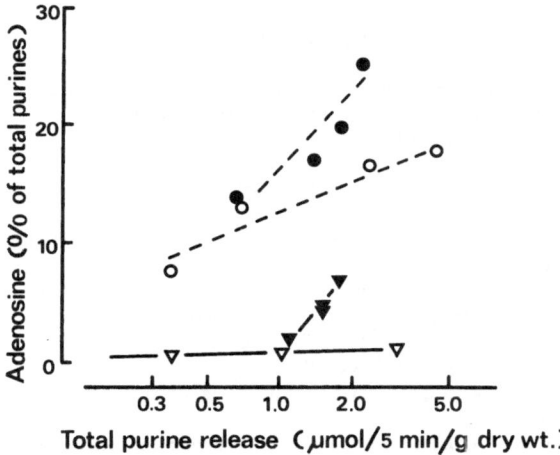

Fig. 2. Adenosine release from adult and newborn hearts during postischemic reperfusion and hypoxia. At similar rates of total purine release, newborn hearts (triangles) release far less adenosine than adult hearts do (circles), both during hypoxia (closed symbols) or during reperfusion (open symbols).

inhibits the normal age-dependent increase in cardiac mitochondrial protein (19,28). Hypoxia, therefore, probably both distorts and slows down normal cardiac development. Animal-model and patient studies show (44) that in case of ventricular hypertrophy especially the endocardial layer of the hypertrophied ventricle has a decreased high-energy phosphate content. This occurs even before the onset of cardioplegia, after which the difference further increases. In a canine model of congenital cyanotic heart disease, created by surgical induction of an atrial-pulmonary shunt and banding of the pulmonary artery, ATP-depletion during cardioplegic arrest is faster than in control animals (39). The full effect is not caused by pulmonary banding alone. After reperfusion, myocardial ATP-content and cardiac function remain depressed in the hypoxemic animals. Cardioplegia in patients proves insufficient, when ischemic time exceeds 85 min. In this study (6), an indirect measure of cardiac ATP (i.e., tissue birefringence) predicts the eventual outcome of surgery. Both low ATP at the onset of cardioplegia and severe ATP-decline during prolonged cardioplegia correlate with negative outcome of surgery as reflected by early death or severely low cardiac output.

Final Conclusions

Studies on postnatal development of structure, function and metabolism of the heart contribute to the general knowledge of the normal adult and newborn heart. Furthermore, certain aspects of cardiac pathology (e.g., hypertrophy, thyrotoxic heart) should be viewed upon as manifestations of abnormal development. The exact knowledge of normal cardiac development is therefore also a prerequisite for the understanding of pathological development.

We have not discussed differences in catecholamine status and reactivity in adult and newborn hemodynamics nor the possibly different sensitivities to calcium or acidosis. We feel, however, that by looking at the difference between synthesis-rate and breakdown-rate of ATP, both adult and newborn cardiac tolerance to ischemia can be further investigated. Such studies in relevant models and in the clinical setting are needed for pediatric cardiac patients to benefit, as well as for our general understanding of cardiac pathophysiology and development.

Acknowledgements

The author is grateful to Ms. C.E. Zandbergen-Visser for secretarial assistance and to Dr. J.W. de Jong for his critical review of the manuscript. He is supported by a grant from the Dutch Heart Foundation.

References

1. Achterberg PW, Harmsen E, De Jong JW: Adenosine deaminase inhibition and myocardial purine release during normoxia and ischaemia. Cardiovasc Res 19:593-598, 1985
2. Achterberg PW, Nieukoop AS, Schoutsen B, De Jong JW: Different ATP-catabolism in the reperfused adult and newborn rat heart. Am J Physiol (accepted for publication)
3. Baldwin KM, Cooke DA, Cheadle, WG: Enzyme alterations in neonatal heart muscle during development. J Mol Cell Cardiol 9:651-660, 1977
4. Barrie SE, Harris P: Myocardial enzyme activities in guinea pigs during development: Am J Physiol 233:H707-H710, 1977
5. Boucek RJ, Shelton ME, Artman M, Landou E: Myocellular calcium regulation by the sarcolemmal membrane in the adult and immature rabbit heart. Basic Res Cardiol 80:316-325, 1985
6. Bull C, Cooper J, Stark J: Cardioplegic protection of the child's heart. J Thorac Cardiovasc Surg 88:287-293, 1984
7. Bull C, Stark J: Cardioplegia in pediatric cardiac surgery. Repair in the first year of life. In: Engelman RM, Levitsky S, eds: A textbook of clinical cardioplegia. New York: Futura, 1982:349-363

8. Buss DD, Hennemann WW, Posner P: Maturation of coronary responsiveness to exogenous adenosine in the rabbit. Basic Res Cardiol 82:290-296, 1987

9. Chambers DE, Parks DA, Patterson, DA, Roy R, McCord JM, Yoshida S, Parmley LF, Downey JM: Xanthine oxidase as a source of free radical damage in myocardial ischemia. J Mol Cell Cardiol 17:145-152, 1985

10. Chiu RCJ, Bindon W: Why are newborn hearts vulnerable to global ischemia? The "lactate hypothesis". Circulation 74:II-133, 1986 (Abstr)

11. David H, Behrisch D, Vivar Flores OD: Postnatal development of myocardial cells after oxygen deficiency in utero. Pathol Res Pract 179:370-376, 1985

12. De Jong JW, Achterberg PW: Developmental differences in myocardial ATP metabolism. Basic Res Cardiol 82, Suppl 2:121-126, 1987

13. Del Nido PJ, Mickle DAG, Williams WG, Wilson GJ, Benson LN Coles JG, Trusler GA: Evidence of myocardial free radiacal injury during elective repair of Tetralogy of Fallot (TOF). Circulation 74:II-77, 1986 (Abstr)

14. Downing SE, Chen V: Dissociation of adenosine from metabolic regulation of coronary flow in the lamb. Am J Physiol 251:H40-H46, 1986

15. Frank JS, Rich TL: Calcium depletion and repletion in rat heart. Age-dependent changes in the sarcolemma. Am J Physiol 245:H343-H353, 1983

16. Friedman WF: The intrinsic physiologic properties of the developing heart. Progr Cardiovasc Dis 15:87-111, 1972

17. Gelli MG, Enhörning G, Hultman E, Bergström J: Glucose infusion in the pregnant rabbit and its effect on glycogen content and activity of foetal heart under anoxia. Acta Paediatr Scand 57:209-214, 1968

18. Hallidie-Smith KA: Myocardial ischemia in infancy. In: Maseri A, ed: Myocardial function in man, ischaemia, failure, cardiomyopathies. Hammersmith Cardiology Workshop Series, Vol 2. New York: Raven Press, 1985:183-186

19. Hallman M, Mäenpää P, Hassinen I: Levels of cytochromes in heart, liver, kidney and brain in the developing rat. Experientia 28:1408-1410, 1972

20. Hammond B, Kontos HR, Hess ML: Oxygen radicals in the adult respiratory distress syndrome, in myocardial ischemia and reperfusion injury, and in cerebral vascular damage. Can J Physiol Pharmacol 63:173-178, 1985

21. Hoerter J: Changes in the sensitivity to hypoxia and glucose deprivation in the isolated perfused rabbit heart during perinatal development. Pflügers Arch 363:1-6, 1976

22. Hoerter JA, Opie LH: Perinatal changes in glycolytic function in response to hypoxia in the incubated or perfused rat heart. Biol Neonate 33:141-161, 1978

23. Hökegard KH, Eriksson BO, Kjellmer I, Magno R, Rosén KG: Myocardial metabolism in relation to electrocardiographic changes and cardiac function during graded hypoxia in the fetal lamb. Acta Physiol Scand 113:1-7, 1981

24. Hollenberg M, Honbo N, Samorodin AJ: Effects of hypoxia on cardiac growth in neonatal rat. Am J Physiol 231:1445-1450, 1976

25. Hopkins SF, McCutcheon EP, Wekstein DR: Postnatal changes in rat ventricular function. Circ Res 32:685-691, 1973

26. Jarmakani JM, Nagatomo T, Nakazawa M, Langer GA: Effect of hypoxia on myocardial high-energy phosphates in the neonatal mammalian heart. Am J Physiol 235:H475-H481, 1978

27. Jarmakani JM, Nakazawa M, Nagatomo T, Langer GA: Effect of hypoxia on mechanical function in the neonatal mammalian heart. Am J Physiol 235:H469-H474, 1978

28. Kinnula VL, Hassinen I: Effect of hypoxia on mitochondrial mass and cytochrome concentrations in rat heart and liver during postnatal development. Acta Physiol Scand 99:462-466, 1977

29. Kohlschütter A, Hausdorf G: Primary (genetic) cardiomyopathies. A survey of possible disorders and guidelines for diagnosis. Eur J Pediatr 145:454-459, 1986

30. Mainwaring RD, Mentzer RM, Ely SW, Rubio R, Berne RM: The role of adenosine in the regulation of coronary blood flow in newborn lambs. Surgery 98:540-545, 1985

31. Mills E, Smith PG: Mechanisms of adrenergic control of blood pressure in developing rats. Am J Physiol 250:R188-R192, 1986

32. Nakanishi T, Jarmakani JM: Developmental changes in myocardial mechanical function and subcellular organelles. Am J Physiol 246:H615-H625, 1984

33. Neely JR, Grotyohann LW: Role of glycolytic products in damage to ischemic myocardium.

Dissociation of adenosine triphosphate levels and recovery of function of reperfused ischemic hearts. Circ Res 55:816-824, 1984

34. Neely JR, Liedtke AJ, Whitmer JT, Rovetto MJ: Relationship between coronary flow and adenosine triphosphate production from glycolysis and oxidative metabolism. In: Roy PE, Harris P, eds: Recent advances in studies on cardiac structure and metabolism, Vol 8 (The cardiac sarcoplasm). Baltimore: Univ Park Press, 1975:301-321

35. Rosén KG, Isaksson O: Alterations in fetal heart rate and ECG correlated to glycogen, creatine phosphate and ATP levels during graded hypoxia. Biol Neonate 30:17-24, 1976

36. Schoutsen B, De Jong JW: Age-dependent increase in xanthine oxidoreductase in various heart cell types. Circ Res 61:604-607, 1987

37. Schoutsen B, De Jong JW, Harmsen E, De Tombe PP, Achterberg PW: Myocardial xanthine oxidase/dehydrogenase. Biochim Biophys Acta 762:519-524, 1983

38. Seguchi M, Harding JA, Jarmakani JM: Developmental changes in the function of sarcoplasmic reticulum. J Mol Cell Cardiol 18:189-195, 1986

39. Silverman NA, Kohler J, Levitsky S, Pavel DG, Fang RB, Feinberg H: Chronic hypoxemia depresses global ventricular function and predisposes to the depletion of high-energy phosphates during cardioplegic arrest: Implications for surgical repair of cyanotic congenital heart defects. Ann Thorac Surg 37:304-308, 1984

40. Toma BS, Wangler RD, DeWitt DF, Sparks HV: Effect of development on coronary vasodilator reserve in the isolated guinea pig heart. Circ Res 57:538-544, 1985

41. Uchino J, Tsuboi KK: Actin accumulation in developing rat muscle. Am J Physiol 219:154-158, 1970

42. Warshaw JB: Cellular energy metabolism during fetal development. IV. Fatty acid activation, acyl transfer and fatty acid oxidation during development of the chick and rat. Devel Biol 28:537-544, 1972

43. Warshaw JB, Terry ML: Cellular energy metabolism during fetal development. II. Fatty acid oxidation by the developing heart. J Cell Biol 44:354-360, 1970

44. Wechsler AS: Deficiencies of cardioplegia. The hypertrophied ventricle. In: Engelman RM, Levitsky S, eds: A textbook of clinical cardioplegia. New York: Futura, 1982:381-390

45. Wittnich C, Chiu RCJ: Is neonatal myocardium more or less vulnerable to ischemic injury? Circ Res 72, Suppl III:361, 1985 (Abstr)

46. Young HH, Shimizu T, Nishioka K, Nakanishi T, Jarmakani JM: Effect of hypoxia and reoxygenation on mitochondrial function in neonatal myocardium. Am J Physiol 245:H998-H1006, 1983

Chapter 3

Principles of Fuel Metabolism in Heart Muscle

H. Taegtmeyer, Division of Cardiology, Department of Medicine,
The University of Texas Medical School at Houston,
Houston, Texas, U.S.A.

The functions of metabolism in the heart are few in number and can be understood without detailed knowledge of elaborate chemical formulas. The primary function of the metabolism of energy-providing substrates is the rapid synthesis of ATP from ADP and inorganic phosphate. Under normal conditions, heart muscle covers its energy needs through the oxidation of glucose, fatty acids, ketone bodies or, to a lesser extent, amino acids. This chapter reviews the mechanisms of metabolic control and fuel selection, as they relate to the function of the heart. Particular emphasis is placed on the interaction between substrates and the control of the citric acid cycle, the final common pathway for all substrates providing reducing equivalents for the respiratory chain.

1. Heart Muscle as Site of Energy Transformation

The purpose of this chapter is to provide a brief outline of cardiac fuel metabolism. To begin with, fuel metabolism cannot be separated from energy metabolism, hence the inclusion of this chapter in a book on Myocardial Energy Metabolism. Secondly, a review on cardiac fuel metabolism seems timely in light of the renewed interest in intermediary metabolism generated by recent techniques using tracer compounds labeled with positron emitting radioisotopes, or by magnetic resonance spectroscopy to assess tissue function and metabolism in the intact heart.

Compared with the vast biochemical literature such a review is necessarily superficial and fragmented. Fortunately, the functions of metabolism in the heart are few in number and can be understood without detailed knowledge of elaborate chemical formulas or reactions. The main purpose of cardiac metabolism is to provide energy for contraction, ion transport, biosynthesis and degradation. The conversion of chemical energy into mechanical energy is termed energy transformation. Like any other living tissue, heart muscle captures and utilizes chemical energy in the form of ATP. The tissue content of ATP is normally around 20 umol/g dry weight. At an O_2 consumption of 4.5 mmol/min/g dry wt. (67) and a P/O ratio of around 3, heart muscle utilizes 1400 umol ATP/min/g dry wt. or 23 umol ATP/s/g dry wt. This calculation shows that under steady-state conditions ATP has to be regenerated as rapidly as it is broken down. The overall concept of this "ATP cycle" is depicted in Fig. 1.

The sources of ATP in heart muscle are oxidative phosphorylation of ADP in the respiratory chain and, to a lesser extent, phosphorylation of ADP either by substrate level phosphorylation in the glycolytic pathway and the citric acid cycle, or by the action of creatine kinase. Since oxidative phosphorylation is quantitatively the main source of ATP and because the key enzymes of oxidative metabolism are located in mitochondria, myocardial cells are well endowed with organelles.

Oxidative phosphorylation of ADP depends on the production of reducing equivalents (protons) and the passage of electrons along the respiratory chain. Since reducing equivalents are produced in the course of substrate catabolism to acetyl-CoA and subsequent oxidation of acetyl-CoA in the citric acid cycle, it is convenient to group the main energy-providing reactions into three stages (Fig. 2). The first stage comprises all reactions leading to acetyl-CoA, the second stage the oxidation of acetyl-CoA in the citric acid cycle (resulting in liberation of CO_2 and reducing equivalents), and the third stage consists of the flow of the electrons down the respiratory chain leading to release of free energy, which is conserved in the energy-rich phosphate bond of ATP.

2. Some Early Experiments

A major impetus for research in cardiac metabolism came from the work of Richard Bing and his co-workers who measured arterio-venous differences of substrates and products across the heart (see ref. 3 for review).

Using coronary sinus sampling, it was shown that the human heart uses the three foodstuffs carbohydrates, fatty acids and (under certain circumstances also) amino acids to varying degrees for its energy production. The mammalian heart can therefore be likened to an omnivore. Bing's work suggested that the arterial concentration of *carbohydrates* and the presence of insulin affect theire relative myocardial usage. The total aerobic metabolism of *glucose, lactate* and *pyruvate* did, however, not account for the total oxygen consumption of the heart, and it was quickly recognized that oxidation of *fatty acids* and *ketone bodies* made up the considerable balance. Myocardial utilization of fatty acids was particularly high during starvation. While oxidation of ketone bodies accounted for approximately 5% of the total myocardial oxygen consumption in normal individuals, the diabetic dog heart utilized a consistently larger quantity of these fuels (70). The utilization of ketone bodies seemed to be governed by their arterial concentration and the quantity of carbohydrate available (4). Bing et al. (4) further suggested that both the human heart and the dog heart may even extract considerable amounts of *amino acids* from the blood when their plasma levels rise to greater than physiological values. This ability of the heart to utilize various substrates can be regarded as an "important factor of safety" (3) in the maintenance of an adequate fuel supply for an organ of vital importance for the survival of the total organism.

3. Principles of Metabolic Control

It is a characteristic property of living cells to permit complex chemical reactions to proceed quickly at relatively low temperatures and concentrations. This efficient energy transfer in cells occurs through enzyme-catalyzed reactions and is probably the result of a long process of evolutionary selection (1).

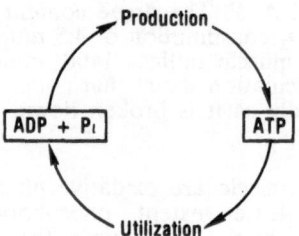

Fig. 1. The cycle of cardiac energy metabolism (ATP cycle). Breakdown of fuels leads to the capture of energy in the form of ATP. Hydrolysis of ATP and ADP and inorganic phosphate (P_i) provides the energy for contraction, ion transport, as well as synthesis and degradation of biomolecules. Stores of ATP in the tissue are very small and must therefore be continuously replenished by phosphorylation of ADP.

Activities of enzymes are influenced by temperature, pH, and other factors, such as allosteric activators or inhibitors. The progress of any enzyme-catalyzed reaction can be determined by measuring either product formation or substrate utilization under optimal environmental conditions (Fig. 3). The hyperbolic configuration of enzyme-activity curves shows that the rate of the reaction diminishes with time and reaches its end-point asymptomatically, which is due to a fall in the concentration of substrate as well as a number of other environmental factors.

In most instances substrate availability (as well as product removal) and enzyme activity are the main factors which control the rate of a biochemical reaction. While this is certainly true for the many reactions which are at *near-equilibrium*, the same does not apply to those enzymes whose activity is modified by factors other than the pathway substrate and which catalyze *non-equilibrium* reactions. Biochemists identify non-equilibrium or regulatory reations in the first instance by measuring tissue concentrations of products and substrates, by calculating their ratio (mass action ratio) and by comparing this value to the equilibrium constant K_{eq} of the reaction. If both are similar, the reaction is at near-equilibrium; if they differ significantly, the reaction is at non-equilibrium. The same principle is expressed in the "crossover theorem" by Chance and Williams (9), which simply states that when a pathway is inhibited at a specific reaction, the substrate concentration will increase and the product concentration will decrease. The next step biochemists have taken in the process of identification of a non-equilibrium or regulatory enzyme is testing the enzyme activity in vitro. Such detailed analysis is desirable because regulatory enzymes set the pace for an entire sequence of reactions in a metabolic pathway, as will be discussed below.

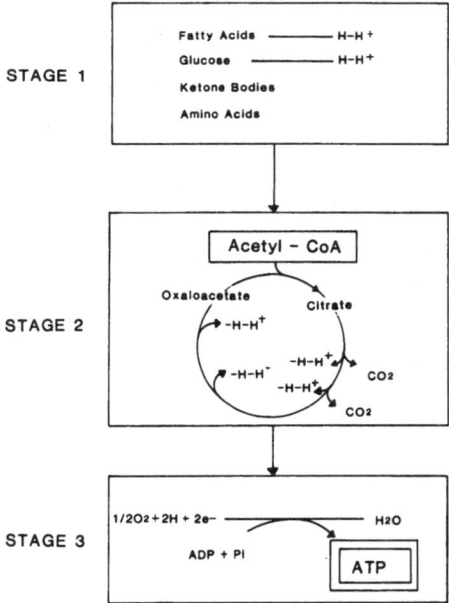

Fig. 2. Stages of catabolic reaction in cardiac energy metabolism. Energy-providing substrates are converted to acetyl-CoA, which is, in turn, oxidized in the citric acid cycle. At many stages in their catabolism, the substrates generate reducing equivalents in the form of $NADH + H^+$ and/or $FADH_2$. Transfer of their electrons to the electron carrying chain and their reaction with molecular oxygen leads to synthesis of ATP.

There are two major classes of regulatory enzymes: allosteric (or non-covalently regulated) enzymes and covalently regulated enzymes. The allosteric theory proposes that an enzyme possesses at least two spatially distinct binding sites on the protein molecule, one is termed the catalytic site, the other is termed the regulatory site. A regulator binds to the regulatory site and modulates the activity of the enzyme. Often the product of a reaction or of a metabolic pathway acts as regulator (feedback regulator).

The second class of regulatory enzymes is modulated through interconversion of their active and inactive forms by covalent modification. Examples of covalently modified enzymes are the enzymes of glycogen metabolism and the pyruvate dehydrogenase complex, both of which undergo phosphorylation and dephosphorylation.

4. Control of Metabolic Pathways

The broad objective of the study of enzyme mechanisms in vivo is the understanding of events which link metabolism to function. The full importance of enzymatic reactions can therefore only be understood when they are viewed in the context of a metabolic pathway. A metabolic pathway can be defined as a series of enzyme-catalyzed reactions which convert a substrate into a product.

Prevailing ideas on the operational control of metabolic pathways are probably too simplistic when one considers the complex conditions inside the cell. An important concept to be considered is the *concept of flux*, i.e., the flow of matter along a series of reactions. Biochemical maps of a series of reactions provide no more informations than a road map. The real test is the journey itself. According to Newsholme and Leech (42), it is useful to think of a metabolic pathway as a series of reactions initiated and terminated by a flux-generating reaction. The concept of a "pacemaker" reaction for a metabolic pathway was first proposed by Krebs (25). More recently Newsholme and Crabtree (41) have defined a "flux-generating step" as any non-equilibrium reaction saturated with pathway substrate to which all other reactions in the pathway must respond in order to achieve a steady state. As it is understood today, a change in the activity of the regulatory enzyme (or transport step, or work load, for that matter) would cause a change in flux through the pathway as long as the enzyme is saturated with its pathway substrate. Newsholme and Crabtree (41) have termed this process "external regulation" and the control of enzyme activity via changes in the pathway substrate concentration as "internal regulation". The near-equilibrium reactions of the pathway are very sensitive to changes in concentrations of substrate, and a small increase in pathway substrate concentration would stimulate the activity of the enzyme sufficiently to accommodate new rates of flux.

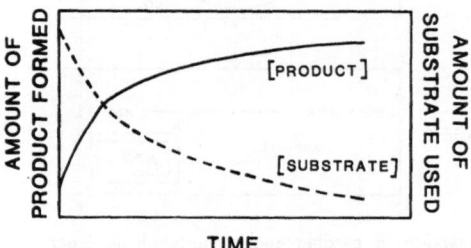

TIME

Fig. 3. Progress of an enzyme-catalyzed reaction. Progress of the reaction is in this example monitored by the appearance of a product (solid line) or the utilization of a substrate (broken line) with time (on the abscissa). The reaction assumes the form of a hyperbola. The velocity of the reaction is measured in the initial, linear portion of the curve.

A traditional approach in studying cardiac metabolism is the analysis of metabolites extracted from rapidly frozen tissue. As discussed earlier, accumulation of metabolites proximal to a metabolic "block" may reveal the presence of a regulatory enzyme. Accumulation of metabolites only suggests, but does not prove, inhibition of flux through a specific step in a pathway. For example, the steady state content of some key metabolites of cardiac energy metabolism may vary considerably depending on the dietary state and substrate supply. It is of note in this regard that there is no correlation between intermediary metabolite levels and cardiac performance. For example, with glucose as the only exogenous substrate, the performance of the heart is excellent, but acetyl-CoA (an important intermediate of oxidative glucose metabolism) is below the detectable range, because the compound is oxidized as fast as it is produced. By contrast, when ketone bodies are the only substrate available to the heart, acetyl-CoA accumulates, but performance and O_2 consumption are low, because of decreased rates of oxidation in the tricarboxylic acid cycle (64). This observation provides an excellent example for the general concept that flux through energy-providing pathways, and not necessarily the steady-state concentrations of metabolites, determines the functional state of the tissue.

5. Metabolism of Energy Providing Substrates

A. Carbohydrates

Carbohydrates of major importance as substrates for cardiac energy metabolism are glucose, glycogen, lactate and pyruvate. Glucose is taken up by the heart-muscle cell, phosphorylated and metabolized as glucose 6-phosphate by several distinct enzyme systems to meet the needs of the cell. These systems include: a) degradation via the Embden-Meyerhof pathway (also termed the glycolytic pathway if it entails metabolism of glucose to lactate only), b) conversion of glycogen which serves as an intracellular glucose store, and c) metabolism via the pentose-phosphate pathway which yields ribose and NADPH. The latter pathway is quantitatively of lesser importance in heart muscle than the former two pathways.

In keeping with the scheme presented in Fig. 2, there are three energy-yielding stages of glucose metabolism: the Embden-Meyerhof pathway leading to pyruvate and acetyl-CoA, the citric acid cycle and the respiratory chain. Each stage is regulated by its own set of controls so that overall flux through the pathway proceeds at a rate just sufficient to satisfy the beat-to-beat needs of the heart for the pathway product, ATP. For example, the relative concentrations of ADP in the cytosol control not only the rate of electron transport and oxidative phosphorylation, but also the turnover rate of the citric acid cycle, the rate of pyruvate oxidation and glucose utilization. Thus, whenever hydrolysis of ATP is increased, a whole sequence of enzyme-catalyzed reactions and transport mechanisms is set into motion.

Glucose transport and phosphorylation. Glucose uptake into heart muscle occurs via a stereo-specific passive membrane transport system that displays saturation kinetics and counter transport (35). This system equilibrates intracellular and extracellular concentrations of glucose but does not move sugar against a concentration gradient. The stereo-specificity of the carrier for sugars of the C-1 chain conformation is not matched by the same degree of selectivity.

There are various factors which influence glucose transport by changing either K_m or V_{max} of the carrier system. These include hormones, such as insulin, growth hormone, epinephrine and cortisol (for review, see ref. 40), anoxia (34), extracellular glucose concentration and cardiac work (38,39,67). In normal heart muscle, cardiac work, the availability of alternative substrates, as well as plasma insulin and glucose concentrations, act in concert as the most important factors regulating glucose uptake.

When hearts are perfused at very low work loads, insulin nearly doubles glucose utilization and lactate production (34,39,67), and during maximal insulin stimulation net transport exceeds the rate of phosphorylation (34). This finding is in good agreement with recent in vivo studies which have estimated glucose transport with [^{3}H]-2-deoxyglucose (20).

Phosphorylation of glucose by hexokinase becomes rate limiting for glycolysis at high rates of glucose transport. Rates of glucose phosphorylation measured in vitro are more than twice as high as the maximal rates of glucose utilization by heart at physiological work load and with glucose as the only substrate (41). The intracellular glucose concentration of heart muscle rises with starvation, diabetes, oxidation of fatty acids, ketone bodies or pyruvate, which indicates inhibition of the phosphorylation step. This is most likely due to accumulation of glucose 6-phosphate, because of inhibition of the phosphofructokinase reaction (see below).

Glycogen formation and breakdown. Glycogen metabolism is without question the biochemical pathway which has contributed more to the understanding of enzyme regulation and the molecular basis of hormone action than any other known system of metabolic control (10). In this system the first example of the control of enzyme activity by an allosteric regulator (activation of phosphorylase by AMP) was described (13), enzyme regulation by reversible covalent modification (phosphorylation) was discovered (24), and the molecular basis of hormone action was elucidated through the discovery of cAMP (53).

Synthesis and degradation of glycogen occur in two separate pathways. The combined effects of protein phosphorylation and dephosphorylation on glycogen synthase and phosphorylase provide an interlocking system by which homones (such as epinephrine) and mechanical activity (through Ca^{2+}) can control the net flux of glucose 1-P into and out of the intracellular glycogen stores. Epinephrine-induced cAMP formation promotes protein phosphorylation and simultaneously inhibits glycogen synthesis while stimulating glycogen breakdown, whereas stimulation of protein dephosphorylation shifts the balance toward glycogen synthesis. Although the interaction of the enzymes of glycogen breakdown and synthesis is more complex than this simple mechanism suggests (for details, see ref. 11), the concept is relatively easy to grasp.

Glycolysis. The catabolism of glucose 6-phosphate in the glycolytic pathway exhibits two major control sites: phosphofructokinase and glyceraldehyde 3-P dehydrogenase.

Allosteric control of phosphofructokinase (PFK) provides for large changes in catalytic activity of the enzyme which consists of six subunits. Phosphofructokinase activity is *increased* by ADP, AMP, Pi, fructose 1,6-bisP (only the alpha anomer), by cAMP, and by ammonium ions. Activity of the enzyme is *decreased* by intracellular acidosis, by ATP, phosphocreatine and citrate. The allosteric control of phosphofructokinase by citrate plays a role in the control of the selection of fuels by decreasing rates of glucose metabolism in the presence of fatty acids, ketone bodies, or pyruvate (51).

The oxidation of glyceraldehyde 3-phosphate to 1,3-diP-glycerate couples reduction of NAD^+ to the formation of a diphosphotriose. It is an energy-conserving step leading to the formation of ATP. The reaction is thought to be at near equilibrium; yet it was found that under conditions of O_2 deprivation or high rates of cardiac work (when PFK becomes strongly activated), glycolysis is controlled further down in the glycolytic pathway (23,55). In vitro the enzyme is strongly inhibited by its products 3-phospho-D-glyceroyl-P and, especially, NADH (33). According to Kobayashi and Neely (23), even in hearts well supplied with oxygen, maximal rates of glucose breakdown are limited by flux through glyceraldehyde 3-P dehydrogenase which is under the restraint of an increasing cytosolic [NADH]/[NAD] as the work load of the heart increases.

Pyruvate Oxidation. In heart muscle pyruvate may either be reduced to lactate (which completes the glycolytic pathway), transaminated to alanine, carboxylated to oxaloacetate or malate, or, most importantly, oxidized to acetyl-CoA. Lactate and most of the alanine are formed in the cytosol by near-equilibrium reactions and both metabolites may be washed out from the cell. In well-oxygenated, working heart muscle, the bulk of pyruvate enters, however, the mitochondrion.

The inner mitochondrial membrane represents a barrier to charged molecules. Specific transport mechanisms are required to enable transport of specific metabolites through the lipid bilayer, and during the past two decades an impressive number of carrier systems have been demonstrated for the transport of both anions and protons (for review, see ref. 29). A specific carrier exists for the transport of pyruvate into the mitochondrial matrix, as it has been most clearly demonstrated with the inhibitor 4-hydroxy-alpha-cyanocinnamate (for review, see ref. 15).

Inside the mitochondrial matrix, pyruvate may be either decarboxylated to acetyl-CoA or carboxylated to oxaloacetate. Although there is a considerable amount of pyruvate carboxylase activity present in heart muscle (59), most of pyruvate is decarboxylated. Oxidative decarboxylation of pyruvate assumes a central position in the regulation of fuel supply to the heart. A system of intricate control mechanisms governs both activation and inactivation of the pyruvate dehydrogenase (PDH) enzyme complex.

The conversion of pyruvate to acetyl-CoA requires the sequential action of three different enzymes: pyruvate decarboxylase, dihydrolipoyl transacetylase, and dihydrolipoyl dehydrogenase. The reaction also requires five different coenzymes or prosthetic groups. These are thiamine pyrophosphate, lipoic acid, CoASH, FAD and NAD. These enzymes and coenzymes are organized into a multienzyme cluster. The overall MW of the PDH complex is about 6 million dalton, i.e., the size of a ribosome. The PDH complex is attached to the inner side of the inner mitochondrial membrane.

PDH is also regulated by covalent modification through a *phosphorylation* and *dephosphorylation* cycle. Multisite phosphorylations provide an indirect means by which the rate and extent of the kinase reactions may regulate reactivation of the PDH complex by the phosphatase. Like most mammalian tissues, heart muscle possesses both active (dephosphorylated) and inactive (phosphorylated) dehydrogenase. The total activity of both together is approximately 30 U/g dry wt. at 30°C (48), of which normally about 20% is in the active form. The relative amount of active PDH increases with an increase in work load. Active PDH may decline to only 1-5% of total PDH during starvation or in alloxan diabetes (22), i.e., when non-carbohydrate substrates become the main fuel for respiration. Both kinase and phosphate phosphatase are probably active in vivo, and the relative proportion of active PDH must therefore be dependent on the relative activities of kinase and phosphatase as well as the intramitochondrial concentration of the effectors of these enzymes.

Most of the known effectors of the phosphorylation/dephosphorylation cycle are also effectors of the kinase reaction. Of the metabolite pairs ATP/ADP, acetyl-CoA/CoASH, NADH/NAD$^+$ and lactate/pyruvate, the first component either activates the kinase or serves as substrate, while the second component inhibits the enzyme. Ca^{2+} and Mg^{2+} both inhibit the kinase and activate the phosphatase reaction, i.e., lead to activation of PDH. The effects of oxidation of fatty acids or ketone bodies are likely to be mediated by the increase in acetyl-CoA they induce (22), and the primary effects of the inhibition/inactivation of PDH by fatty acids or ketone bodies is the acetyl-CoA/CoASH ratio (44). Conversely, an increase in cardiac work may inhibit the PDH kinase due to a fall in NADH, acetyl-CoA and ATP, leading to activation of PDH (23).

B. Fatty Acids

The importance of fatty acids as fuel for respiration and energy metabolism in the heart is easy to understand. First, fatty acids carry considerably more reducing equivalents per g molecular weight than glucose. Secondly, the energy stores of fatty acids in the body in the form of triglycerides (i.e., fatty acids esterified with glycerol) quantitatively by far exceed the energy stores of glucose (i.e., glycogen). Thus, triglycerides constitute the main fuel reserve in mammals. In times of food deprivation, fatty acids are mobilized and serve as the preferred fuel for respiration in most tissues (with the notable exception of nerve tissue and red blood cells).

The mechanism of fatty acid oxidation by beta-oxidation was discovered by Knoop in 1904 long before the glycolytic pathway and the citric acid cycle were known. Despite this early start and despite the importance of fatty acids as fuel for oxidative ATP production in heart muscle and many other tissues, the rate-controlling steps of fatty acid metabolism are not as well understood as those of carbohydrate metabolism.

The heart can utilize both complex lipids from the blood (14) as well as albumin-bound free fatty acids (60). In well-oxygenated heart muscle, the rate of utilization seems to depend chiefly on their plasma concentration and on the work load of the heart (36,45).

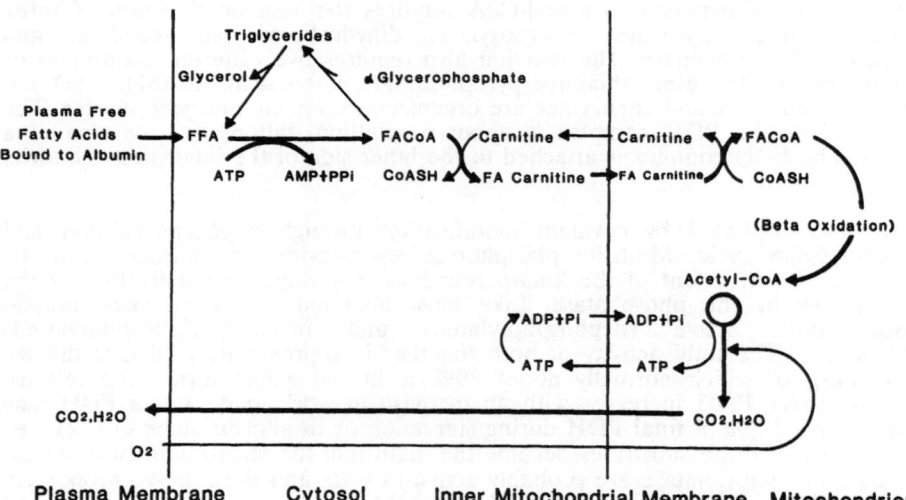

Fig. 4. Main pathways of long-chain fatty acid metabolism in heart muscle. Long-chain free fatty acids (FFA) are activated at the outside of the inner mitochondrial membrane and transported across the inner mitochondrial membrane by the carnitine-acyltransferase system. Inside the mitochondrion they undergo beta-oxidation to acetyl-CoA, which is oxidized in the citric acid cycle. FFA activated in the cytosol, but not transported into the mitochondrion, may be esterified with alpha-glycerophosphate to form triglycerides. Heavy lines indicate the main pathway of long-chain fatty acid metabolism in heart muscle. Fatty acids of short chain length do not utilize the carnitine-acyltransferase system and undergo beta-oxidation directly.

An overview of fatty acid oxidation in heart muscle is shown in Fig. 4. The pathway starts with liberation of fatty acids from triglycerides and/or albumin bound transport of free fatty acids (FFA) in the blood, and it ends with the entry of acetyl-CoA into the citric acid cycle. FFA of any chain length are able to cross the plasma membrane, and are activated on the outer mitochondrial membrane by esterificiation with CoASH to form fatty acyl-CoA. Metabolism of long-chain FFA continues with fatty-acyl transfer to carnitine, transport into the mitochondria in exchange for carnitine, reesterification with CoASH, beta-oxidation and, finally, oxidation of acetyl-CoA. A certain amount of FFA taken up by heart muscle is esterified with glycerol in the cytosol to form triglycerides. Thus, the fatty acyl-CoA formed on the outer mitochondrial membrane has two possible fates, either oxidation via the citric acid cycle in the mitochondria or conversion to triglycerides in the cytosol. Which pathway is taken depends on the rate of transfer of long-chain fatty acyl-CoA across the inner mitochondrial membrane and on the rate of esterification in the cytosol. Increased net triglyceride synthesis by the heart has been observed with starvation, diabetes and also with ischemia. It is not known whether this is the result of increased rates of esterification or a decreased rate of lipolysis.

Before the acyl group can be oxidized, it has to be moved from the outer mitochondrial membrane into the mitochondrion. Transfer of long-chain fatty acyl-CoA into mitochondria requires a three-step membrane transport process which is the rate-controlling step for long-chain fatty acid oxidation. The first step of the sequence is the transfer of the acyl group from CoA to carnitine (Fig. 4), catalyzed by the enzyme carnitine acyl-CoA transferase I (CAT I). The carnitine-acylcarnitine translocase system has been partially characterized in isolated heart mitochondria (46) and in the perfused heart (21). An exchange reaction moves acyl-carnitine across the inner mitochondrial membrane in a stoichiometric exchange for carnitine. In the liver, and most likely also in heart muscle, CAT I is the regulatory site for fatty acid oxidation with high concentrations of malonyl-CoA acting as inhibitor of the enzyme during the carbohydrate-fed state (30,31). Malonyl-CoA arises from carboxylation of acetyl-CoA in the cytosol. The presence of this compound in heart muscle has been demonstrated (32). The carnitine-acyl unit traverses through the inner mitochondrial membrane and is transferred to CoASH inside the mitochondrial matrix by a second carnitine acyl-CoA transferase (CAT II) located at the inner surface of the inner mitochondrial membrane.

Once inside the mitochondria, acyl-CoA is committed to oxidation by the system of beta-oxidation. The beta-oxidation pathway is controlled by fluctuations in the concentrations of substrates (acyl-CoA, NAD^+, and FAD^+; ref. 7), or, expressed in more physiological terms, by work load and oxygen supply of the heart. It is not yet clear how the rate of oxidation of FFA in the intact heart is related to citric acid cycle activity and the rate of oxidative phosphorylation.

C. Ketone Bodies

The ketone bodies, acetoacetate and beta-hydroxybutyrate, are produced by the liver, when supply of FFA is abundant and exceeds the liver's capacity for their complete oxidation in the citric acid cycle. Ketone bodies, like carbohydrates and fatty acids, share in the supply of acetyl-CoA for oxidation in the citric acid cycle and for energy production in heart muscle (72). Compared with the complexity of glucose and long-chain fatty acid metabolism, the metabolism of ketone bodies requires only a few steps, none of which appear to be subject to any regulation.

A unique feature of ketone body metabolism is the activation of acetoacetate to acetoacetyl-CoA by succinyl-CoA, which itself is a citric acid cycle intermediate. A few facts have emerged with respect to the regulation of ketone body metabolism in the heart. The enzymes degrading beta-hydroxybutyrate and acetoaceate to acetyl-

CoA, beta-hydroxybutyrate dehydrogenase, 3-oxoacid-CoA transferase and acetoacetyl-CoA thiolase, are of comparatively high activity in heart muscle (see refs. 2,71). Ketone bodies, when present in relatively high concentrations, spare glucose utilization by the heart. The ability of the heart to oxidize ketone bodies may constitute a buffer against qualitative changes in substrate supply when the relative proportions of oxidizable fuel are shifting as a result of changes in the dietary or hormonal state of the body (56). Even in physiological situations, such as short-term starvation or exercise, ketone body concentrations in the plasma can rise by up to 50-fold (65).

It is of interest to note that ketone bodies, when present as the only exogenous substrate for rat heart, do not sustain the full work output of the heart, and their rate of oxidation in the citric acid cycle is insufficient to meet the energy needs of the heart (64,67). This relative inhibition of the citric acid cycle occurs at the level of 2-oxoglutarate dehydrogenase. It is most likely related to sequestration of the enzyme's cofactor CoASH, which is bound in the form of its thioesters, acetyl-CoA, and as acetoacetyl-CoA, and thus not available for the reaction in the citric acid cycle (64). Whether or not this inhibition of the citric acid cycle is important in clinical situations such as diabetic cardiomyopathy, is a matter of speculation (68).

D. Amino Acids

Amino acids are not only the "building blocks" of myocardial proteins, but, through their metabolism, they are also an integral part of myocardial energy metabolism. It cannot be emphasized too strongly that each amino acid is different from the others with respect to control of its metabolism. Many amino acids are either degraded or synthesized by heart muscle (e.g., glutamate, glutamine, aspartate, alanine, leucine) while others are not metabolized at all (e.g., phenylalanine, tyrosine).

In general, transamination is the first step of amino acid degradation in heart muscle. One of the functions of amino-transferases in heart muscle (for comparative data, see ref. 26) is the provision of carbon skeletons for the citric acid cycle. The amino acids aspartate and glutamate also play an important role in the transfer of reducing equivalents across the mitochondrial membrane for oxidation of cytosolic NADH by the mitochondrial electron transport chain (6), as shown in Fig. 5. Current evidence suggests that in heart muscle, and similarly in liver, reducing equivalents are indirectly carried into the mitochondrial compartment by metabolic anions of the malate-aspartate cycle (57,58). In the malate-aspartate cycle, cytosolic NADH, which

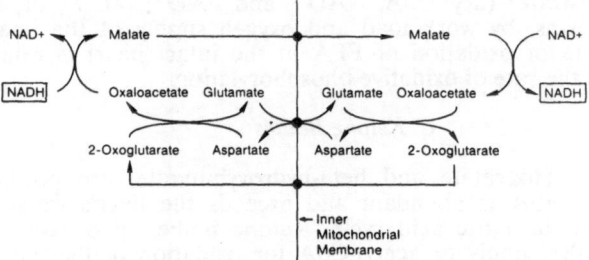

Fig. 5. The malate-aspartate shuttle, an important pathway of cytosolic NADH oxidation. Cytosolic NADH arises from the glyceraldehyde 3-P dehydrogenase reaction and can be reoxidized by either the alpha-glycerophosphate dehydrogenase, the malate dehydrogenase or lactate dehydrogenase reaction. The first two reactions are concerned with the transfer of reducing equivalents into the mitochondrial matrix. Quantitatively the most important mechanism for transfer of reducing equivalents is the malate-aspartate shuttle which involves transamination of aspartate and glutamate, as depicted.

cannot cross the inner mitochondrial membrane, is oxidized by the reduction of oxaloacetate to malate. Malate enters the mitochondrion and is oxidized to oxaloacetate, which is transaminated with glutamate to form 2-oxoglutarate and aspartate. Aspartate and 2-oxoglutarate leave the mitochondrion. Transamination of these metabolites regenerates oxaloacetate and glutamate in the cytosol, the net effect being a transfer of hydrogen ions across the mitochondrial membrane. A second postulated function for the malate-aspartate shuttle is the delivery of inter-mediates (malate) to the citric acic cycle, acting as a modulator of cycle activity.

Myocardial amino acid metabolism during hypoxia, ischemia and reperfusion has been the subject of a number of studies in the intact and isolated heart muscle (5,66,69). The amino acid alanine was found to be, like lactate, an end product of anaerobic glucose breakdown, arising through transamination of glutamate in the reaction:

$$\text{glutamate} + \text{pyruvate} \longrightarrow \text{2-oxoglutarate} + \text{alanine}.$$

This reaction is not the end of the pathway. While alanine leaves the cell, 2-oxoglutarate is further metabolized and decarboxylated to succinate via substrate level phosphorylation in the citric acid cycle (63).

6. Competition and Selection of Energy-Providing Fuels

While the heart plays only a minor role in the overall fuel balance of the whole body, it obeys the same rules of metabolic regulation as other tissues do. Specifically, it has been known for a good numner of years that in the fasting animal the energy requirements of the heart are met largely by the oxidation of fatty acids with an respiratory quotient (R.Q.) of 0.7 (18), and that after carbohydrate feeding the heart switches its preference to glucose with an increase of the R.Q. to about 0.9 to 1.0 (19).

It is possible to make a number of assumptions from data available in the literature. As already stated, under aerobic conditions the heart is able to derive energy from a variety of fuel sources such as FFA, ketone bodies, glucose, lactate, and even certain amino acids. Depending on the physiological state of the individual, only plasma glucose levels are relatively constant, whereas levels of lactate, FFA and ketone bodies may vary over a wide range. Thus, it is not surprising that the fuel for aerobic energy production in heart muscle varies with the plasma concentration of individual substrates.

Heart muscle itself is also capable of contributing to the fuels of respiration by releasing both FFA and glucose from their stores of triglycerides and glycogen. The most striking example of endogenous fuel utilization occurs with ischemia, when glycogen is broken down to lactate (see, e.g., ref. 37). Thus, with regard to endogenous fuel consumption, we have the situation that during normoxia and normal coronary flow, exogenous fuels provide the bulk of the fuel for respiration, whereas with ischemia and restricted coronary flow, the endogenous fuel glycogen contributes a large portion of anaerobic energy production through metabolism to pyruvate, lactate and alanine.

Fuel selection by normoxic heart muscle continues to be an intriguing question. The *first factor* which affects the utilization of an individual substrate in a mixture of substrates is its plasma concentration. Glucose concentrations are relatively constant, whereas concentrations of fatty acids, ketone bodies or lactate can fluctuate. Accordingly, FFA have been identified as the preferred fuel for respiration by the heart especially when their plasma concentration was high after an overnight fast (54). Likewise, lactate has been identified as a preferred fuel by the heart when lactate concentrations were high (16).

A *second factor* which determines selection of fuels by the heart is the ease of access of fuels to the enzymes of oxidative metabolism in the mitochondrion. There are fewer control steps involved in the metabolism of lactate or acetoacetate to acetyl-CoA than in the metabolism of either fatty acids or glucose to acetyl-CoA. Thus, when present in sufficient concentrations, both lactate or ketone bodies may contribute a large amount of acetyl-CoA for oxidation in the citric acid cycle.

A *third factor* which determines the selection of fuels by the heart is its O_2 supply and the specific P/O ratio (i.e., moles of ADP phosphorylated per mole of O_2 consumed) of fuels. For instance, one mole of glucose, when oxidized, uses 6 moles of O_2 and yields 36 moles of ATP (P/O ratio = 3.0), whereas the oxidation of one mole of palmitate utilizes 23 moles of O_2 and yields 129 moles of ATP (P/O ratio = 2.8). Thus, glucose produces about 8% more ATP per mole of oxygen than do FFA. Although this may not seem a very large difference, an improved P/O ratio may become important for the heart which is extracting very large amounts of oxygen. In situations of limited O_2 reserve, oxidation of glucose can provide a greater yield of ATP. Consequently, the heart may then prefer glucose over FFA as fuel for respiration. Inhibition of the oxidation of a long-chain fatty acid derivative (palmitoylcarnitine) by pyruvate has recently been found in rat-heart mitochondria (8). Suppression of FFA oxidation by glucose and/or pyruvate is different from the situation encountered in true ischemia, where the O_2 supply is completely insufficient to meet the energy needs of cells and rates of anaerobic glycolysis are necessarily increased.

A *fourth factor* which determines the selection of fuels is the interaction of *intermediary metabolites* arising in the course of degradation of energy-providing substrates with regulatory steps in the pathway of energy conversion of substrates. Such regulatory steps have been described earlier in the discussion of glucose and fatty acid metabolism. For example, the mechanism by which FFA suppress glucose utilization has been delineated in the perfused heart. Addition of fatty acids, ketone bodies or pyruvate cause a major depression in glucose utilization through coordinated inhibition at the four control sites: glucose transport, hexokinase, phosphofructokinase, and pyruvate dehydrogenase by mechanisms already discussed. The opposite question, i.e., how FFA are replaced by glucose as an energy source when the starved organism is refed, has not yet been addressed. In recent experiments with a perfused working heart preparation, capable of sustaining a near-physiological work output in vitro for more than 1 h, glucose (5 mM) suppressed oleate (1 mM) removal by 35% (67). These effects of glucose on FFA metabolism are relatively large and of special interest in light of measurable amounts of the carnitine acyltransferase inhibitor malonyl-CoA in heart muscle from fed animals and its decline in heart muscle from starved animals (32). The exact mechanism of glucose inhibition of FFA metabolism is still unknown.

The *fifth factor* which determines the selection of fuels is *hormone levels*. Plasma insulin concentration, and probably the insulin/glucagon ratio, play a major role in the fuel preferences of tissues. Insulin regulates glucose entry into muscle, adipose tissue and other cells. Likewise, the rate of adipose tissue lipolysis is under hormonal control. Randle et al. (50) have integrated these observations in the glucose-fatty acids cycle: glucose oxidation is inhibited by FFA when FFA are elevated (low insulin, fasting), and FFA release is inhibited by glucose uptake (high insulin, fed).

7. Control of the Citric Acid Cycle

The function of the citric acid cycle in the heart is to oxidize the acetyl-group of acetyl-CoA to CO_2 in the process of generating reducing equivalents. Reduction of the electron carriers NAD^+ and FAD requires addition of H_2O, as can be easily seen from the net reaction of the cycle

$$CH_3 - CO \sim SCoA + 3 H_2O \longrightarrow 2 CO_2 + 4 [2H] + CoASH,$$

where [2H] represents a pair of reducing equivalents removed to reduce an electron carrier. It is reasonable to assume that the H_2O added to the cycle arises from reduction of molecular O_2 in the respiratory chain. The production of reducing equivalents from H_2O is probably the most important function of the citric acid cycle, in addition to the other functions such as the generation of intermediates like citrate (the metabolic signal which regulates phosphofructokinase) and the disposition of the products of carbohydrate, FFA and amino acid metabolism.

There are three non-equilibrium enzymes in the citric acid cycle: citrate synthase, isocitrate dehydrogenase (both the NAD^+- and the $NADP^+$-dependent isozyme) and 2-oxoglutarate dehydrogenase. Under physiological conditions, oxidation of acetyl-CoA in the citric acid cycle is directly related to the rate of ATP production by the oxidative phosphorylation of ADP in the respiratory chain. Not surprisingly, in isolated working rat hearts perfused with glucose as the only substrate, the tissue content of acetyl-CoA is so low that it cannot be measured (64). Hence, the citric acid cycle, as the main source for reducing equivalents, must be under a strict control between supply and demand, because the rate of acetyl-CoA production and oxidation has to respond instantly to a wide range of energy needs.

Control is exercised by the tight relationship between electron transport in the respiratory chain and ATP synthesis. The major determinants of mitochondrial respiratory rate are extramitochondrial [ATP]/[ADP][P_i], intramitochondrial [NAD^+]/[NADH] and O_2 tension. If ATP synthesis cannot occur (e.g., when the [ATP]/[ADP] ratio is high), the reducing equivalents NADH and $FADH_2$ will not be oxidized by the respiratory chain. A very simple mechanism of control of the citric acid cycle is, therefore, that the lack of oxidized cofactors (NAD^+ and FAD) will slow down those reactions of the citric acid cycle which depend on coupled oxidation and reduction. This feedback mechanism prevents unnecessary utilization of fuels when ATP supplies are adequate.

The enzymes of the citric acid cycle are located in the mitochondrial matrix or inner mitochondrial membrane in close proximity to the enzymes of the respiratory chain. However, unlike the respiratory chain or the glycolytic pathway, the citric acid cycle is not a linear pathway. According to Baldwin and Krebs (1), the advantage of a cyclic process is that it can offer opportunities for regulation which a linear pathway cannot provide. Unfortunately, methods of conventional metabolic analysis (such as measuring accumulation of metabolites) frequently fail to elucidate control mechanisms in a cyclic process. Furthermore, understanding of the control of the citric acid cycle is also complicated by lack of knowledge about metabolite concentrations in mitochondria.

Earlier investigators (see, e.g., refs. 27,28) have pointed out that the synthesis of citrate, catalyzed by citrate synthase in the reaction

$$oxaloacetate + acetyl\text{-}CoA + H_2O \longrightarrow citrate + CoASH,$$

must be a rate-limiting step, because subsequent metabolites in the citric acid cycle do not normally accumulate. This means that they are removed as rapidly as they are produced or that their rate of removal is limited by their rate of supply. In other words, the citrate synthase reaction is a non-equilibrium reaction which initiates the flux to which other participants in the pathway must respond in order to achieve a steady state. The assumption that citrate synthase is rate limiting has been supported by the discovery of several feedback inhibitors of the enzyme (see ref. 62 for review). Citrate synthase is also inhibited by several products of the pathway which citrate synthase initiates, such as NADH, ATP, or succinyl-CoA. However, for a number of reasons it is unlikely that the in vitro findings also account for the regulation of citrate synthase in the intact, functioning heart muscle. Regulation of citrate synthase by NADH may become clearer when one considers the citric acid cycle as a whole. The overall result of the citric acid cycle (or the oxidation of acetate by O_2) is the obligatory phosphorylation of ADP by the respiratory chain.

The key reactions for the cycle cause a reduction of the coenzymes of the respiratory chain. When the relevant products (three $NADH_2$ and one reduced flavoprotein) enter the respiratory chain, H^+ atoms are transferred from the coenzymes to molecular O_2, coupled with the synthesis of ATP. Thus, the *primary* end products of the cycle reactions are $NADH_2$ and reduced flavoprotein, and the *final* product of the cycle and its associated reactions as a whole is ATP.

The obligatory coupling of the oxidation of reduced coenzymes with the phosphorylation of ADP (i.e., synthesis of ATP from ADP and P_i) means that the [ATP] can indirectly regulate the dehydrogenase steps by controlling the [ADP] and [P_i]. This type of control would apply to all mitochondrial dehydrogenases; but it would appear that it cannot regulate the initiating step of the cycle, the citrate synthase reaction, because this is not a dehydrogenase reaction. Thus, citrate synthase seems to be the only important enzyme in the cycle not directly regulated by the availability of ADP and P_i for oxidative phosphorylation.

Certain regulatory mechanisms exist within the cycle when the heart switches from the oxidation of glucose to the oxidation of FFA or ketone bodies. Randle et al. (49) first drew attention to the fact that during the oxidation of acetate plus glucose the cycle transiently operates in two spans, the first span controlled by citrate synthase (and the availability of oxaloacetate), and the second span controlled by 2-oxoglutarate dehydrogenase. This enzyme is a multi-enzyme complex which bears a certain similarity to pyruvate dehydrogenase in its components and reaction sequence. It is also inhibited by its products, succinyl-CoA and NADH (17), and it requires Ca^{2+}, but, unlike PDH, it is not regulated by a phosphorylation-dephosphorylation cycle. Regulation of 2-oxoglutarate dehydrogenase is thought to be of crucial importance in the accumulation of citrate, induced in rat heart by the oxidation of FFA and ketone bodies (52).

The experiments of Randle et al. (49) have shown that flux through citrate synthase is strictly substrate dependent. As more acetyl-CoA becomes available, more oxaloacetate is formed by transamination of aspartate with 2-oxoglutarate. As a result, the tissue content of aspartate falls and the tissue content of glutamate rises. More recent experiments by Hassinen's group have shown that at least part of the oxaloacetate is also derived from carboxylation of pyruvate (43,47).

Regulation of flux through 2-oxoglutarate dehydrogenase is complex and still not well understood. In addittion to product inhibition, there are at least two additional factors which need to be taken into account: a) the overall activity of the enzyme, which was found to be directly correlated with flux through the citric acid cycle (12), and b) availability of free CoASH, a cofactor for the reaction (64).

8. Summary and Physiological Implications

The effects of cardiac work on cardiac metabolism are mediated by a feedback system which keeps energy production in step with energy utilization on a beat-to-beat basis. The effects of cardiac work have already been considered in conjunction with individual fuels for respiration. These may be summarized as follows: Within a physiological range, an increase in the work load of the heart leads to: a) an increase in work output, b) an increase in O_2 consumption, c) an increase in citric acid cycle turnover, d) an increase in flux through the Embden-Meyerhoff pathway, and/or e) an increase in beta-oxidation of fatty acids. The integration of this sequence of events is still not completely understood, but it is assumed that the decline in the phosphorylation state of adenine nucleotides, which raises the extra-mitochondrial [ADP] as a consequence of increased ATP utilization, is the central event.

The phasic mechanical activity of the heart necessitates a continuous supply of nutrients and O_2 for energy production. Heart muscle possesses the ability to choose between a number of different energy-providing substrates circulating in the blood stream, but the mechanisms of fuel selection are not entirely clear as yet. Balance calculations indicate that the energy *value* in the form of reducing equivalents is greater for glucose than for fatty acids, because of the greater P/O ratio of the former.

The selection of fuels rests within a number of regulatory enzymes whose kinetic properties, as discussed earlier, are controlled by factors other than substrate concentrations and product removal. There is good reason to assume that a few regulatory enzymes are responsible for alterations in flux through any metabolic pathway and thus provide for the most economical selection of fuel in a given physiological situation.

To date there are few, if any, examples of in situ measurements of regulatory enzymes relative to flux through the pathway. A rare example of the integrative control of metabolism, neural activity and cardiac function has been reported by Skinner et al. (61). The authors collected multiple freeze-biopsy samples from the myocardium of the conscious pig and demonstrated an increase in activated phosphorylase with repeated stressful situations. The increase declined with time, as the animals became used to the situation. Data like these are, however, few, and it is hoped that more will be learned on the integrative control of cardiac metabolism in the future. It needs to be stressed once more that control of flux through a metabolic pathway cannot be achieved by any single enzyme acting in isolation and that regulation is a shared property of many different intra- and extracellular effectors.

Acknowledgement

The author is the recipient of a USPHS Research Career Development Award (NHLBI, No. 5 K04 HL01246).

References

1. Baldwin S, Krebs HA: The evolution of metabolic cycles. Nature 291:381-382, 1981
2. Beis A, Sammit VA, Newsholme EA: Activities of 3-hydroxybutyrate dehydrogenase, 3-oxoacid CoA-transferase and acetoacetyl-CoA thiolase in relation to ketone body utilization in muscles fron vertebrates and invertebrates. Eur J Biochem 104:209-215, 1980
3. Bing RJ: The metabolism of the heart. Harvey Lectures 50:27-70, 1955
4. Bing RJ, Siegel A, Ungar I, Gilbert M: Metabolism of the human heart. 2. Studies on fat, ketone and amino acid metabolism. Am J Med 16:504-515, 1954
5. Bittl JA, Shine KI: Protection of ischemic rabbit myocardium by glutamic acid. Am J Physiol 245:H406-H412, 1983
6. Borst P: Hydrogen transport and transport of metabolites. In: Karlson P, ed: Funktionelle and morphologische Organisation der Zelle. Berlin: Springer Verlag, 1963:137-158
7. Bremer J, Wojtczak AB: Factors controlling the role of fatty acid beta-oxidation in rat liver mitochondria. Biochim Biophys Acta 280:515-530, 1972
8. Brosnan JT, Reid K: Inhibition of palmitoylcarnitine oxidation by pyruvate in rat heart mitochondria. Metabolism 34:588-593, 1985
9. Chance B, Williams GR: Respiratory enzymes in oxidative phosphorylation. J Biol Chem 217:382-397, 1955
10. Cohen P: Control of enzyme activity. London: Chapman & Hall, 1976

11. Cohen P: The hormonal control of glycogen metabolism in mammalian muscle by multisite phosphorylation. Biochem Soc Trans 7:459-480, 1979
12. Cooney GJ, Taegtmeyer H, Newsholme EA: Tricarboxylic acid cycle flux and enzyme activities in the isolated working rat heart. Biochem J 200:701-703, 1981
13. Cori GT, Colowick SP, Cori CF: The action of nucleotides in the disruptive phosphorylation of glycogen. J Biol Chem 123:381-389, 1938
14. Cruickshank EWH, Kosterlitz HW: The utilization of fat by the aglycaemic mammalian heart. J Physiol (Lond) 99:208-223, 1941
15. Denton RM, Halestrap AP: Regulations of pyruvate metabolism in mammalian tissues. Essays Biochem 15:37-77, 1979
16. Drake AJ, Haines JR, Noble MIM: Preferential uptake of lactate by the normal myocardium in dogs. Cardiovasc Res 14:65-72, 1980
17. Garland PB: Some kinetic properties of pig heart oxoglutarate dehydrogenase that provide a basis for metabolic control of the enzyme activity. Biochem J 92:10C-11C, 1964
18. Goodale WT, Hackel DB: Myocardial carbohydrate metabolism in normal dogs, with effects of hyperglycemia and starvation. Circ Res 1:509-517, 1953
19. Goodale WT, Olson RE, Hackel DB: The effects of fasting and diabetes mellitus on myocardial metabolism in man. Am J Med 27:212-220, 1959
20. Hom FG, Goodner CJ, Berrie MA: A (3H)2-deoxyglucose method for comparing rates of glucose metabolism and insulin repsonses among rat tissues in vivo. Diabetes 33:141-152, 1984
21. Idell-Wenger JA, Neely JR: Regulation of uptake and metabolism of fatty acids by muscle. In: Dietschry JM, Gotto AM Jr, Omtko JA, eds: Disturbances in lipid and lipoprotein metabolism. Bethesda: American Physiological Society, 1978:269-284
22. Kerbey AL, Randle PJ, Cooper RH, Whitehouse S, Pask HT, Denton RM: Regulation of pyruvate dehydrogenase in rat heart. Biochem J 154:327-348, 1976
23. Kobayashi K, Neely JR: Control of maximum rates of glycolysis in rat cardiac muscle. Circ Res 44:166-175, 1979
24. Krebs EG, Fischer EH: The phosphorylase b to a converting enzyme of rabbit skeletal muscle. Biochim Biophys Acta 20:150-157, 1956
25. Krebs HA: Control of metabolic processes. Endeavour 16:125-132, 1957
26. Krebs HA: Some aspects of the regulation of fuel supply in omnivorous animals. Adv Enzyme Regul 10:397-420, 1972
27. Krebs HA, Lowenstein JM: The tricarboxylic acid cycle. In: Greenberg DM, ed: Metabolic pathways. Vol 1. New York: Acad Press, 1960:129-202
28. LaNoue KF, Bryla J, Williamson JR: Feedback interactions in the control of citric acid cycle activity in rat heart mitochondria. J Biol Chem 247:667-679, 1972
29. LaNoue KF, Schoolwerth AC: Metabolite transport in mitochondria. Annu Rev Biochem 48:871-922, 1979
30. McGarry JD, Foster DW: Regulation of hepatic fatty acid oxidation and ketone body production. Annu Rev Biochem 49:395-420, 1980
31. McGarry JD, Leatherman GF, Foster DW: Carnitine palmitoyl transferase I. The site of inhibition of hepatic fatty acid oxidation by malonyl-CoA. J Biol Chem 253:4128-4136, 1978
32. McGarry JD, Mills SE, Long CS, Foster DW: Observation on the affinity for carnitine and malonyl-CoA sensitivity of carnitine palmitoyl transferase I in animal and human tissues. Demonstration of the presence of malonyl-CoA in nonhepatic tissues of the rat. Biochem J 214:21-28, 1983
33. Mochizuki S, Neely JR: Control of glyceraldehyde-3-phosphate dehydrogenase in cardiac muscle. J Mol Cell Cardiol 11:221-236, 1979
34. Morgan HE, Henderson MJ, Regen DM, Park CR: Regulation of glucose uptake in muscle. 1. The effects of insulin and anoxia on glucose transport and phosphorylation in the isolated, perfused heart of normal rat. J Biol Chem 236:253-261, 1961
35. Morgan HE, Neely JR: Insulin and membrane transport. In: Steiner DF, Freinkel N, eds: Handbook of physiology. Washington: American Physiological Society, 1972:323-331
36. Neely JR, Bowman RH, Morgan HE: Effects of ventricular pressure development and palmitate on glucose transport. Am J Physiol 216:804-811, 1969
37. Neely JR, Grotyohann LW: Role of glycolytic products in damage to ischemic myocardium. Dissociation of ATP levels and recovery of function of reperfused ischemic hearts. Circ Res 55:816-824, 1984

38. Neely JR, Liebermeister H, Battersby EJ, Morgan HE: Effeacts of pressure development on oxygen consumption by isolated rat heart. Am J Physiol 212:804-814, 1967
39. Neely JR, Liebermeister H, Morgan HE: Effect of pressure development on membrane transport of glucose in isolated rat heart. Am J Physiol 212:815-822, 1967
40. Neely JR, Morgan HE: Relationship between carbohydrate and lipid metabolism and the energy balance of heart muscle. Annu Rev Physiol 36:413-459, 1974
41. Newsholme EA, Crabtree B: Theoretical principles in the approach to control of metabolic pathways and their application to glycolysis in muscle. J Mol Cell Cardiol 11:839-856, 1979
42. Newsholme EA, Leech AR: Biochemistry for the medial sciences. Chichester: Wiley, 1983
43. Nuutinen EM, Peuhkurinen KJ, Pietilainen EP, Hiltunen JK, Hassinen IE: Elimination and replenishment of tricarboxylic acid cycle intermediates in myocardium. Biochem J 194:867-875, 1981
44. Olson MS, Dennis SC, DeBuysere MS, Padma A: The regulation of pyruvate dehydrogenase in the isolated perfused rat heart. J Biol Chem 253:7369-7375, 1978
45. Oram JF, Bennetch SL, Neely JR: Regulation of fatty acid utilization in isolated perfused rat hearts. J Biol Chem 248:5299-5309, 1973
46. Pande SV, Parvin R: Pyruvate and acetoacetate transport in mitochondria: a reappraisal. J Biol Chem 253:1565-1573, 1978
47. Peuhkurinen KJ, Hassinen IE: Pyruvate carboxylation as an anaplerotic mechanism in the isolated perfused rat heart. Biochem J 202:67-76, 1982
48. Randle PJ: Regulation of glycolysis and pyruvate oxidation in cardiac muscle. Circ Res 38, Suppl I: I8-I12, 1976
49. Randle PJ, England PJ, Denton RM: Control of the tricarboxylate cycle and its interactions in rat heart. Biochem J 117:677-695, 1970
50. Randle PJ, Garland PB, Hales CN, Newsholme EA: The glucose fatty-acid cycle. Its role in insulin sensitivity and the metabolic disturbances of diabetes mellitus. Lancet 1:785-789, 1963
51. Randle PJ, Garland PB, Hales CN, Newsholme EA, Denton RM, Pogson CI: Interactions of metabolism and the physiological role of insulin. Rec Progr Hormone Res 22:1-41, 1966
52. Randle PJ, Tubbs PK: Carbohydrate and fatty acid metabolism. In: Berne RM, Sperelakis N, Geiger SR, eds: The handbook of physiology. Vol I, Sect 1. The cardiovascular system. Bethesda: American Physiological Society, 1979:805-844
53. Robison GA, Butcher RW, Sutherland EW: Cyclic AMP. London: Acad Press, 1971
54. Rothlin ME, Bing R: Extraction and release of individual free fatty acids by the heart and fat deposits. J Clin Invest 40:1380-1385, 1961
55. Rovetto MJ, Lamerton WF, Neely JR: Mechanisms of glycolytic inhibition in ischemic rat hearts. Circ Res 37:742-751, 1975
56. Rudolph W, Maas D, Ritcher J, Hasinger F, Hoffman H, Dohm P: Ueber die Bedeutung von Acetoacetat und beta-Hydroxybutyrat im Stoffwechsel des menschlichen Herzens. Klin Wschr 43:445-451, 1965
57. Safer B, Smith CM, Williamson JR: Control of the transport of reducing equivalents across the mitochondrial membrane in perfused rat heart. J Mol Cell Cardiol 2:111-124, 1971
58. Safer B, Williamson JR: Mitochondrial-cytosolic interactions in perfused rat heart. J Biol Chem 248:2570-2579, 1973
59. Scrutton MC: Regulation of metabolism in complex organisms. In: Kent PW, ed: New approaches to genetics. London: Oriel Press, 1978:157-178
60. Shipp JC, Opie LH, Challoner D: Fatty acid and glucose metabolism in the perfused heart. Nature 189:1018-1019, 1961
61. Skinner JE, Bedes SD, Entman ML: Psychological stress activates phosphorylase in the heart of conscious pig without increasing heart rate and blood pressure. Proc Natl Acad Sci (USA) 80:4513-4517, 1983
62. Srere PA: Controls of citrate synthase activity. Life Sci 15:1695-1710, 1974
63. Taegtmeyer H: Metabolic responses to cardiac hypoxia. Increased production of succinate by rabbit papillary muscles. Circ Res 43:808-815, 1978
64. Taegtmeyer H: On the inability of ketone bodies to serve as the only energy providing substrate for rat heart at physiological work load. Basic Res Cardiol 78:435-450, 1983
65. Taegtmeyer H: Six blind men explore an elephant: Aspects of fuel metabolism and the control of tricarboxylic acid cycle activity in heart muscle. Basic Res Cardiol 79:322-336, 1984

H. Taegtmeyer

66. Taegtmeyer H, Ferguson AG, Lesch M: Protein degradation and amino acid metabolism in autolyzing rabbit myocardium. Exp Mol Pathol 26:52-62, 1977
67. Taegtmeyer H, Hems R, Krebs HA: Utilization of energy-providing substrates in the isolated working rat heart. Biochem J 186:701-711, 1980
68. Taegtmeyer H, Passmore JM: Defective energy metabolism of the heart in diabetes. Lancet 1: 139-141, 1985
69. Taegtmeyer H, Peterson MB, Ragavan VV, Ferguson AG, Lesch M: De novo alanine synthesis in isolated oxygen deprived rabbit myocardium. J Biol Chem 252:5010-5018, 1977
70. Ungar I, Gilbert M, Siegel A, Blain JM, Bing RJ: Studies on myocardial metabolism. IV. Myocardial metabolism in diabetes. Am J Med 18:385-396, 1955
71. Williamson DH, Bates MW, Page MA, Krebs HA: Activities of enzymes involved in acetoacetate utilization in adult mammalian tissues. Biochem J 121:41-47, 1971
72. Williamson JR, Krebs HA: Acetoacetate as fuel of respiration in the perfused rat heart. Biochem J 80:540-542, 1961

Chapter 4

Carnitine and Cardiac Energy Supply

R. Ferrari, Chair of Cardiology,
University of Brescia, Brescia, Italy

The purpose of this survey is to briefly consider some metabolic aspects of the human heart. The pathways of lipid utilization for energy production have received special attention. The function of L-carnitine in mitochondrial fatty acid oxidation and its role in myocardial energy production is analysed. Mention is also made of the effects of L-carnitine on myocardial metabolism of patients with coronary artery disease, either at rest or during pacing-induced myocardial ischaemia.

Introduction

The energy needed for cardiac contraction comes from the breakdown of chemical substances, the substrates. In the heart this process occurs mainly aerobically, i.e., substrates are combined with oxygen to form water, carbon dioxide and energy. The amount of oxygen consumed is often used as a measure of the total energy available. The ultimate fate of the liberated energy is to take the form of heat. However, before this happens completely, part of the energy is used for contraction, a process during which the heart may perform external work, as is usually the case in the intact body.

Substrate Metabolism of Human Heart

Knowledge of the fuels of the human heart started with the introduction of coronary sinus catherization by Bing and his associates in 1947. The chemical composition of arterial blood entering the heart was compared with that of coronary sinus blood leaving the heart. From such studies it was established that glucose, lactate and fatty acids are the heart's major sources of energy (3,4). It has also been recognized that the myocardium is able to utilize such fuels as pyruvate, acetate, ketone bodies and amino acids, but the normal circulating levels are too low for them to be considered as major sources of energy, even when the external supply is raised (8). The hearts can also use, under certain circumstances, its internal energy stores such as glycogen and lipid (32).

A cruder estimate of the type of substrate used by the heart can be achieved by the respiratory quotient, which is calculated by comparing the rate of oxygen uptake with the rate of production of carbon dioxide. A respiratory quotient near to one implies oxidation of glucose and/or lactate, whereas a lower value implies fatty acid oxidation. Because the myocardial respiratory quotient was frequently low, early workers were alerted to the importance of the role of lipids as the major myocardial fuels (24). However, there is still some controversy about which is the preferred myocardial substrate.

Substrate Preference

It is often stated that lipids are the preferred myocardial substrates (3,23,33). However, most studies of myocardial utilization of unesterified fatty acids (NEFA) in humans have been undertaken in resting fasting subjects. In this state the circulating NEFA levels are high (over 1000 uEq/L). Thus, they may provide more than 60% of the myocardial energy requirements (2,4,27). But if the energy requirements are satisfied by the supply of carbohydrates, NEFA extraction will be decreased (1). For example, if there is a rise in the arterial lactate, as it occurs during exercise, there is also an increase in the myocardial utilization of lactate (2,4,21,28). Thus, as it is also shown in Fig. 1, myocardial substrate preference is critically dependent on the availability of substrates in the circulation at the moment of the study (12). It is therefore crucial that substrate utilization is studied when all the substrates are available in adequate concentration. Such studies show that in the heart carbohydrates and lipids are concerned predominantly with the provision of ATP for contraction. Oxidation of carbohydrate substrates accounts for up to 35% of oxygen consumption. Extraction of NEFA may account for about 70% of the oxygen consumption in the post-absorptive state. The human heart, like that of the rat, exhibits a predilection for fatty acids as respiratory fuel (3).

As the heart derives most of its energy from oxidative metabolism, enthalpy changes can be taken to be equivalent to the oxygen consumption. The amount of heat liberated per gram of fat oxidized about (9 kcal/g) is more that twice that per gram of carbohydrate about (4 kcal/g). Nevertheless, the enthalpies are quite similar, when expressed per liter of oxygen consumed (fat 4.69 kcal/g O_2; carbohydrate 5.05 kcal/L O_2). This is because more oxygen is required to oxidize a gram of fat than a gram of carbohydrate.

Pathways of Substrate Utilization and Energy Production

The overall process of energy production involves the complete oxidation of substrates to carbon dioxide and water (Fig. 2). Energy (as ATP) can be generated in the myocardium by two metabolic pathways, glycolysis (anaerobic) and oxidative phosphorylation (aerobic). Under normal conditions the majority (90%) of ATP is produced by the latter pathway (20).

Before a carbohydrate (such as glucose) can enter the tricarboxylic acid cycle, it must be broken down via the glycolytic pathway. Therefore oxidation of glucose is dependent on the enzymes of anaerobic glycolysis. Lactate, the other major carbohydrate fuel, is dependent on the activity of the lactate dehydrogenase complex.

Although the heart is able to synthesize structural lipids for incorporation into the membranes, the most likely fate of a fatty acid molecule taken up by the heart is oxidation. The first step in fatty acid oxidation is their activation to form acyl-coenzyme A derivatives. The enzymes first concerned with the use of fatty acids

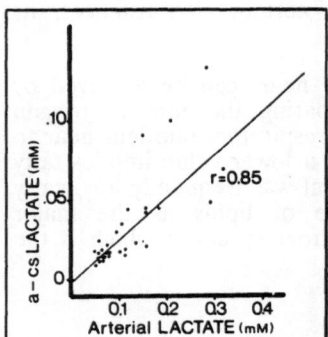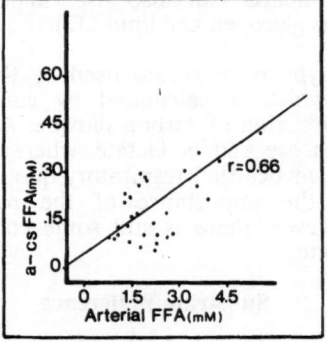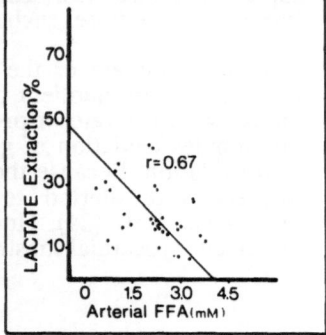

Fig. 1. Relationship between arterial substrate concentration and myocardial arteriovenous difference or extraction of free fatty acids (FFA) and lactate. a = arterial; cs = coronary sinus. Data were obtained in catheterized patients (12).

are therefore the acyl-coenzyme A synthetases which exist in the heart with different fatty acid specificities and locations (16). Short-chain fatty acids (C_8 or less) are oxidized by heart mitrochondria in the absence of external coenzyme A or of L-carnitine, the role of which is discussed later in this chapter. This is because the relevant acyl-coenzyme A synthetases are in the same compartment (mitochondrial matrix) as the enzymes of beta-oxidation and the citrate cycle. However, for the oxidation of long-chain acids, this activation must be followed by the transfer of long-chain acyl-coenzyme A into the interior of the mitochondria.

It is rather curious that the long-chain fatty acids that are physiologically really important, are converted to their coenzyme A derivatives on the wrong side of the mitochondrial inner membrane, which is impermeable to coenzyme A. This is particularly strange because the activation of fatty acid is accomplished by two distinct enzymes that share the same cytosolic coenzyme A pool, which is very small. One of these enzymes, representing 80% of the total activity, is attached to the mitochondrial outer membrane, and the other to the endoplasmic reticulum (7).

Function of L-Carnitine in Mitochondrial Fatty Acid Oxidation

Carnitine (L-3-hydroxy-4-trimethylaminobutyric acid) was discovered in 1905 as an abundant constituent of muscle (17). Its metabolic function remained completely obscure for over 50 years. Only in 1959 Fritz (15) showed that carnitine increased long-chain fatty acid oxidation in liver and heart.

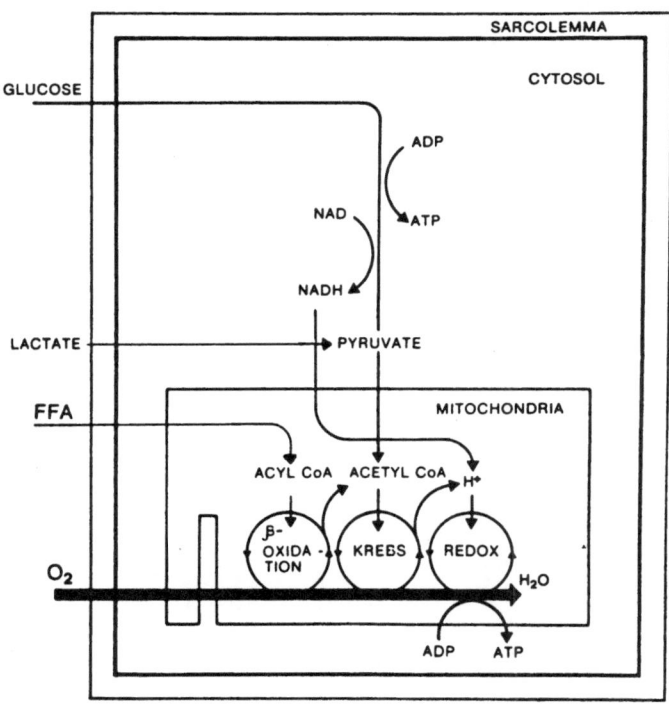

Fig. 2. Schematic representation of anaerobic metabolism in the myocardium.

Carnitine is synthetized almost exclusively in the liver, from lysine and methionine (5). Bohmer et al. (6) have found that heart cells in culture possess a specific carnitine uptake mechanism. The plasma concentration of carnitine is about 30 uM; that in the heart is about 100-fold higher. In more recent years, it has been suggested (18) that: a) L-carnitine, the naturally occurring form, is able to stimulate fatty acid oxidation by tissue homogenates; b) fatty acyl esters of carnitine are useful substrates for mitochondria; c) these compounds are formed by enzyme-catalyzed reversible acyl transfer from coenzyme A.

Later it has been recognized that two distinct classes of carnitine acyltransferases exist, normally referred to as carnitine acetyl- and palmitoyltransferases. Little is known about the heart carnitine palmitoyltransferases; however, inhibition experiments with purified liver enzymes have suggested that in the mitochondria they exist in two forms with considerably different properties. One of these transferases, the outer or A enzyme, is accessible to extramitochondrial coenzyme A, while the inner, or B enzyme, is within the inner membrane, and it uses the coenzyme A of the matrix (Fig. 3). The overall steps of free fatty acid oxidation can be summarized as follows (Fig. 3): a) extramitochondrial formation of acyl-coenzyme A; b) transfer of the acyl-group to carnitine by the outer transferase; c) delivery of the carnitine ester to the inner transferase; d) transfer of the acyl-group to the matrix coenzyme A; e) beta-oxidation; f) finally, a carnitine acyl-carnitine translocase system transfers intramitochondrial carnitine to outside the mitochondria. Experiments with heart mitochondria have demonstrated that penetration of the inner mitochondrial membrane by acyl-carnitine occurs by the acyl-carnitine ⟷ carnitine exchangers or translocases. They are analogous to the ADP ⟷ ATP exchange.

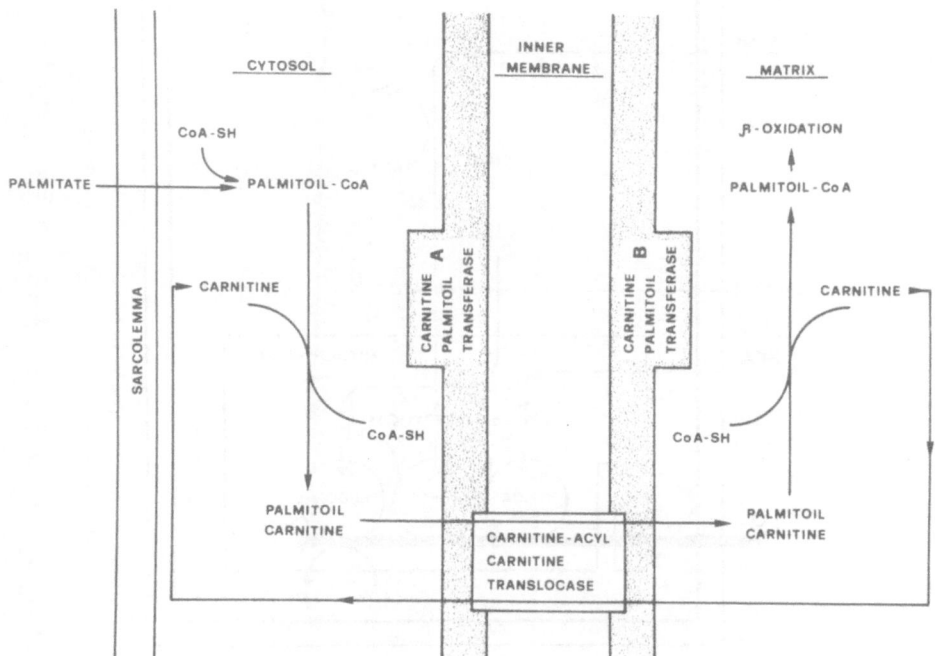

Fig. 3. Schematic representation of the L-carnitine cycle.

The process exhibits 1:1 stoichiometry, so that the total carnitine content of mitochondria remains unchanged. This content is about 3 nmol/mg protein in isolated mitochondria, corresponding to a matrix concentration of 2-4 mM, similar to the concentration in the cytostol (25).

Role of L-Carnitine in Myocardial Energy Production

From the previous discussion, the role of L-carnitine in energy production from fatty acid oxidation is obvious: L-carnitine takes part in the shuttle mechanism whereby long-chain fatty acids are transformed to acyl-carnitine derivatives and transported across the inner mitochondrial membrane. This is impermeable to long-chain fatty acids and to their coenzyme A esters. Once across the membrane, the acyl-carnitine is reconverted to carnitine and to acyl-coenzyme A, which undergoes beta-oxidation. This role of L-carnitine is essential in heart free fatty acid metabolism. Interestingly, the myopathic-type of carnitine deficiency is associated with cardiomyopathy as well as specific carnitine deficiency in the heart (10). In such cases FFA have always been shown to accumulate in the myocardium, with impaired long-chain fatty acid oxidation by muscle homogenates. These alterations can be corrected by exogenous administration of L-carnitine.

From Fig. 4 it also appears that L-carnitine might have a role not only in energy production, but also, indirectly, in energy delivery from the mitochondria to the cytosol. In fact long-chain acyl-coenzyme A is known to inhibit mitochondrial adenine-nucleotide translocase (26,29-31), the enzyme responsible for the one to one exchange of ATP and ADP (19). There is evidence that one of the most important effects of L-carnitine administration is the reduction of long-chain acyl-coenzyme A in heart tissue, resulting in a reduction of the inhibition of the adenine-nucleotide translocase. In this way, L-carnitine allows ATP to be formed from long-chain acyl-coenzyme A and transported from the mitochondria to the cytosol.

It is often assumed that L-carnitine has also a role to permit the import of acetyl-coenzyme A by mitochondria as with the long-chain derivatives. However, doubts are threefold: a) acetyl-coenzyme A is generated within the mitochondrial matrix; b) isolated mitochondria scarsely use external acetyl-coenzyme A, even in the presence of carnitine; c) isolated mitochondria apparently contain only one form of carnitine acetyltransferase (9).

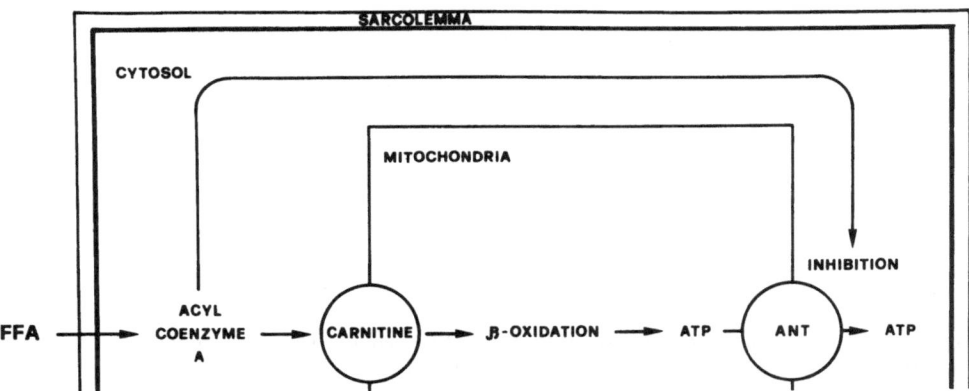

Fig. 4. Schematic representation of the effect of L-carnitine on adenine-nucleotide translocase (ANT).

Effects of L-Carnitine on Myocardial Metabolism of Patients
at Rest and during Pacing-Induced Ischaemia

We have investigated the influence of L-carnitine, the necessary cofactor for mitochondrial transport of long-chain fatty acids (11-14), on myocardial carbohydrate and lipid metabolism in 16 patients with coronary artery disease. The patients were also subjected to two periods of rapid coronary sinus pacing (Fig. 5). They received a central venous infusion of L-carnitine (40 mg/kg) in 50 ml at 200 mg/min prior to the second pacing period. The two pacing periods were separated by a 45-min interval (12). The haemodynamic data at rest, reported in Table 1, show that acute administration of L-carnitine did not significantly change the parameters

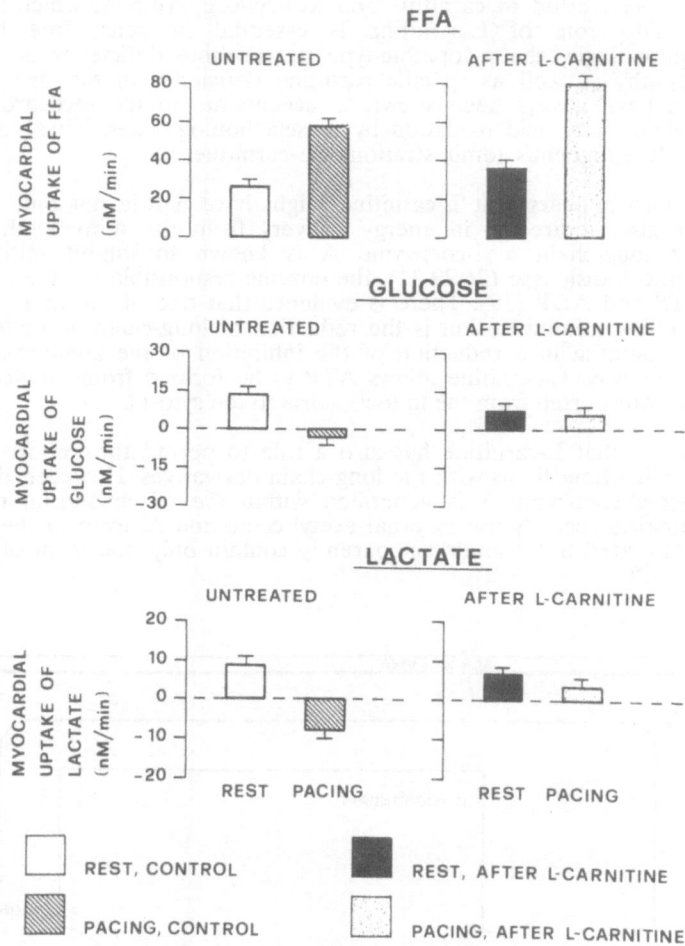

Fig. 5. Effect of L-carnitine on myocardial uptake of FFA, glucose and lactate of 16 patients with coronary artery disease, either at rest or during pacing. Data represent the mean ± S.E. L-carnitine significantly (P<0.001, analysis of variance) modified the behaviour of myocardial uptake of glucose and lactate during the second pacing period, when lactate and glucose production was converted into extraction.

measured, including total coronary flow. In addition L-carnitine failed to modify the changes in haemodynamic parameters induced by atrial pacing, except for the left ventricular end-diastolic pressure which was reduced from 22.1 ± 8.2 to 16.7 ± 48 mmHg (P < 0.05, ref. 13).

Figure 5 shows that the same concentration of L-carnitine induced important changes in myocardial substrate uptake, either at rest or after pacing. At rest, L-carnitine induced a small increase in myocardial uptake of free fatty acids (FFA), concomitant with a decrease of glucose uptake and no major changes in lactate uptake. During pacing, lacatate production was converted to extraction. An increase in the uptake of FFA and of glucose, though small, was demonstrated. All these alterations are due to the metabolic properties of L-carnitine, as coronary flow during atrial pacing was unchanged.

There is no simple explanation for these effects of L-carnitine. During ischaemia, the reduced availability of oxygen leads to a decrease in mitochondrial electron transport, which in turn causes an accumulation of NADH and long-chain acyl-coenzyme A. An increased NADH/NAD ratio and acyl-coenzyme A/coenzyme A ratio may inhibit the activity of pyruvate dehydrogenase (PDH), the enzyme which regulates the entry of pyruvate into the citric acid cycle. Under these conditions pyruvate is preferentially converted to lactate with consequent lactate production. L-carnitine can reduce lactate production under ischaemic conditions either by activating PDH directly or by provoking a decrease in the acetyl-coenzyme A/coenzyme A ratio, as a consequence of acetyl removal from coenzyme A, mediated by carnitine coenzyme A acetyltransferase. Therefore, the utilization of pyruvate and consequently of lactate in the oxidative pathway, induced by L-carnitine, may be attributed to enhanced PDH activity, rather than to an increase in oxygen availability induced by L-carnitine.

Another possible explanation for the effects of L-carnitine on lactate metabolism is an indirect effect on phosphofructokinase activity (PFK), the enzyme regulating the rate of anaerobic glycolysis. The activity of this enzyme is inhibited by a high cytosolic ATP concentration, whilst it is stimulated by low cytosolic ATP concentration. Under ischaemic conditions the ATP concentration in the cytosol decreases either as a result of reduced oxidative metabolism or because the activity of adenine-nucleotide translocase is inhibited by long-chain acyl-coenzyme A. This causes sequestration of ATP into the mitochondrial matrix with a decrease of cytosolic ATP and a consequent stimulation of PFK. By removing long-chain acyl-coenzyme A, carnitine can also prevent its inhibitory action on adenine-nucleotide translocase, thus improving the ATP transfer to the cytosolic compartment. The increased cytosolic ATP levels might decrease PFK activity and lactate production.

Table 1. Effect of L-carnitine on haemodynamic variables in cardiac patients at rest

Variable	Before	After L-carnitine
Heart rate (beats/min)	79 ± 4	76 ± 3
Mean aortic systolic pressure (mmHg)	146 ± 4	144 ± 6
Mean aortic diastolic pressure (mmHg)	76 ± 6	73 ± 2
Pulmonary artery pressure (mmHg)	18 ± 1	18.2 ± 3
Cardiac output (litre/min)	5.9 ± 0.7	5.9 ± 0.6
Coronary sinus blood flow (ml/min)	127 ± 14	129 ± 12

See legend to Fig. 5. No statistically significant differences were observed.

It is possible that the increased PDH and the decreased PFK activity are concurrently involved in the reduction of lactate formation induced by L-carnitine administration. These data show the complexity of lactate metabolism that can be positively affected by interventions like with L-carnitine, a compound without haemodynamic effects, acting primarily on lipid metabolism.

References

1. Ballard FB, Danforth WH, Naegles S, Bing RJ: Myocardial metabolism of fatty acids. J Clin Invest 39:717-723, 1960
2. Bing RJ: The metabolism of the heart. Harvey Lecture Series 50:27-70, 1954
3. Bing RJ: Cardiac metabolism. Physiol Rev 45:171-213, 1965
4. Bing RJ, Siegel A, Vitale A, Balboni F, Sparks E, Taeschler M, Klapper M, Edwards S: Metabolic studies on the human heart in vivo. Am J Med 15:284-296, 1953
5. Bohmer T: Conversion of butyrobetaine to carnitine in the rat in vivo. Biochim Biophys Acta 343:551-557, 1974
6. Bohmer T, Eiklid K, Jonssen J. Carnitine uptake into human heart cells in culture. Biochim Biophys Acta 465:627-633, 1977
7. De Jong JW, Hülsmann WC: A comparative study of palmitoyl-coenzyme A synthetase activity in rat-liver, heart and gut mitochondrial and microsomal preparations. Biochim Biophys Acta 197:127-135, 1970
8. Drake AJ: Substrate utilization in the myocardium. Basic Res Cardiol 77:1-11, 1982
9. Edwards YH, Chase JFA, Edwards MR, Tubbs PK: Carnitine acetyltransferase: the question of multiple forms. Eur J Biochem 46:209-215, 1974
10. Engel AG: Possible causes and effects of carnitine deficiency in man. In: Frenkel RA, McGarry JD, eds: Carnitine biosynthesis, metabolism and functions. New York: Acad Press, 1980:271-285
11. Ferrari R, Agnoletti G, Ciampalini G, Albertini A, Visioli O: Coronary sinus lactate release as an index of myocardial anaerobiosis: effects of intervention. Adv Cardiol 33:115-126, 1986
12. Ferrari R, Bolognesi E, Raddino R: Reproducibility of metabolical parameters during successive coronary sinus pacing in patients known to have coronary artery disease. Int J Cardiol 9:231-234, 1985
13. Ferrari R, Cucchini F, Di Lisa F, Raddino R, Bolognesi R, Visioli O: The effects of L-carnitine on myocardial metabolism of patients with coronary artery disease. Clin Trials J 21:41-58, 1984
14. Ferrari R, Cucchini F, Visioli O: The metabolical effects of L-carnitine in angina pectoris. Int J Cardiol 5:213-216, 1984
15. Fritz IB: Action of carnitine on long chain fatty acid oxidation by liver. Am J Physiol 197:297-304, 1959
16. Groot PHE, Scholte HR, Hülsmann WC: Fatty acid activation: specificity, localization and function. Adv Lipid Res 14:75-126, 1976
17. Gulewitsch W, Krimberg R: Zur Kentniss der Extraktivstoffe der Musckeln. Ueber das Carnitin. Z Physiol Chem 45:326-330, 1905
18. Hoppel CL: Carnitine palmitoyltransferase and transport of free fatty acids. In: Martonosi A, ed: The enzymes of biological membranes. Vol 2. New York: Plenum, 1976:119-143
19. Klingenberg M, Scherer B, Stengel-Rutkowski L, Buchholz M, Grebe K: Experimental demonstrations of the reorientating (mobile) carrier mechanism exemplified by the mitochondrial adenine nucleotide translocator. In: Azzone GF, ed: Mechanisms in bioenergetics. New York: Acad Press, 1973:257-284
20. Kobayashi K, Neely JR: Control of maximum rates of glycolysis in rat cardiac musle. Circ Res 44:166-175, 1979
21. Kuel J, Doll E, Steim H, Fleer U, Reindell H: Ueber den Stoffwechsel des menschlichen Herzens III. Der oxydative Stoffwechsel des menschlichem Herzens unter verschiedenden Arbeitsbedingungen. Pflügers Arch 282:43-53, 1965
22. Margaria R, Cerretelli P, Aghemo P, Sassi G: Energy cost of running. J Appl Physiol 18:367-370, 1963
23. Neely JR, Morgan HE: Relationship between carbohydrate and lipid metabolism and the energy balance of heart muscle. Annu Rev Physiol 36:413-459, 1974
24. Opie LH: Metabolism of the heart in health and disease. II. Am J Cardiol 77:100-122, 1969

25. Oram JE, Bennetch SL, Neely JR: Regulation of fatty acid utilization in isolated perfused rat hearts. J Biol Chem 248:5299-5309, 1973
26. Pande SV, Blanchaer MC: Reversible inhibition of mitochondrial adenosine diphosphate phosphorylation by long chain acyl CoA esters. J Biol Chem 246:402-411, 1971
27. Rothlin ME, Bing RJ: Extraction and release of individual free fatty acids by the heart and fat depots. J Clin Invest 40:1380-1386, 1961
28. Shepard RJ, Allen C, Benade AJA, Davies CTM, Di Prampero PE, Hedman R, Merriman JE, Myhre K, Simmons R: The maximum oxygen intake. An international reference standard in cardio-respiratory fitness. Bull WHO 38:757-764, 1968
29. Shrago E: Myocardial adenine nucleotide translocase. J Mol Cell Cardiol 8:497-504, 1976
30. Shrago E, Shug AL, Sul H, Bittar N, Folts JD: Control of energy production in myocardial ischemia. Circ Res 39, Suppl:75-79, 1976
31. Shug AL, Shrago E, Bittar N, Folts JD, Koke JR: Acyl-CoA inhibition of adenine nucleotide translocation in ischemic myocardium. Am J Physiol 228:689-692, 1975
32. Van der Vusse GJ, Reneman RS: Glycogen and lipids (endogenous substrates). In: Drake-Holland AJ, Noble MIM, eds: Cardiac metabolism. Chichester: Wiley, 1983:215-237
33. Zierler KL: Fatty acids as substrates for heart and skeletal muscle. Circ Res 38:459-463, 1976

Chapter 5

Adenine Nucleotides, Purine Metabolism and Myocardial Function

P.W. Achterberg, Cardiochemical Laboratory, Thoraxcenter,
Erasmus University Rotterdam, The Netherlands

During cardiac contraction there is a very rapid breakdown and synthesis of the high-energy phosphates ATP and creatine phosphate. Whenever a disbalance exists between synthesis and breakdown of ATP a series of enzymes will act to catabolize AMP to adenosine, inosine, (hypo)xanthine and urate which will leave the heart. This will eventually result in decreased adenine nucleotide and ATP levels. A number of observations are discussed in order to relate the function of the heart and the metabolism of ATP and its catabolites. Good evidence is still lacking to conclude that ATP-content and heart function are either directly related or non-related. Major problems appear to be adequate definition of heart function and the fact that the ATP-catabolite adenosine exerts several modulatory effects on heart function. For these and other reasons further studies on purine (nucleotide) metabolism and heart function are needed.

Introduction

The continuous pumping of blood by the heart requires a large amount of (bio)chemical energy in the form of adenosine triphosphate (ATP) in order to drive the relaxation of the myofibrillar contractile proteins, to maintain cationic homeostasis and to activate intermediary metabolism and synthesis of high molecular structural and functional cell components. The normal ATP-content of cardiac tissue falls around 20-30 umol per g dry wt. for most mammalian species studied. In addition the heart will contain between 3 and 5 umol ADP per g dry wt., most of which is probably protein-bound, and an even smaller amount of AMP (0.3-1.0 umol/g dry wt.). In well-oxygenated hearts ATP therefore makes up 80 to 90% of total adenine nucleotides, if we do not include the nicotinamide adenine dinucleotides. This apparently constant amount of cardiac ATP is in fact the result of a continuous, fast breakdown and synthesis of ATP. The latter has been estimated between 150 and 600 umoles of ATP synthetized per g dry wt. per minute in isolated rat hearts (28). This means that between 5 and 10% of total ATP can be used (and resynthetized) during a single contraction. Only oxidative (oxygen dependent) phosphorylation in the cardiac mitochondria is able to provide such a high ATP-synthesis rate. Apart from ATP, the heart contains relatively large amounts (25-40 umol/g dry wt.) of another high-energy phosphate, i.e., creatine phosphate, which is also continually turned over to creatine, with formation of ATP from ADP, the first-formed ATP-catabolite. When the formation of ATP is inhibited, either by lack of oxygen, hypoxia, poisoning of oxidative phosphorylation, or ischemia, the ATP-content in the heart will decline, ADP and AMP contents will rise, and AMP will be broken down. A long chain of enzymatically mediated events will lead to the formation of adenosine, inosine, hypoxanthine and, occasionally, xanthine and urate (1-3,17). These purine nucleosides (adenosine, inosine) and oxypurines can pass the cell membrane relatively easy and will be washed out off the heart. Fig. 1 shows the time-course of the ATP-breakdown cascade in autolyzing rat heart as determined by Gerlach and co-workers (13,15) more than 20 years ago.

J.W. de Jong (Ed.), *Myocardial Energy Metabolism*, Martinus Nijhoff Publishers, Dordrecht/Boston/Lancaster, 1988

If reoxygenation occurs after a period of ischemia and adequate oxidative metabolism is restored, the heart will be left with a decreased amount of adenine nucleotides. We suggest that adequate reperfusion of non-irreversibly damaged hearts will result in ADP- and AMP-contents which are no longer increased. If still increased, however, AMP-catabolism and purine release will rapidly lead to further adenine nucleotide loss and ATP-decrease, which, beyond a certain level, is incompatible with adequate cellular function and will be associated with cell death (26). During hypoxia, increased work-load or hypoperfusion, the inadequate oxygen supply to the heart can be increased by coronary vasodilation, which is at least partially mediated by the ATP-catabolite adenosine (7). The supply-demand ratio for oxygen, effected by the balance between formation and breakdown of ATP, can thus be considered as the trigger for adenosine-mediated coronary vasodilation (2,38). From the above considerations alone, it is obvious that there exists an intricate and complex relationship between myocardial ATP-content and turnover, purine metabolism and heart function, which will be further discussed in the following paragraphs.

ATP-Content and Myocardial Function

During adequate supply of oxygen and substrates, the heart is capable of increasing its work-performance and ATP-turnover rate several fold. This will not be accompanied by major changes in myocardial ATP-content. In addition, when cardiac

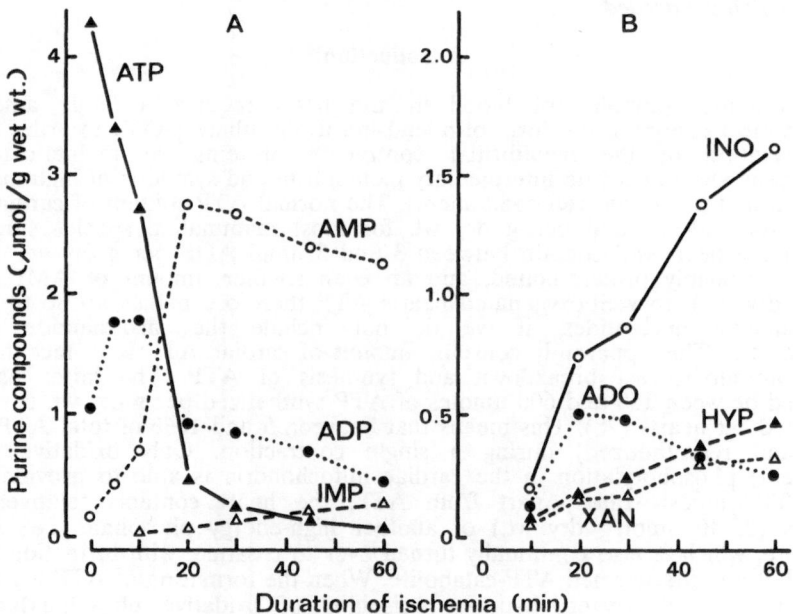

Fig. 1. Time-course of adenine nucleotide degradation (A) and accumulation of purine nucleosides and oxypurines (B) in autolyzing rat heart. The data were redrawn from Deuticke and Gerlach (13). Abbreviations: INO = inosine, ADO = adenosine, HYP = hypoxanthine, XAN = xanthine. Note the very fast ATP-decline over the first 20 min and the simultaneous rapid rise in AMP, adenosine and inosine.

work-performance is fully arrested in the presence of oxygen and substrate (e.g., by giving high doses of calcium antagonists or providing zero calcium) again normal ATP-values will be found. Superficially, this would appear to contradict any relation at all between ATP-content and heart function. Still, it has been found that myocardial ATP-content during hypoxia of varying degrees of severity correlates with myocardial function (see Fig. 2). Other studies, however, failed to observe significant decline in ATP-contents during hypoxia, while cardiac function was already severely depressed (8). The latter NMR-study did show, however, a correlation between ATP-turnover rate via the creatine kinase reaction and cardiac function, expressed as the rate-pressure product. It should be noted that NMR measurements of ATP-content are not very sensitive.

Decreased myocardial ATP-contents after short and reversible ischemia in vivo are restored very slowly (10,23,35,40). A similar slow restoration is found for the decreased post-ischemic function (10,27). Myocardial ATP-content after reperfusion is directly and sometimes linearly (20) related to the ATP-content at the end of ischemia. During hypoxia, ischemia or during inadequate reperfusion, metabolic alterations, other than decreased ATP-content, are thought to contribute to depression of function or to the alteration of the ATP-function relationship. Acidosis (lactate accumulation) or intracellular build-up of calcium or phosphate have been suggested as causative factors, together with fatty acids or fatty acid derivatives. In general it can be expected that factors which inhibit ATP production rate, will lead to larger disturbances of ATP-content than factors which inhibit ATP-consumption, i.e., contractile function.

Indirect evidence for the importance of cardiac ATP-content in maintaining cardiac function comes from studies on the protective effects of calcium-antagonists against ischemic damage. Calcium antagonists (diltiazem, nifedipine, nisoldipine, bepridil) have a dose-dependent negative inotropic action on the heart. When given before the onset of ischemia, there is a dose-dependent inhibition on ischemic ATP-breakdown as reflected by the decreased release of ATP-catabolites (purines) during ischemia (11,16,19). Work performance before the onset of low-flow ischemia is directly related to purine release during ischemia (12). Under these conditions (low-flow ischemia, glucose as substrate), it is most probably the formation of ATP which is restricted. Therefore reduction of ATP-breakdown is the palliative measure of choice.

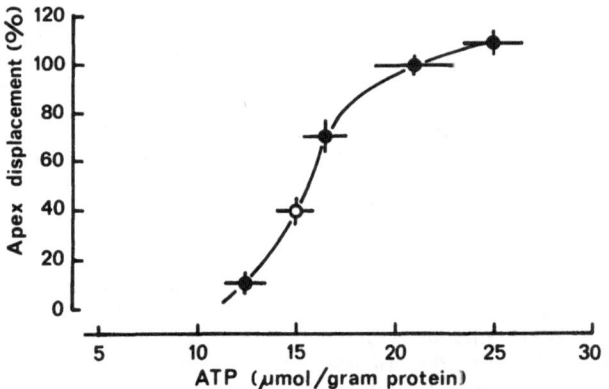

Fig. 2. Relation between ATP-content and contractility in hypoxic rat heart. ATP-contents were determined in isolated rat hearts, subjected to varying degrees of hypoxia (●) or ischemia (o). Myocardial function is given as percentage of prehypoxic function. Apex displacement is used as a measure of contractility. Data were redrawn from Stam and De Jong (39).

In a number of studies, good correlations have been observed between ATP-content and myocardial function after ischemia and reperfusion (22,24,25,30,34,41). The actual curves, however, which describe the ATP-function relationship (Fig. 3) are different in these studies. After pre-ischemic conditioning of hearts by 10 min anoxia, Neely and Grotyohann (31) found a very steep relationship, i.e., fast decline of function over a short range of ATP-contents. They concluded that there is no relation between ATP-content and heart function. Whether AMP and ADP contents returned to normal levels in these anoxia-pretreated, reperfused hearts, was not reported (31). Non pretreated, reperfused hearts in the same study still had enormously elevated AMP-levels. Because evidence has been gathered (3,9) that adenine nucleotide breakdown to purines is regulated by cytosolic AMP-content, hearts with strongly elevated AMP-contents are bound to loose adenine nucleotides (and thus ATP) at a fast rate. Hearts of cardiomyopathic hamsters contain severely depressed amounts of ATP at the moment when signs of heart-failure in vivo are apparent. The protective effect of high oral doses of calcium antagonists on these hearts coincides with the finding of normalized ATP-values (42).

The discrepancies observed in various reports on the correlation between ATP and function (Fig. 3) could be related to the manner in which cardiac function is determined or expressed. For instance, developed tension or pressure of the heart is directly related to the applied or existing preload. It is repeatedly found that post-ischemic, reperfused hearts exhibit increased preload (beginning contracture?). Function determined at this increased preload should be expressed in relation to the control function at similarly high preload, and not, as is often done, in relation to function (developed pressure) at a lower preload. It has on occasion been reported that developed pressure of isolated hearts during early post-ischemic reperfusion was actually higher than the pre-ischemic control value (29). In addition, changes in heart rate may also cause changes in the inotropic status of the heart. In general, it proves extremely difficult to obtain relevant measures of cardiac function (14). We feel that most controversies on the relationship between heart function and metabolic parameters can be reduced to problems related to adequate determination of function or inotropic status of the hearts investigated. Another factor which tends to obscure good interpretation of data is the fact that differences may exist in both ATP-content and function in the outer and inner layers of the ischemic and reperfused heart (33). NMR-measurements and biochemical determinations of ATP in whole heart homogenates will be inadequate if such differences exist. We conclude that it is not yet fully clear how the dynamic relationship between ATP-content or ATP-turnover on the one hand and contractile activity or contractile potential on the other is regulated. A major contribution to the regulatory dynamics appears to be effected through operation of the creatine kinase reaction. In this respect it should be considered that gradients of ATP and ADP may exist between mitochondria and the site of the myofibrillar, contractile ATP-ase, with the creatine kinase equilibrium functioning to lessen the steepness of ATP- and ADP-gradients. The fact that the flux through creatine kinase is at least severalfold higher than the ATP-formation rate could point into that direction. Evidence for the existence of ATP-gradients has been found in hypoxic liver cells (5), which have far lower rates of ATP-turnover than the heart.

It is often argued that the cellular ATP-concentration cannot directly influence contractility, because the K_m of the contractile ATP-ases will always be far lower than estimated ATP-concentrations. It should be noted, however, that ordinary Michaelis-Menten kinetics do not apply when enzyme concentrations are extremely high, such as at the myofibrillar ATP-ase sites. Again the possible existence of ATP-gradients could strongly influence the actual ATP-concentration at the ATP-ase site.

Myocardial ATP-contents in physiologically normal hearts tend to be constant and depleted ATP-contents will be increased, albeit slowly, during reperfusion (10,23,35,40). It appears therefore that normal ATP-content is a prerequisite for normal cardiac metabolism and function. ATP is a known modulator and substrate for a host of cardiac enzymes (e.g., kinases, ATP-ases), which execute a number of essential reactions. The precise sites of action, which will be critically influenced by a decreased cytosolic or mitochondrial ATP-concentration, are therefore very difficult to identify. It would be interesting in this respect to investigate the metabolic properties of non-irreversibly damaged, but partially ATP-depleted hearts at various work loads.

One important implication of a direct relationship between ATP-content, heart function and metabolism may be that an already ATP-depleted heart is bound to completely loose its function or enter a stage of irreversible damage, if ATP-levels have not been restored in the inter-ischemic period. Studies have been performed (18,36), in which short periods of repeated ischemia/reperfusion were compared with a long period of ischemia of similar total duration. It appeared, however, that short-during ischemia and reperfusion might have a protective effect on ATP-decline during a next ischemic period (18,36). It is not clear, whether the observed decreased heart function, the observed "overshoot" of creatine phosphate or the combination of both provides the temporarily protection against ATP-decline. It is tempting to conclude that creatine phosphate overshoot reflects an intact capability

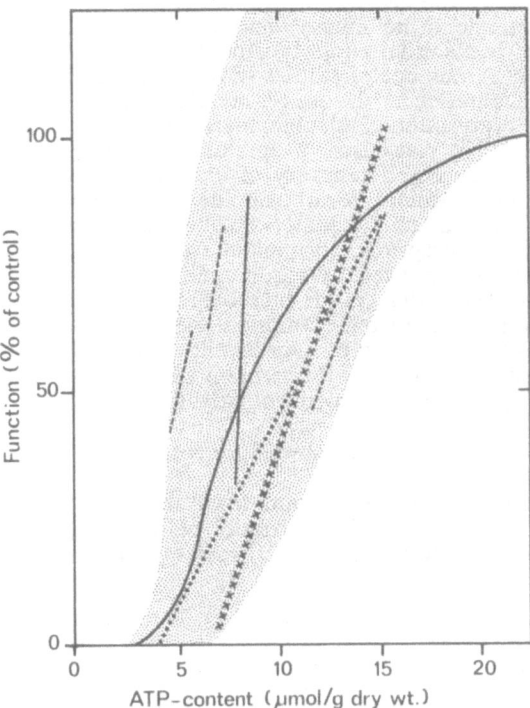

Fig. 3. Myocardial ATP-content and function in reperfused isolated rat heart preparations. Data were taken from publications by different authors. (-----) ref. 22; (——) ref. 31; (••••) ref. 30; (xxxxx) ref. 41.

of the heart to synthetize but not to use ATP (21). Finally, there is no solid evidence that severely ATP-depleted hearts will continue to perform in the normal to maximal range of work, which can be observed in non ATP-depleted hearts. In the next paragraph we shall further elaborate on the effects of ATP-catabolites, with the emphasis on adenosine, on regulation of cardiac function.

Purine Metabolism and Function of the Heart

Adenosine release from the heart is increased, whenever a disbalance exists between formation and breakdown of ATP. This can be either caused by increased work-load, decreased oxygen supply, mitochondrial poisoning, catecholamine stress or substrate-induced (acetate) breakdown of ATP to the direct adenosine precursor AMP (1,3). The physiological effects of adenosine may be looked upon as controlling the balance between supply and demand of oxygen or energy (32,38). Adenosine is a coronary vasodilator, also in man, and many coronary vasodilatory drugs (dipyridamole, dilazep, hexobendine, papaverine) presumably act by enhancing adenosine effects through interference with the normally fast transport and catabolism of adenosine.

In addition, adenosine reduces heart rate and slows conduction through the atrioventricular node (6) and anti-catecholaminergic effects of adenosine have been noticed (37). Several beneficial effects on the ischemic heart have also been reported for the adenosine catabolite inosine (4). Both vasodilation and positive inotropic effects by inosine have been noticed in animal and human studies (see Chapter 26). Inosine effects may well be caused by modification or attenuation of adenosine effects or of the availability of adenosine at the site of its receptors. Both adenosine (via adenosine kinase) and inosine can be reincorporated in the myocardial adenine nucleotide pool. Inosine is first catabolized to hypoxanthine before incorporation. A number of studies have indicated the feasibility of increased adenine nucleotide formation in the post-ischemic heart by giving purine compounds (adenosine, inosine, hypoxanthine, adenine, 5-amino-4-imidazolecarboxamide riboside; see Chapters 1,10,26). No consensus of opinion exists as yet, however, whether this will actually result in increased recovery of cardiac function (18). Enhanced ATP-restoration by 5-amino-4-imidazolecarboxamide riboside coincides with decreased cardiac function for which no adequate explanation has as yet been provided. As outlined in a previous paragraph, this might well be caused by the difficulties in determining the relevant indices of cardiac function. The study of the intricate relationship between myocardial purine metabolism, function and general metabolism of the heart deserves much further attention. Attempts to speed up restoration of post-ischemic ATP-content might be worthwhile, even if ATP-concentration is not directly related to cardiac function. Restoration of ATP-regulated metabolism of the post-ischemic heart remains a goal worthy of pursuit.

Acknowledgements

P.W. Achterberg is supported by a grant from the Dutch Heart Foundation. The author is grateful for the secretarial assistance of Mrs. M.J. Kanters-Stam and Mrs. C.E. Zandbergen-Visser, and for the critical review of the manuscript by Dr. J.W. de Jong.

References

1. Achterberg PW: Formation and breakdown of adenosine in the heart. Investigations on myocardial purine metabolism. Acad thesis, Erasmus University Rotterdam, 1986
2. Achterberg PW, Harmsen E, De Jong JW: Adenosine deaminase inhibition and myocardial purine release during normoxia and ischemia. Cardiovasc Res 19:593-598, 1985
3. Achterberg PW, Stroeve RJ, De Jong JW: Myocardial adenosine cycling rates during normoxia and under conditions of stimulated purine release. Biochem J 235:13-17, 1986

4. Aviado DM: Inosine: a naturally occurring cardiotonic agent. J Pharmacol (Paris) 14, Suppl III:47-71, 1983
5. Aw TY, Jones DP: ATP concentration gradients in cytosol of liver cells during hypoxia. Am J Physiol 249:C385-C392, 1986
6. Belhassen P, Pelleg A: Electrophysiologic effects of adenosine triphosphate and adenosine on the mammalian heart: clinical and experimental aspects. J Am Coll Cardiol 4:414-424, 1984
7. Berne RM: Cardiac nucleotides in hypoxia: Possible role in regulation of coronary blood flow. Am J Physiol 204:317-322, 1963
8. Bittl JA, Balschi JA, Ingwall JS: Contractile failure and high-energy phosphate turnover during hypoxia: ^{31}P-NMR surface coil studies in living rat. Circ Res 60:871-878, 1987
9. Bünger R, Soboll S: Cytosolic adenylates and adenosine release in perfused working heart. Eur J Biochem 159:111-115, 1986
10. DeBoer LWV, Ingwall JS, Kloner RA, Braunwald E: Prolonged derangements of canine myocardial purine metabolism after a brief coronary artery occlusion not associated with anatomic evidence of necrosis. Proc Natl Acad Sci USA 77:5471-5475, 1980
11. De Jong JW, Harmsen E, De Tombe PP, Keijzer E: Nifedipine reduces adenine nucleotide breakdown in ischemic rat heart. Eur J Pharmacol 81:89-96, 1982
12. De Jong JW, Huizer T, Tijssen JPG: Energy conservation by nisoldipine in ischemic heart. Br J Pharmacol 83:943-949, 1984
13. Deuticke B, Gerlach E: Abbau freier Nucleotide in Herz, Skeletmuskel, Gehirn, und Leber der Ratte bei Sauerstoffmangel. Pflügers Arch 292:239-254, 1966
14. Elzinga G, Westerhof N: How to quantify pump function of the heart. The value of variables derived from measurements on isolated muscle. Circ Res 44:303-308, 1979
15. Gerlach E, Deuticke B, Dreisbach RH: Der Nucleotid-Abbau im Herzmuskel bei Sauerstoffmangel und seine mögliche Bedeutung für die Coronardurchblutung. Naturwiss 50:228-229, 1963
16. Harmsen E, De Tombe PP, De Jong JW: Synergistic effect of nifedipine and propranolol on adenosine (catabolite) release from ischemic rat heart. Eur J Pharmacol 90:401-409, 1983
17. Harmsen E, De Tombe PP, De Jong JW, Achterberg PW: Enhanced ATP and GTP synthesis from hypoxanthine or inosine after myocardial ischemia. Am J Physiol 246:H37-H43, 1984
18. Hoffmeister HM, Mauser M, Schaper W: Repeated short periods of regional myocardial ischemia: effect on local function and high energy phosphate levels. Basic Res Cardiol 81:361-372, 1986
19. Huizer T, De Jong JW, Achterberg PW: Protection by bepridil against myocardial ATP-catabolism is probably due to negative inotropy. J Cardiovasc Pharmacol 10:55-61, 1987
20. Ichihara K, Abiko Y: Effects of diltiazem and propranolol on irreversibility of ischemic cardiac function and metabolism in the isolated perfused rat heart. J Cardiovasc Pharmacol 5:745-751, 1983
21. Ichihara K, Abiko Y: Rebound recovery of myocardial creatine phosphate with reperfusion after ischemia. Am Heart J 108:1594-1597, 1984
22. Ichihara K, Neely JR: Recovery of ventricular function in reperfused ischemic hearts exposed to fatty acids. Am J Physiol 249:H492-H497, 1985
23. Isselhard W, Lauterjung KL, Witte J, Bau T, Hübner G, Giersberg O, Heugel E, Hirt HJ: Metabolic and structural recovery of left ventricular canine myocardium from regional complete ischemia. Eur Surg Res 7:136-155, 1975
24. Jarmakani JM, Nagatomo T, Nakazawa M, Langer GA: Effect of hypoxia on myocardial high-energy phosphates in the neonatal mammalian heart. Am J Physiol 235:H474-H481, 1978
25. Jarmakani JM, Nakazawa M, Nagatomo T, Langer GA: Effect of hypoxia on mechanical function in the neonatal mammalian heart. Am J Physiol 235:H469-H474, 1978
26. Jennings RB, Steenbergen C: Nucleotide metabolism and cellular damage in myocardial ischemia. Annu Rev Physiol 47:727-749, 1985
27. Kloner RA, DeBoer LWV, Darsee JR, Ingwall JS, Hale S, Tumas J, Braunwald E: Prolonged abnormalities of myocardium salvaged by reperfusion. Am J Physiol 241:H591-H599, 1981
28. Kobayashi K, Neely JR: Control of maximum rates of glycolysis in rat cardiac muscle. Circ Res 44:166-175, 1979
29. Lipasti JA, Alanen KA, Eskola JW, Nevalainen TJ: Ischaemic contracture in isolated rat heart: reversible or irreversible myocardial injury? Exp Pathol 28:89-95, 1985
30. Nayler WG: Protection of the myocardium against postischemic reperfusion damage. The combined effect of hypothermia and nifedipine. J Thorac Cardiovasc Surg 84:897-905, 1982

31. Neely JR, Grotyohann LW: Role of glycolytic products in damage to ischemic myocardium. Dissociation of adenosine triphosphate levels and recovery function of reperfused ischemic hearts. Circ Res 55:816-824, 1984
32. Newby AC: Adenosine and the concept of "retaliatory metabolites". Trends Biochem Sci 9:42-44, 1984
33. Prinzen FW, Arts T, Van der Vusse GJ, Coumans WA, Reneman RS: Gradients in fiber shortening and metabolism across ischemic left ventricular wall. Am J Physiol 250:H255-H264, 1986
34. Reibel DK, Rovetto MJ: Myocardial ATP synthesis and mechanical function following oxygen deficiency. Am J Physiol 234:H620-H624, 1978
35. Reimer KA, Hill ML, Jennings RB: Prolonged depletion of ATP and of the adenine nucleotide pool due to delayed resynthesis of adenine nucleotides following reversible myocardial ischemic injury in dogs. J Mol Cell Cardiol 13:229-239, 1981
36. Reimer KA, Murry CE, Yamasawa I, Hill ML, Jennings RB: Four brief periods of myocardial ischemia cause no cumulative ATP loss or necrosis. Am J Physiol 251:H1306-H1315, 1986
37. Schrader J, Baumann G, Gerlach E: Adenosine as inhibitor of myocardial effects of catecholamines. Pflügers Arch 372:29-35, 1977
38. Sparks HV, Bardenheuer H: Regulation of adenosine formation by the heart. Circ Res 58:193-201, 1986
39. Stam H, De Jong JW: Sephadex-induced reduction of coronary flow in the isolated rat heart: A model for ischemic heart disease. J Mol Cell Cardiol 9:633-650, 1977
40. Swain JL, Sabina RL, McHale PA, Greenfield JC, Holmes EW: Prolonged myocardial nucleotide depletion after brief ischemia in the open-chest dog. Am J Physiol 242:H818-H826, 1982
41. Watts JA, Koch CD, LaNoue KF: Effects of Ca^{2+} and heart function after ischemia. Am J Physiol 238:H909-H916, 1980
42. Wikman-Coffelt J, Sievers R, Parmley WW, Jasmin G: Verapamil preserves adenine nucleotide pool in cardiomyopathic Syrian hamster. Am J Physiol 250:H22-H28, 1986

Chapter 6

On the Role of Myocardial AMP-Deaminase

A.C. Skladanowski, Department of Biochemistry,
Academic Medical School Gdansk, Gdansk, Poland

In this chapter the AMP-deaminase activity in hearts of different species is compared with that in skeletal muscle. Some features of the well-characterized skeletal muscle enzyme, like binding to myosin filaments and deficiencies in humans, are described. Experimental models to study the role of cardiac AMP-deaminase, i.e., isolated enzyme, cardiomyocytes and isolated perfused heart, are discussed. Substrate saturation kinetics and regulatory effects of nucleotides, phosphate, and activated fatty acids are summarized; their significance for in vivo regulation considered. In addition the behavior of AMP-deaminase under various adenylate energy charges and its implications for the regulation of enzyme in ischemic tissue is examined. Myocardial AMP-deaminase is shown to be a tissue-specific isoenzyme, with different kinetic and regulatory forms. Developmental changes of AMP-deaminase and other enzymes of adenylate metabolism enzymes in myocardial cells are reviewed. The flux through the pathway catalyzed by AMP-deaminase in normoxic and ischemic heart is analyzed. A probable function of the reaction - preservation of purine nucleotides inside the myocardial cell - is proposed.

Introduction

Adenosine 5'-monophosphate is the first ьonenergetic catabolite of ATP-breakdown in contracting muscle. It is also one of the precursor metabolites for the restoration of high-energy phosphates. The large, time-dependent changes of AMP levels in ischemic/reperfused heart (58) suggest an efficiently regulated system for its transformation. The adenylate kinase equilibrium is responsible for AMP-conversion to the di- and triphosphate level. AMP can be either dephosphorylated to adenosine or deaminated to IMP. The former reaction is involved in the main stream of adenine nucleotide catabolism in the heart. It produces physiologically active adenosine (see Chapters 8 and 9). The parallel hydrolysis of the 6-amino group attached to adenine ring in AMP gives IMP. Independently of further degradation, IMP is also the common point between de novo purine nucleotide synthesis and hypoxanthine salvage. Both reactions are part of ATP and GTP biosynthesis, thus being of undoubted importance for energy-metabolite replenishment. The reamination of IMP to AMP, an energy-dependent two-step process, is involved in AMP deamination in the purine nucleotide cycle (37).

AMP-deaminase (AMP-aminohydrolase, AMP-D; EC 3.5.4.6) catalyzes the hydrolysis of AMP to IMP and ammonia. It is found in the majority of cells and organs of all species studied so far (79). The reaction is irreversible. The enzyme isolated from vertebrates is highly specific towards 5'-AMP, with only slight reactivity with 2'-deoxy-5'-AMP (61). AMP-deaminase is exclusively localized in the cytosolic compartment. Only one report indicates also its location in the mitochondria (22).

The physiological role of AMP-deaminase is still unclear, despite its general occurrence. This holds even for skeletal muscle, where its activity is one to two orders of magnitude higher than in other tissues, including heart muscle (52). Nevertheless, the activity changes during myoblast differentiation (40) as well as the metabolite flux through the AMP-deamination pathway (see, e.g., refs. 1,2,39) indicate the significance of AMP-deaminase in heart purine nucleotide catabolism. This contribution may be lower than in skeletal muscle. The aim of this chapter is to discuss these potencies and to suggest a physiological role of AMP-deaminase in heart tissue.

Skeletal and Heart Muscle AMP-Deaminase

Tissue Activities

Purzycka (52) proved AMP-deaminase to be present in rat-heart muscle. This author revised the hypothesis of Conway and Cooke (13): compulsory dephosphorylation prior to deamination of AMP in heart ventricle, auricle being an exception in that sequence of adenylic acid breakdown. Nakatsu and Drummond (41) compared the activity of adenylate deaminase in heart of several species. They ranked them from nondetectable in dog heart, through rabbit and pigeon heart with 0.75 U/g wet wt., rat heart with 1.8 U/g, to the exceptionally high activity of 7 U/g in turtle heart. The activities of heart AMP-deaminase reported by other authors varied from 0.15 U/g wet wt. (rabbit, ref. 74), through about 0.75 U/g (man, ref. 47), to about 2.4 U/g (rat, ref. 76), depending also on the assay conditions.

Comparative studies revealed a wide spectrum of skeletal muscle AMP-D to heart AMP-D activity ratios in tissue extracts. These extend from 26 (pig, ref. 23), through 55 (man, ref. 47), to about 260 (rabbit heart and white skeletal muscle, ref. 45). If one disregards the assay conditions, the activity of heart AMP-deaminase is about two orders of magnitude lower than that of skeletal muscle. However, the enzyme activity in many tissues could be either overestimated due to superimposed activity of both 5'-nucleotidase and adenosine deaminase or underestimated because of nonoptimal conditions. The former can be avoided by the use of inhibitors. The enzyme activity in heart extracts is not inhibited by endogenous low-molecular compounds, but it is in homogenized skeletal muscle (23).

AMP-deaminase activity is much more pronounced in white type muscles in rabbit; it exceeds 6-10 times the activity in red muscles (55). A histochemical technique applied to hamster skeletal muscle demonstrated the strongest staining for the enzyme in the gastrocnemius fast, glycolytic fibers, and only weak staining in the slow, oxidative fibers (21). Thus well-oxygenated muscles (red skeletal muscle and heart, with oxidative energy production prevailing over glycolysis) clearly contain a lower capacity to deaminate AMP directly than the glycolysis-dependent muscles do.

Binding to Myosin Filament

Essential in the physiological role of AMP-deaminase catalysis could be enzyme regulation during muscle contraction. The formation of complexes between proteins of the contractile apparatus and the enzymes involved in muscle energy formation is well-known [e.g., creatine kinase (77), adenylate kinase (57)]. The interaction which exists between AMP-deaminase and myofibrils is sufficiently strong (K_d = 10^{-10} M) to get more than 99% of the binding sites of the enzyme occupied in vivo (14). The binding is remarkably increased in skeletal muscle contracting by electrical stimulation of the sciatic nerve (66). This treatment does not change the total enzyme activity. However, it increases the muscle ammonia and IMP content. Phosphate (< 10 mM) and GTP, which are inhibitors of AMP-deaminase, abolish the binding completely (44).

The purified enzyme binds to rat-muscle myosin in a molar ratio 1:3, mainly with the light meromyosin portion of the myosine molecule (67). However, the exclusive binding with the heavy meromyosin part or with subfragment 2 has also been reported for the rabbit enzyme (3). Myosin-bound AMP-deaminase is more sensitive to regulation by nucleotides than is the free enzyme (68). On the other hand, adenylosuccinate synthetase from rat muscle, the second enzyme of the purine nucleotide cycle, interacts with F-actin. This is not observed with AMP-deaminase (48). During contraction, when myosin (to which AMP-deaminase binds) slides along actin (to which adenylosuccinate synthetase binds), the cyclic operation of the purine nucleotide cycle could be triggered (67). Contradictory results have been published about the isozymic specificity of deaminase binding with myosin. Slight interference exists between rat muscle myosin and cardiac AMP-deaminase on the one hand (67). Efficient binding of heart enzyme to myosin is reported on the other (44). Both heart and skeletal muscle AMP-D bind by about 70% in that case (44).

The physiological role of the interaction adenylate deaminase-thick filament is unclear. The extent of interaction could be regulated by pH and phosphate (5). Transitions could occur in the range measured for exercising skeletal or ischemic heart muscle. The association could vary with the exercise-recovery cycle or with ischemic-postischemic changes in the heart. Thus the process may have significance in the metabolism of contractile tissues.

Enzyme Deficiencies

Myoadenylate deaminase deficiency is the most common of the known enzyme defects in muscle (for review, see ref. 17). Two forms occur, one being a primary type inherited as a complete gene block. The other is a secondary form consequent to muscle damage from other diseases. In the last form, the residual activity is 1-10% of normal; the other muscle enzymes are also depleted. Neuromuscular dysfunctions are more often associated with the secondary form. The complete lack of skeletal muscle AMP-D isozyme is accompanied by normal AMP-D levels in other tissues.

Among many cases, two related individuals have been described with dilated cardiomyopathy combined with skeletal myopathy, characterized by type I fiber atrophy (62). The metabolic consequences of AMP-D deficiency are mostly treated as connected with the disruption of the purine nucleotide cycle and unability of the muscle to restore high-energy phosphates deposit from adenylate catabolites after exercise (59) or turning off the one from the anaplerotic processes involved (63).

Models to Study Cardiac AMP-Deaminase

Different experimental models addressed the question whether the low activity of cardiac AMP-deaminase correlates with its minor role in heart adenylate catabolism. It is of particular interest to compare the properties of isolated AMP-deaminase with those of AMP-D in situ. Many authors isolated and characterized cardiac AMP-deaminase from various species. All purification procedures include a very efficacious phosphocellulose chromatographic step, first used by Smiley et al. (72) for rabbit skeletal muscle enzyme. The specific activities of the obtained highly purified AMP-D preparations were: 30 [pig (23)], 37 [duck (50)], 40-50 [rat (42)] and 30-50 U/mg protein [cow (Skladanowski, unpublished data)]. Only one was apparently homogeneous according to the criterion of SDS-electrophoresis (23). The molecular weights were estimated only for beef- and duck-heart AMP-deaminase. The former appeared to be a tetrameric structure with mol. wt. 161,000 D (71). The duck-heart enzyme contained two homotetrameric protein molecules of different isozymes, mol. wt. 190,000 and 240,000 D (50).

Isolated cardiomyocytes allow us to estimate AMP-deaminase's share in the whole adenylate metabolism. Preparations of isolated adult ventricular cells are now available in which the morphological and metabolic features of intact heart tissues are well preserved (e.g., refs. 11,16,25,33, and Chapter 15). Such myocyte preparations offer a convenient way to study nucleotide synthesis and degradation under conditions existing in the myocardial cell. In addition, we can look at the intracellular compartmentation of various components. Myocytes lose ATP and adenylates as a function of time during anaerobic incubation in the absence of glucose (25), the presence of uncouplers and glycolysis inhibitors (20), or the presence of 2-deoxyglucose and oligomycin (39). Measurement of ATP-catabolites in a cell homogenate, which can be done by HPLC, may reveal AMP-deaminase activity in situ.

The activities of various pathways may be gauged in intact cells of primary cardiomyocyte culture under physiological conditions, employing labelled precursors (80). This approach allows us to design a model of synthesis - degradation pathways and estimate the importance of different reactions, for example those catalyzed by AMP-deaminase. Its activity estimated in cardiomyocyte extract equals to 0.078 U/mg protein. It is of the order found in human fibroblasts and 8-fold lower than in skeletal muscle myotubes (80). Developmental changes of cardiac AMP-deaminase could be observed with embryonic heart cells (myoblasts), differentiated to myocytes, using selective cultivation conditions (40). The extensive change in the amount of AMP-deaminase, occurring during cell transformation, could be a promising approach to estimate its role in this kind of cell.

Important conclusions on the role of AMP-deamination in the myocardium have been derived from experiments with perfused isolated heart. In the Langendorff preparation (36), most commonly used, decreased perfusion or changed gas composition (from O_2 to N_2) initiates net ATP-catabolism. Catabolic enzymes could be activated. Tissue IMP changes during ischemia could indicate enhanced AMP-D activity, the sole pathway able to produce IMP during a relatively short time. Perfusion with substrates for ATP-resynthesis (adenine, adenosine, inosine, hypoxanthine, etc.) could establish AMP-deaminase's share in anabolic routes (see Chapter 26). Additionally, inclusion of enzyme inhibitors [like erythro-9-(2-hydroxy-3-nonyl)adenine (EHNA) for adenosine deaminase] and analysis of the degradation products in the perfusate allow us to revaluate two deamination steps existing in adenylate metabolism: adenosine and AMP-deamination. A specific, membrane-permeable AMP-deaminase inhibitor is not available. Therefore, direct insight in the AMP-D activity in the whole organ is impossible.

Preliminary data in human-heart biopsies, collected during open-heart surgery and analyzed with HPLC, demonstrate distinct IMP production, which correlates with decreased ATP-levels (Smolenski & Skladanowski, unpublished results).

Kinetic Properties and Regulation of Cardiac AMP-Deaminase

Substrate saturation kinetics of AMP-D under optimal conditions of ionic strength and pH ($\geq$100 mM KCl, pH 6.5) reveals a slightly sigmoidal curve, suggesting allosteric regulation. However, the Hill coefficient is reported sometimes as close to 1 (4,23), whereas the majority of values obtained are higher: 1.3 [rat (30)], 1.3 [duck (51)], 1.33 [man (32)], 1.7 [pigeon (29)], 2.0 [cow (70)], 2.35 [pig (53)]. Cooperativity between AMP-binding sites presumably exists, as the calculated number of functional protomers according to the model of Monod et al. is 2 (29, 30).

The half-saturation constant for heart AMP-D under optimal K^+ (or Na^+) concentrations and in the absence of other effectors is higher than that for skeletal muscle AMP-D (for exception, see ref. 42). It equals to: 1.86 mM [duck (51)], 2.11 mM [hen (31)], 2.78 mM [pigeon (29)], 3.04 mM [rat (30)], 3.1 mM [rabbit (4)], 3.8 mM [pig (23)], 3.9 mM [cow (70)], and 5.0 mM [pig (53)].

Purified myocardial AMP-deaminase is regulated by nucleotides, inorganic and organic phosphates, activated fatty acids, and alkali metal ions. The isolated enzyme is unstable and practically inactive in the absence of metal cations (Skladanowski, unpublished results). In the presence of at least 100 mM K^+ or Na^+, confirmed as saturating, AMP-D from beef heart is regulated by nucleotides and phosphate (Skladanowski, unpublished data) in a way corresponding to a "K_M-type" of enzyme, whereas a mixed "K_M,V_{max}-type" of regulation is observed at low ion concentrations (70).

Activation by ATP and ADP

Subphysiological concentrations of ATP (0.5 mM) and near-physiological concentrations of ADP (0.5 - 1.5 mM) are able to activate maximally the majority of heart AMP-D preparations studied so far. The activation constants for ATP and ADP, calculated for rat-heart AMP-D at 0.75 mM AMP concentration, were 0.037 and 0.095 mM, respectively (30). $K_{0.5}$ was decreased from 3.04 to about 0.6 mM in the presence of 0.5 mM ATP, and to about 1 mM in the presence of 1.5 mM ADP (30). Rabbit-heart AMP-D was activated up to 300% by 0.05 mM ATP at 0.1 mM AMP concentration, but only to 30% by 0.05 mM ADP (4). ATP-Mg^{2+} was shown to be an even stronger activator (4); at a physiological Mg^{2+} concentration, a part of ATP exists as a magnesium complex (19). ATP and ADP effects are further enhanced (about 100% with ATP, and 50% with ADP) in the presence of artificial phospholipid membranes (53). This is not true for skeletal muscle AMP-deaminase. Chung and Bridger (12) reported a stronger stimulation of rabbit-heart AMP-D by ADP than by ATP. This dissimilarity is probably caused by a chelating effect of citrate.

Inhibition by Phosphate and GTP

Inorganic phosphate, phosphocreatine and several other organic phosphate compounds inhibit the cardiac enzyme analogous to skeletal muscle AMP-deaminase. I_{50} values are calculated to be about 0.5 mM and 2.0 mM for phosphocreatine and P_i inhibition of rabbit-heart AMP-D at 0.1 mM AMP, respectively (4). The binding of the phosphate to the enzyme molecule shifts the substrate saturation curve to a distinct sigmoid (4), but ATP reverses this effect instantly (28).

GTP is known as the most effective inhibitor of AMP-D from skeletal muscle (7). However, its effect on rat-heart AMP-deaminase is more complex. Activation is even observed at low GTP concentrations (30). Rabbit-heart AMP-D is inhibited by about 25% at 0.1 mM AMP in the presence of 5 uM GTP. The activation by micromolar concentrations of ADP and ATP is abolished (4). The inhibition of duck-heart AMP-D by various GTP concentrations is biphasic. It is probably caused by a different sensitivity of the two existing isoforms (51).

Inhibition by Activated Fatty Acids

Palmitoyl-CoA and stearyl-CoA strongly inhibit the activity of AMP-deaminase from beef (69), hen (31) and pigeon (29) heart at concentrations as low as 5 uM. Both acyl-CoA's inhibit well below their critical micelle concentrations (73). Many enzymes are inhibited by activated fatty acids (78). For some the effect is due to ligand-induced irreversible dissociation of polymeric enzymes (34). The inhibition of heart AMP-D could be of physiological importance during ischemia when an increase in activated fatty acids level in the tissue is observed (6).

Regulation by Adenylate Energy Charge

The response of AMP-deaminase to variations in adenylate energy charge ([ATP] + 0.5[ADP])/([ATP] + [ADP] + [AMP]) reflects best its behavior in a dynamically changing intracellular environment. The activities of pig-heart AMP-D (23) and the beef-heart enzyme (Skladanowski, unpublished results) are suppressed at energy charge values in the physiological range (0.8 - 0.9). A decrease of this parameter, occurring during ischemia (56), could deinhibit AMP-D and be responsible for IMP accumulation in myocardium (refs. 20,26; Smolenski & Skladanowski, unpublished data). On the other hand, rabbit-heart AMP-D (74) and the rat-heart enzyme (Skladanowski, unpublished results) are unaffected by energy charges below 0.9. In contrast, the skeletal muscle enzyme increases gradually in activity, when the energy charge decreases from 1.0 to 0.0 (74). Rabbit- and rat-heart AMP-deaminase always seem to be in the activated state.

Is Cardiac AMP-Deaminase regulated in Vivo?

According to Burger and Lowenstein (10), AMP deamination dominates AMP dephosphorylation due to the cardiac ATP level. However, during accelerated adenylate catabolism, the heart produces considerable amounts of adenosine, and little IMP (58).

The activation of cardiac AMP-D by ATP seems to be insignificant for in vivo regulation. We disregard the question how 5'-nucleotidase, responsible for adenosine production, is affected by ATP (27,38). The fall of ATP-level during ischemia may not exceed 60% (26). Even then AMP-D remains saturated by ATP. Inhibitory effects by GTP as well as P_i are most likely permanently suppressed by ATP-activation (4). Therefore, an ischemia-induced P_i-increase (24) probably does not play a significant role in heart AMP deamination. This thesis is confirmed by the fact that 2 mM P_i does not change hen-heart AMP-D substrate saturation kinetics in the presence of 5 mM ATP (28). ADP is unlikely to regulate AMP-D activity in the muscle cell, because it preferentially binds to actin filaments.

The key role in AMP-D regulation in heart therefore seems to be played by the level and accessibility of substrate. The free AMP concentration in guinea-pig heart is well below 1 uM (8). This has implications for both cytosolic 5'-nucleotidase and adenylate deaminase. Since the $K_{0.5}$ for AMP of each enzyme appears to be considerable higher than its free cytosolic concentration, the catalytic properties of either enzyme are unlikely to be rate-controlling. As mentioned before, AMP dephosphorylation prevails of deamination. Substrate availability for these potentially competing enzymes is therefore probably limited by intracellular compartmentation.

In conclusion none of the properties of heart AMP-D seems to be more significant than the other for in vivo regulation of this enzyme. More complex experiments on the cumulative effects of effectors with micromolar AMP concentrations could settle this problem.

Development of Heart AMP-Deaminase

Changes in properties of skeletal muscle AMP-D during growth are due to isozymic pattern differentiation (61). According to Ogasawara et al. (43), rat-heart tissue possesses one type of AMP-deaminase isozyme (denoted C). This belongs to the isozyme family: homoprotomeric A (skeletal muscle), B (liver), C (heart), and heteroprotomeric hybrids composed of the B and C type subunits. On the other hand, three types of isozymes are found in human heart. Two have the properties of the erythrocytic forms E_1 and E_2, and one those of the hepatic form L (47). Two forms differing in kinetic and regulatory properties, have also been detected in beef heart (Skladanowski, unpublished results).

Developmental changes occur in the amount of isozymes in rat heart, with a shift from five isoenzymes to only one adult form (46). In most tissues (including heart), AMP-D activities are higher in mature cells or organs than in the early postnatal stage (46). Chicken-heart AMP-D activity increases about four times between 18 days of postnatal development and adulthood (54).

Striking differences in adenylate metabolism are encountered in rat-heart myoblasts, differentiating to myocytes. The 14-fold increase in AMP-D activity correlates well with a jump in creatine kinase activity (40). These and other enzyme activity changes during development are presented in Table 1.

AMP-deaminase may be important for the conservation of purine nucleotides through the operation of purine nucleotide cycle. This also holds for 5'-nucleotidase, operating through vasodilation by adenosine (or inosine). The ratio AMP-D : 5'-nucleotidase changes during differentiation from 0.11 to 0.48. This indicates that, in rat-heart myocytes, the interconversion of purine nucleotides is mainly from AMP to IMP, not inversely (40). The same is true for rat skeletal muscle cells (81).

The increase of enzymes responsible for ATP sparing and purine ring conservation in myocytes correlates temporarily with an increase of the ATP demand of and use by contracting myocytes.

Heart AMP-Deaminase and AMP Catabolism

During oxygen deficiency cytosolic ADP cannot be rephosphorylated by creatine phosphate. ADP is then used more effectively by myokinase than by creatine kinase and mitochondrial oxidative phosphorylation. Judged from AMP-binding equilibria constants, AMP levels increase from about 100 nM during normoxia to about 800 nM in ischemic hearts (8). This change occurs in a strikingly lower range than that suggested by total tissue AMP analysis: 100-200 uM (9). The change in free AMP is several fold larger, and thus could be a better stimulus for cytoplasmic enzymes, converting AMP. Under these conditions AMP becomes available for adenylate deaminase and 5'-nucleotidase, either for further degradation to IMP or to the coronary vasodilator adenosine, respectively. In fact, several studies demonstrate a rise in myocardial IMP level during ischemia (15,26,49) and especially during anoxia (18). Also in cardiomyocytes, subjected to conditions stimulating ATP catabolism, IMP distinctly increases (20,39). However, the superiority of the AMP → adenosine route over that of AMP → IMP in ischemic heart adenylate catabolism is not questioned. It cannot be explained by kinetic differences between AMP-D and cytosolic 5'-nucleotidase, because these enzymes, isolated from rat heart (27,30), have very similar properties. Nevertheless, the total enzyme activity of 5'-nucleotidase is about 15 times higher than that of AMP-deaminase (60).

The question remains: What are the further fates of accumulated IMP? Part is most likely hydrolyzed to inosine, transported from the myocytes to the endothelial cells, and degraded to uric acid. The rest establishes a pool of nondiffusable nucleotides, ready for recovery of depleted adenylates by the purine nucleotide cycle. The presence and role of this cycle in heart musle is still controversial (75). Time-delayed variation in flux through different parts of the cycle – AMP-deaminase during

Table 1. Changes of enzyme activities in heart adenylate metabolism during cell differentiation and development

Enzyme	Increase	Reference
Creatine kinase	10-12x	(35,40)
AMP-deaminase	14x	(40)
Adenosine deaminase	1.4x	(40)
5'-Nucleotidase	3.1x	(40)
Purine nucleoside phosphorylase	1.7x	(40)
Myokinase	ca. 3x	(35)
Xanthine oxidase	manifold	(64)
Adenosine kinase	1.4x	(40)
HGPRT	1.7x	(40)
APRT	1.6x	(40)

ischemia and adenylosuccinate synthetase with adenylosuccinase during recovery – could be an explanation. The size of the remaining IMP pool must depend on the rate of cytosolic dephosphorylation by 5'-nucleotidase. This reaction is inhibited by increased P_i and by decreased energy charge (27). Thus the preservation of IMP would be better during more severe ischemia.

The importance of AMP-deamination in AMP catabolism during either normoxia or ischemia is indirectly confirmed by the use of EHNA, a strong adenosine deaminase inhibitor, during rat-heart perfusion with adenosine (2), or 5'-amino-5'-deoxyadenosine (an adenosine kinase inhibitor) combined with EHNA during guinea-pig heart perfusion (65). EHNA increases the proportion of adenosine released, without changing total purine release in normoxia from 7 (without inhibitor) to 18%, and from 15 (without inhibitor) to 60% of total purine release during ischemia. It suggests that purine formation through AMP → IMP → inosine, bypassing the inhibited step is significant under both conditions (2). The decrease of intramyocardial adenosine without change of total purine release after homocysteine infusion during ischemia also indicates that purine formation could occur partly via IMP under ischemic conditions (1).

Concluding Remarks

1. Myocardium contains AMP-deaminase activity which is about two orders of magnitude lower than that in skeletal muscle. The enzyme content rises sharply in the myocardial cell during development, together with other enzymes responsible for ATP maintenance.

2. Isolated AMP-deaminase is activated by ATP, ADP, K^+ or Na^+ ions, and inhibited by GTP, P_i, activated fatty acids. The activation by ATP is enhanced by phospholipid bilayers. All nucleotide effects are modulated by Mg^{2+}. The reaction rate in vivo is mainly regulated by availability of the substrate AMP, which is normally present in submicromolar amounts. Changes of enzyme effectors are probably insignificant for in vivo regulation of AMP-deaminase in view of the prevailing ATP activation.

3. AMP deamination is responsible for a considerable part of the purine catabolites released from ischemic heart and probably for the entire purine efflux during normoxia.

4. The postulated physiological role for cardiac AMP-deaminase is preservation of the purine ring inside myocytes (as nondiffusable IMP) during increased ATP catabolism. Salvaged IMP could later be the source for fast adenylate pool restoration.

References

1. Achterberg PW, De Tombe PP, Harmsen E, De Jong JW: Myocardial S-adenosylhomocysteine hydrolase is important for adenosine production during normoxia. Biochim Biophys Acta 840:393-400, 1985
2. Achterberg PW, Harmsen E, De Jong JW: Adenosine deaminase inhibition and myocardial purine release during normoxia and ischemia. Cardiovasc Res 19:593-598, 1985

3. Ashby B, Frieden C: Interactions of AMP-aminohydrolase with myosin and its subfragments. J Biol Chem 252:1869-1872, 1977

4. Barsacchi R, Ranieri-Raggi M, Bergamini C, Raggi A: Adenylate metabolism in the heart. Regulatory properties of rabbit cardiac adenylate deaminase. Biochem J 182:361-366, 1979

5. Barshop BA, Frieden C: Analysis of the interaction of rabbit skeletal muscle adenylate deaminase with myosin subfragments. A kinetically regulated system. J Biol Chem 259:60-66, 1984

6. Bittar N, Shug AL, Koke JR, Folts JD, Shrago ES: Inhibited adenine nucleotide translocation in mitochondria isolated from ischemic myocardium. Recent Adv Stud Cardiac Struct Metab 7:137-143, 1976

7. Brody TG, Costello JF: Activation and inhibition of AMP deaminase by GTP and ATP. Biochim Biophys Acta 350:455-460, 1974

8. Bünger R: Thermodynamic state of cytosolic adenylates in guinea pig myocardium. Energy-linked adaptive changes in free adenylates and purine nucleoside release. In: Gerlach E, Becker BF, eds: Topics and perspectives in adenosine research. Berlin: Springer Verlag, 1987:223-235

9. Bünger R, Soboll S: Cytosolic adenylates and adenosine release in perfused working heart. Comparison of whole tissue with cytosolic non-aqueous fractionation analyses. Eur J Biochem 159:203-213, 1986

10. Burger R, Lowenstein JM: Adenylate deaminase. III. Regulation of deamination pathways in extracts of rat heart and lung. J Biol Chem 242:5281-5288, 1967

11. Cheung JY, Thompson IG, Bonventre JV: Effects of extracellular calcium removal and anoxia on isolated rat myocytes. Am J Physiol 243:C184-C190, 1982

12. Chung L, Bridger WA: Activation of rabbit cardiac AMP aminohydrolase by ADP. A component of a mechanism guarding against ATP depletion. FEBS Lett 64:338-340, 1976

13. Conway EJ, Cooke R: LIX. The deaminase of adenosine and adenylic acid in blood and tissues. Biochem J 33:479-492, 1939

14. Cooper J, Trinick J: Binding and location of AMP deaminase in rabbit psoas muscle myofibrils. J Mol Biol 177:137-152, 1984

15. Deuticke B, Gerlach E: Abbau freier Nucleotide in Herz, Skeletmuskel, Gehirn und Leber der Ratte bei Sauerstoffmangel. Pflügers Arch 229:239-254, 1966

16. Dow JW, Harding NGL, Powell T: Isolated cardiac myocytes. I. Preparation of adult myocytes and their homology with intact tissue. Cardiovasc Res 15:483-514, 1981

17. Fishbein WN: Myoadenylate deaminase deficiency: Inherited and acquired forms. Biochem Med 33:158-169, 1985

18. Frick GP, Lowenstein JM: Studies of 5'-nucleotidase in the perfused rat heart. J Biol Chem 251:6372-6378, 1976

19. Garfinkel L, Altschuld RA, Garfinkel D: Magnesium in cardiac energy metabolism. J Mol Cell Cardiol 18:1003-1013, 1986

20. Geisbuhler T, Altschuld RA, Trewyn RW, Ansel AZ, Lamka K, Brierley GP: Adenine nucleotide metabolism and compartmentalization in isolated adult rat heart cells. Circ Res 54:536-546, 1984

21. Gilloteaux J: Ader M AMP deaminase histoenzymology in hamster skeletal muscle. Acta Histochem 73:47-51, 1983

22. Greger J, Fabianowska K: Relationship between 5'-nucleotidase, adenosine deaminase, AMP deaminase, ATP-(Mg^{2+})-ase activities and dTMP kinase activity in rat liver mitochondria. Enzyme 24:54-60, 1979

23. Harmsen E, Verwoerd TC, Achterberg PW, De Jong JW: Regulation of porcine heart and skeletal muscle AMP-deaminase by adenylate energy charge. Comp Biochem Physiol 75B:1-3, 1983

24. Hearse DJ, Crome R, Yellon DM, Wyse R: Metabolic and flow correlates to myocardial ischemia. Cardiovasc Res 17:452-458, 1983

25. Hohl C, Ansel A, Altschuld RA, Brierley GP: Contracture of isolated rat heart cells on anaerobic to aerobic transition. Am J Physiol 242:H1022-H1030, 1982

26. Humphrey SM, Holliss DG, Seelye RN: Myocardial adenine pool depletion and recovery of mechanical function following ischemia. Am J Physiol 248:H644-H651, 1985

27. Itoh R, Oka J, Ozasa H: Regulation of heart cytosol 5'-nucleotidase by adenylate energy charge. Biochem J 235:847-851, 1986
28. Kaletha K: Hen heart AMP-deaminase - The combined effect of ATP, ADP and orthophosphate on the enzyme activity. Int J Biochem 16:83-85, 1984
29. Kaletha K, Bogdanowicz S, Raffin J-P: Regulatory properties of pigeon heart muscle AMP deaminase. Biochimie 69:117-123, 1987
30. Kaletha K, Skladanowski A: Regulatory properties of rat heart AMP deaminase. Biochim Biophys Acta 568:80-90, 1979
31. Kaletha K, Skladanowski A: Regulatory properties of 14 day embryo and adult hen heart AMP-deaminase. Int J Biochem 16:75-81, 1984
32. Kaletha K, Skladanowski A, Bogdanowicz S, Zydowo M: Purification and some regulatory properties of human heart adenylate deaminase. Int J Biochem 10:925-929, 1979
33. Kao RL, Christman EW, Luh SL, Kraubs JM, Tylers GF, Williams EH: The effects of insulin and anoxia on the metabolism of isolated mature rat cardiac myocytes: Arch Biochem Biophys 203:587-599, 1980
34. Kawaguchi A, Bloch K: Inhibition of glutamate dehydrogenase and malate dehydrogenase by palmitoyl coenzyme A. J Biol Chem 251:1406-1412, 1976
35. Kimes BW, Brandt BL: Properties of a clonal muscle cell line from rat heart. Exp Cell Res 98:367-381, 1976
36. Langendorff O: Untersuchungen am überlebenden Saugethierherzen. Arch Gesam Physiol 61:291-332, 1895
37. Lowenstein JM: Ammonia production in muscle and other tissues: the purine nucleotide cycle. Physiol Rev 52:382-414, 1972
38. Lowenstein JM, Yu M-K, Naito Y: Regulation of adenosine metabolism by 5'-nucleotidases. In: Berne RM, Rall TW, Rubio R, eds: Regulatory function of adenosine. The Hague: Nijhoff Publ, 1983:117-132
39. Meghji P, Holmquist CA, Newby AC: Adenosine formation and release from neonatal-rat heart cells in culture. Biochem J 229:799-805, 1985
40. Müller MM, Rumpold H, Schopf G, Zilla P: Changes of purine metabolism during differentiation of rat heart myoblasts. In: Nyhan WL, Thompson LF, Watts RWE, eds: Purine and pyrimidine metabolism in man. New York: Plenum Press, 1986:475-484
41. Nakatsu K, Drummond GI: Adenylate metabolism and adenosine formation in the heart. Am J Physiol 223:1119-1127, 1972
42. Ogasawara N, Goto H, Watanabe T: Isozymes of rat AMP deaminase. Biochim Biophys Acta 403:530-537, 1975
43. Ogasawara N, Goto H, Watanabe T: Isozymes of AMP deaminase. In: Müller MM, Kaiser E, Seegmiller JE, eds: Purine metabolism in man - II. Regulation of pathways and enzyme defects. New York: Plenum Press, 1977:212-222
44. Ogasawara N, Goto H, Yamada Y: Effects of various ligands on interaction of AMP deaminase with myosin. Biochim Biophys Acta 524:442-446, 1978
45. Ogasawara N, Goto H, Yamada Y: AMP deaminase isozymes in rabbit red and white muscles and heart. Comp Biochem Physiol 76B:471-473, 1983
46. Ogasawara N, Goto H, Yamada Y, Watanabe T: Distribution of AMP-deaminase isozymes in rat tissues. Eur J Biochim 87:297-304, 1978
47. Ogasawara N, Goto H, Yamada Y, Watanabe T, Asano T: AMP deaminase isozymes in human tissues. Biochim Biophys Acta 714:298-306, 1982
48. Ogawa H, Shiraki H, Matsuda Y, Nakagawa H: Interaction of adenylosuccinate synthetase with F-actin. Eur J Biochem 85:331-337, 1978
49. Parker JC, Smith EE, Jones CE: The role of nucleoside and nucleobase metabolism in myocardial adenine nucleotide regeneration after cardiac arrest. Circ Shock 3:11-20, 1976

50. Pekkel' VA, Kirkel' AZ: Purification and certain physicochemical properties of myocardial adenylate deaminase. Biokhimiya (Engl Transl) 44:1311-1319, 1979
51. Pekkel' VA, Kirkel' AZ, Gorkin VZ: Kinetic and regulatory properties of myocardial adenylate deaminase. Biokhimiya (Engl Transl) 45:290-298, 1980
52. Purzycka J: AMP and adenosine aminohydrolases in rat tissues. Acta Biochim Polon 9:83-93, 1962
53. Purzycka-Preis J, Prus E, Wozniak M, Zydowo M: Modification by liposomes of the adenosine triphosphate-activating effect on adenylate deaminase from pig heart. Biochem J 175:607-612, 1978
54. Purzycka-Preis J, Wrzolkowa T, Pawlak-Byczkowska E, Zydowo M: Developmental changes of AMP-deaminase activity in chick heart muscle. Int J Biochem 6:885-887, 1975
55. Raggi A, Ronca-Testoni S, Ronca G: Muscle AMP aminohydrolase. II. Distribution of AMP aminohydrolase and creatine kinase activities in skeletal muscle. Biochim Biophys Acta 178:619-622, 1969
56. Reimer KA, Murry CE, Yamasawa I, Hill ML, Jennings RB: Four brief periods of myocardial ischemia cause no cumulative ATP loss or necrosis. Am J Physiol 251:H1306-H1315, 1986
57. Richards EG, Chung C-S, Menzel DB, Olcott HS: Chromatography of myosin on diethylaminoethyl-Sephadex. Biochemistry 6:528-540, 1967
58. Rubio R, Berne RM, Dobson JG Jr: Sites of adenosine production in cardiac and skeletal muscle. Am J Physiol 225:938-953, 1973
59. Sabina RL, Swain JL, Olanow CW, Bradley WG, Fishbein WN, DiMauro S, Holmes EW: Myoadenylate deaminase deficiency. Functional and metabolic abnormalities associated with disruption of the purine nucleotide cycle. J Clin Invest 73:720-730, 1984
60. Saleem Y, Niveditha T, Sadasivudu B: AMP deaminase, 5'-nucleotidase and adenosine deaminase in rat myocardial tissue in myocardial infarction and hypothermia. Experientia 38:776-777, 1982
61. Sammons DW, Chilson OP: AMP-deaminase: Stage-specific isoenzymes in differentiating chick muscle. Arch Biochem Biophys 191:561-570, 1978
62. Scholte HR, Busch HFM, Luyt-Houwen IEM: Familial AMP deaminase deficiency with skeletal muscle type I atrophy and fatal cardiomyopathy. J Inher Metab Dis 4:169-170, 1981
63. Scholte HR, Busch HFM, Meijer AEFH, Luyt-Houwen IEM, Vaandrager-Verduin MHM: The role of myoadenylate deaminase deficiency in myopathy. Klin Wochenschr 65, Suppl X:26, 1987 (Abstr)
64. Schoutsen B, De Jong JW: Age-dependent increase in xanthine oxidoreductase differs in various heart cell types. Circ Res 61:604-607, 1987
65. Schrader J: Metabolism of adenosine and sites of production in the heart. In: Berne RM, Rall TW, Rubio R, eds: Regulatory function of adenosine. The Hague: Nijhoff Publ, 1985:133-156
66. Shiraki H, Miyamoto S, Matsuda Y, Momose E, Nakagawa H: Possible correlation between binding of muscle type AMP deaminase to myofibrils and ammoniagenesis in rat skeletal muscle on electrical stimulation. Biochem Biophys Res Commun 100:1099-1103, 1981
67. Shiraki H, Ogawa H, Matsuda Y, Nakagawa H: Interaction of rat muscle AMP deaminase with myosin. I. Biochemical study of the interaction of AMP deaminase and myosin in rat muscle. Biochim Biophys Acta 566:335-344, 1979
68. Shiraki H, Ogawa H, Matsuda Y, Nakagawa H: Interaction of rat muscle AMP deaminase with myosin. II. Modification of the kinetic and regulatory properties of rat muscle AMP deaminase by myosin. Biochim Biophys Acta 566:345-352, 1979
69. Skladanowski A, Kaletha K, Zydowo M: Inhibition of AMP-deaminase from beef heart by palmitoyl and stearyl-CoA. Int J Biochem 9:43-45, 1978
70. Skladanowski A, Kaletha K, Zydowo M: Potassium-dependent regulation by ATP and ADP of AMP-deaminase from beef heart. Int J Biochem 10:177-181, 1979
71. Skladanowski A, Kaletha K, Zydowo M: Hydro- and thermodynamic properties of bovine heart AMP-deaminase. Int J Biochem 13:865-869, 1981
72. Smiley KL, Berry A, Suelter CH: An improved purification, crystallization and some properties of rabbit muscle 5'-adenylic acid deaminase. J Biol Chem 242:2502-2506, 1967

73. Smith L, Powell G: The critical micelle concentration of some physiologically important fatty acyl-coenzyme A's as a function of chain length. Arch Biochem Biophys 244:357-360, 1986
74. Solano C, Coffee CJ: Differential response of AMP-deaminase isozymes to changes in the adenylate energy charge. Biochem Biophys Res Commun 85:564-571, 1978
75. Taegtmeyer H: On the role of the purine nucleotide cycle in the isolated working rat heart. J Mol Cell Cardiol 17:1013-1018, 1985
76. Takala T, Hiltunen JK, Hassinen IE: The mechanism of ammonia production and the effect of mechanical work load on proteolysis and amino acids catabolism in isolated perfused rat heart. Biochem J 192:285-295, 1980
77. Turner DC, Walliman T, Eppenberger HM: A protein that binds specifically to the M-line of skeletal muscle is identified as the muscle form of creatine kinase. Proc Natl Acad Sci (USA) 70:702-705, 1973
78. Weber G, Lea MA, Stamm NB: Sequential feedback inhibition and regulation of liver carbohydrate metabolism through control of enzyme activity. Adv Enzyme Regul 6:101-107, 1968
79. Zielke CL, Suelter CH: purine, Purine nucleoside, and purine nucleotide aminohydrolases. In: Boyer PD, ed: The enzymes. Vol IV - Hydrolysis. Other C-N bonds, phosphate esters. New York: Acad Press, 1971:47-78
80. Zoref-Shani E, Kessler-Icekson G, Wasserman L, Sperling O: Characterization of purine nucleotide metabolism in primary rat cardiomyocyte cultures. Biochim Biophys Acta 804:161-168, 1984
81. Zoref-Shani E, Shainberg A, Sperling O: Alteration in purine nucleotide metabolism during muscle differentation in vitro. Biochem Biophys Res Commun 116:507-512, 1983

Chapter 7

Redox Manipulation of Free Cardiac Adenylates and Purine Nucleoside Release

Reciprocity between Cytosolic Phosphorylation Potential and Reduction-Oxidation State or Free AMP in Perfused Working Heart

R. Bünger, R.T. Mallet and D.A. Hartman, Department of Physiology, School of Medicine, Uniformed Services University of the Health Sciences, Bethesda, Maryland, U.S.A.

The thermodynamic equilibrium relations between free adenylates and redox state in the cytosol were reviewed as applicable to perfused heart. Studies with isolated working guinea-pig hearts examined cytosolic redox state, phosphorylation potential, and the free concentration of AMP ([AMP]) in relation to the coronary venous output of adenylate catabolites. Excess pyruvate (20 mM), relative to 20 mM lactate, produced increases in the phosphorylation potential and total cardiac AMP content, but inorganic phosphate content, [ADP], [AMP], and purine nucleoside release were decreased. Conversely, excess lactate induced the releases of inosine and large amounts of xanthine, the reduced precursor of uric acid. When 15 mM glucose was the sole substrate, lactate output and adenosine plus inosine release changed concordantly during adrenergic stimulation. During low-flow ischemia with adrenergic stimulation, output of purines showed a large but transient increase, was mainly (67% - 87%) in the form of adenosine plus inosine, and was attenuated by cytosolic oxidation due to intracoronary administration of 5 mM pyruvate. Uric acid output accounted for 40% - 50% of total purine release during normoxia, but for not more than 10% during early ischemia. It was concluded that, in perfused working heart, 1) the cytosolic oxidation by pyruvate can increase myocardial energy state, apparently in accordance with a reciprocal relation between the cytosolic $[NADH]/[NAD^+]$ ratio and the phosphorylation potential, 2) a high phosphorylation potential is associated with very low, i.e., submicromolar [AMP] and minimum adenylate catabolite output, and 3) the cytosolic phosphorylation potential may exert control over free [AMP], which appears to function as a major determinant of 5'-nucleotidase flux.

Introduction and Overview

Previously we demonstrated a quantitative association between the rate of myocardial purine nucleoside release and the calculated concentration of free AMP ([AMP]) in the cytosol (6,10,11). Because [AMP], like the free [ADP]/[ATP] ratio, was highly responsive to physiologic changes in myocardial energy balance and/or utilization, free cytosolic AMP was regarded as an integral component of the free extramitochondrial adenylate system. Established characteristics of the free cytosolic adenylate system in working heart are 1) adaptability of the cytosolic phosphorylation potential (ATP potential) to the acute myocardial energy balance over a wide range of workloads, 2) an apparent linear relationship between free cytosolic [ADP] and myocardial oxygen consumption (mitochondrial respiration) in the physiological range, with [ADP] between 10 and 100 uM, and 3) strict compartmentation of AMP between cytosol and mitochondrial matrix, with [AMP] in the upper submicromolar range being highly responsive to the [ADP]/[ATP] ratio. Since, moreover, the [ADP]/[ATP] ratio is a major determinant of the cytosolic phosphorylation potential ([ATP]/([ADP] · [P$_i$]), myocardial energy state and [AMP] are reciprocally related. Net release of adenosine plus inosine appeared to correlate more closely with [AMP] than with total myocardial AMP content (6,10), supporting the possibility that 5'-nucleotidase flux is controlled by the availability of cytosolic AMP. Studies in skeletal muscle, liver, brain and erythrocytes have suggested that cytosolic redox state and ATP potential are, in principle, reciprocally related (27,54). It seems therefore logical to assume that cytosolic reduction potential and [AMP] may change concordantly. This raised the possibility that myocardial release of adenylate catabolites (purine nucleosides and bases) might be responsive to the cytosolic redox state per se, with the phosphorylation potential functioning as the intermediary metabolite mediator system. In fact, the coronary effluent lactate/ pyruvate concentration ratio and the rate of myocardial purine nucleoside output were directly related (6).

In the present study we examined further the relationships between cytosolic oxidation-reduction state, ATP phosphorylation potential, free cytosolic [AMP], and net release of purine nucleosides and bases, using the isolated physiologically performing guinea-pig heart as the experimental model. The cytosolic redox state was externally manipulated by infusion of excess amounts of redox substrates (lactate, pyruvate) under the conditions of normoxia and global, low-flow ischemia with residual oxygen consumption.

Theoretical background

Creatine kinase and myokinase are powerful enzymes in skeletal muscle, heart and brain, i.e., in tissues with potentially high ATP turnover rates (36,38). The enzyme couple is strategically located in the extramitochondrial space (mitochondrial intermembrane + cytosolic compartments), but is absent from the intramitochondrial matrix (23,38,46). The kinase equilibria are only responsive to the thermodynamic ("free") concentrations of ATP, ADP, and AMP ([ATP], [ADP], [AMP]) or to the cytosolic [ATP]/[ADP] ratio. Since the bulk of myocardial ATP is located in the extramitochondrial compartment (10,20,52) and because actomyosin only insignificantly binds ATP (40), the chemically determined overall myocardial ATP content has been taken as a reasonable approximation of [ATP]. Current estimates based on total tissue analysis, non-aqueous or digitonin fractionation techniques place cytosolic [ATP] in the 7 to 12 mM range (10,20,52). These concentrations are principally found in hearts or cardiomyocytes with "energized" mitochondria under respiratory control, since energization of the inner mitochondrial membrane produces a gradient for the [ATP]/[ADP] ratio (higher outside) across this membrane (22,25,26,52). The subcellular distribution of ADP is complex, because actin and also myosin tightly bind ADP (21,40,48) and because the energized ADP-ATP translocase of the inner mitochondrial membrane exchanges extramitochondrial ADP in preference for intramitochondrial ATP (22,25,26). Thus, ADP rephosphorylation by creatine kinase, binding by actomyosin, and the vectorial ADP-ATP translocase during mitochondrial respiratory control act in concert to reduce the extramitochondrial [ADP] to a physiologically permissible minimum. In fact, the creatine kinase equilibrium predicts [ADP] in the 10 to 100 uM range as a function of the physiological state of the heart (6,10).

Free ADP concentrations in the low submillimolar range are consistent with the absence of a definitive ADP signal in ^{31}P-NMR studies in, e.g., heart and skeletal muscle. Consequently, relatively high values between 10,000 - 100,000 M^{-1} of the cytosolic phosphorylation potential are obtained. The value of [ATP]/([ADP]·[P$_i$]), combined with the standard free energy of ATP hydrolysis [$\Delta G°_{ATP}$ = -32.35 kJ/mol under conditions prevailing in vivo (54)], determines the actual free energy change during ATP hydrolysis in the cytosol:

(Eq. 1) $\Delta G_{ATP} = \Delta G°_{ATP} + RT \ln([ADP] \cdot [P_i])/[ATP]$

in which R = gas constant, 8.314 J/°K/mol, and T = absolute temperature, ° Kelvin.

In pressure-overloaded, isolated working guinea-pig heart, the cytosolic phosphorylation potential can fall to near 7,000 M^{-1} yielding ΔG_{ATP} = -55.17 kJ/mol (13.18 kcal/mol); on the other hand, in non-working, empty, beating Langendorff heart, the ATP potential is near or higher than 80,000 M^{-1} [$\Delta G_{ATP} \geq$ -61.45 kJ/mol (14.68 kcal/mol), ref. 10]. The free energy change of cytosolic ATP hydrolysis may be compared with the 52.7 kcal liberated by the mitochondrial respiration chain when a pair of electron passes from intramitochondrial NADH to oxygen to yield H_2O (30). Assuming a coupling ratio of 3 ATP per oxygen atom reduced, the energy stored in cytosolic ATP varies between 39.54 kcal (75% efficiency) at high workloads to 44.04 kcal (83.5% efficiency) at low workloads. The reason for the apparent decrease in cytosolic energy storage efficiency at high workloads (i.e., high rates of ATP turnover) is not clear, but may be related in part to the fact that energy is obligatorily expended when intramitochondrial ATP (with a relatively low energy potential) must be exported at increased rates into the cytosol, which has a relatively higher phosphorylation potential (22,25,26,52).

Free cytosolic ADP cannot be directly measured by current invasive techniques (acid extraction, sonication plus density gradient fractionation, digitonin cell membrane lysis). However, division of the chemically measured cellular ATP concentration by the value of cytosolic [ATP]/[ADP] (derived from the creatine kinase equilibrium equation), yields the mean cytosolic [ADP]. This approach appears to be justified, since most of guinea-pig cardiac creatine kinase (about 89%) may be located in the cytosol (37). There appears to be no substantial ATP concentration gradient between cytosol and nucleus (50) and the intermembrane space of mitochondria is freely permeable to cytosolic adenylates, protons, and creatine compounds. However, the [ADP] obtained by equilibrium equations ignores the possible existence of microenvironmental diffusion gradients (e.g., ref. 34). Possible differences in the chemical activity of water in the various subparenchymal and subcellular compartments are also not taken into consideration.

The myokinase reaction, like the creatine kinase reaction, is reversible. In skeletal muscle, heart, and brain, myokinase activities are at least as high as those of creatine kinase (38). Coronary endothelium also contains creatine phosphate (and hence creatine kinase) and myokinase in substantial quantities (35). The myokinase equilibrium constant is dependent on the hydrogen ion concentration and free Mg^{2+} ([Mg^{2+}]) in the cytosol. Thus, under conditions of low constant net metabolic throughput (relative to the total potential activity) and constant metabolite levels for a sufficiently extended period (metabolic steady state), the myokinase reaction seems likely to reach the thermodynamic near-equilibrium state:

(Eq. 2) $[AMP] = K_{myk} \cdot [ADP]^2/[ATP]$

K_{myk} = 1.12 at pH 7.2, [Mg^{2+}] = 0.9 mM (29).

Even in systems with rather low myokinase activities, e.g., erythrocytes and liver, the enzyme probably operates at near-equilibrium (54). Moreover, it has been argued (49,54) that in brain, where activities of myokinase and creatine kinase are about 30% of those in heart (38), the enzymes are nevertheless sufficiently powerful to catalyze near-equilibrium reactions under appropriate conditions. In isolated working guinea-pig heart perfused for prolonged periods at constant pre- and afterloads, whether examined under the conditions of low, normal or high-energy outputs or during low-flow ischemia, we recently found essentially the same square root-dependencies between thermodynamic [AMP] (myokinase equation) and cytosolic [ADP]/[ATP] (creatine kinase equation). These findings suggest near-equilibrium catalyses by the kinase couple in perfused heart under various steady-state physiological and pathophysiological conditions (6).

Since H^+ is an important reactant of the creatine kinase and because this enzyme is Mg^{2+}-dependent, the value of the expression $[H^+]/K_{cpk}$, an essential term in the creatine kinase equation, is strongly H^+- and Mg^{2+}-dependent (29):

(Eq. 3) $[ATP]/[ADP] = [CrP]/[Cr] \cdot [H^+]/K_{cpk}$

(Eq. 4a) $\log([H^+]/K_{cpk}) = -0.87 \cdot pH + 8.31$; free $[Mg^{2+}] = 0.9$ mM
(Eq. 4b) $\log([H^+]/K_{cpk}) = -0.86 \cdot pH + 8.30$; free $[Mg^{2+}] = 3.6$ mM

in which CrP = creatine phosphate, Cr = creatine.

It is thus critical to obtain independent values for cytosolic $[H^+]$ and $[Mg^{2+}]$, in order to strengthen confidence in data and conclusions derived from myokinase and creatine kinase equilibrium calculations. A number of recent physico-chemical, cell biological and ^{31}P-NMR studies indicate that the free Mg^{2+} concentration in cell water is relatively constant at 0.5 to 1.0 mM in mammalian organs, including heart (15,19,32,43,55). Bicarbonate and CO_2 measurements in tissues and perfusing media as well as ^{31}P-NMR detectable phosphate chemical shifts indicate intracellular pH values near 7.2 in well-perfused and oxygenated tissues (10,15,19,32,54). Using the Henderson-Hasselbalch equation in combination with measured tissue bicarbonate and perfusion fluid $[CO_2]$ we obtained the following operational equation for calculation of intracellular pH in isolated working guinea-pig hearts (6,10):

(Eq. 5) $pH_i = 7.524 \cdot e^{(-0.0008786 \cdot pCO_2)}$, n = 20, r = 0.959

in which pCO_2 = coronary effluent pCO_2 (mmHg) in hearts perfused with a Krebs-Henseleit buffer, pH 7.4 (25 mM bicarbonate), oxygenated with $O_2/CO_2 = 95/5$.

Three powerful cytosolic enzymes, glyceraldehyde-3-phosphate dehydrogenase (GAPDH), 3-phosphoglycerate kinase (PGK), and lactate dehydrogenase (LDH) (41), associate the cytosolic ATP potential with the overall cytosolic oxidation-reduction state, if it is assumed that GAPDH and LDH share free cytosolic $NAD^+(H)$ as a common coenzyme. In theory cytosolic reduction state and ATP potential are reciprocally related according to the following two relationships (27,54):

(Eq. 6) $[ATP]/([ADP] \cdot [P_i]) = [GAP]/[3\text{-}PG] \cdot [Pyr]/[Lac] \cdot K$

in which $K = K_{gapdh} \cdot K_{pgk}/K_{ldh}$; GAP = glyceraldehyde-3-phosphate, 3-PG = 3-phosphoglycerate, Pyr = pyruvate, Lac = lactate; and

(Eq. 7) $[Pyr]/[Lac] = K_{ldh} \cdot [NAD^+]/([NADH] \cdot [H^+])$.

Since the [ADP]/[ATP] ratio appears to serve a dual role as the major determinant of the value of the cytosolic phosphorylation potential (Eq. 3 divided by $[P_i]$) and of [AMP] (Eq. 2), it seems logical to suggest that [AMP] and the free cytosolic $[NADH] \cdot [H^+]/[NAD^+]$ ratio are directly related, provided induced changes in [Pyr]/[Lac] are not compensated by those in [GAP]/[3-PG]. The complexity of the interactions between the cytosolic $NAD^+(H)$-dependent dehydrogenases, phosphoglycerate kinase, and myokinase is increased by the fact that the equilibrium constants of both kinases, especially K_{pgk}, are highly sensitive to $[Mg^{2+}]$ and ionic strength (14,29).

Methods

Hearts were isolated from male albino guinea pigs (350-600 g body wt.) fed a standard chow diet. Isolation, preparation, non-recirculating hemoglobin-free perfusion and physiologic performance characteristics of the isolated, perfused working hearts have been detailed elsewhere (7,9). Hemodynamic parameters [spontaneous heart rate (HR), coronary venous outflow and aortic flow, mean left ventricular filling pressure (P_v), and phasic and mean aortic pressure (P_a)] were continuously monitored using an 8-channel direct recorder (Hewlett Packard). Left ventricular pressure work per min was judged from the heart rate-pressure product expressed as HR · ΔP ($\Delta P = P_a$-P_v) (e.g., refs. 7,12). All hearts were perfused at constant filling pressures between 12 and 16 cmH$_2$O. Perfusion medium was a Krebs-Henseleit bicarbonate buffer (28), pH 7.40-7.45, 38°C, 294 $\pm$ 0.4 mOsm (n = 29), oxygenated with 95% O$_2$:5% CO$_2$; the medium was modified to contain free calcium in the physiologic concentration range between 1.5 - 2.5 mEq/L. The perfusion fluid was supplemented with 5 mM glucose (+ 5 U/L bovine insulin) as the sole energy substrate or in combination with pyruvate or L-lactate as described in the table and Fig. 1. When additional excess Na-pyruvate or Na-lactate were used the NaCl concentration of the medium was reduced accordingly to maintain osmolarity constant. L-Norepinephrine (NE) stock solution (500 uM), stabilized by 2 mg/ml sodium metabisulfite (Winthrop-Breon, New York, NY), was infused directly into the left atrium of the working hearts, yielding arterial perfusate NE concentrations between 0.2-1.1 uM; these NE levels represented near-maximum effective NE doses in the isolated perfused guinea-pig heart (7). Water for perfusion media and analytical tests was freshly prepared from a low-output quartz double-distillation system. To remove bacteria and particulate impurities all solutions and media including those for enzymatic analyses and HPLC were filtered (Millipore filter, 0.22 um) immediately prior to use.

Myocardial O$_2$ uptake was calculated from the continuously monitored arterio-venous O$_2$ concentration difference (Clark-type oxygen electrode) and coronary venous effluent rate. Arterial and coronary venous pO$_2$, pCO$_2$, and pH were measured as previously described (10,11). Intracellular pH was estimated from measured coronary venous pCO$_2$, using Eq. 5 (6).

To measure the volume of the extracellular space, [^{14}C]carboxylate-inulin (15 mg/L sp. act. = 1.5 · 10^6 dpm/mg) was infused for 6 min prior to stop-freezing the hearts. Radioactivity was measured by liquid scintillation counting using a standard cocktail (9). When experiments were terminated by freeze-fixation of the working hearts, a Wollenberger clamp at the temperature of liquid N$_2$ was used. Myocardial metabolites were measured in neutralized perchloric acid tissue extracts, using the following standardized extraction procedure. Frozen tissue was pulverized in a mortar under liquid N$_2$; 1 g pulverized myocardium at the temperature of liquid N$_2$ was added to 4 ml ice cold 0.3 N perchloric acid; this ensured slow acidic thawing of tissue powder at temperatures below 0°C. The crude ice cold homogenate was stirred (660 rpm) for 1 min using a precooled (0°C) Teflon rod of a diameter 2 mm smaller than that of the plastic test tube (inner diameter = 14 mm); during stirring the test tube was placed in an ice/water mixture. Subsequent centrifugation (12,000 g, Sorvall RC-5) at 0 - 4°C for 10 min was followed by gentle pH-controlled neutralization of the supernatant fluid to a slightly acidic pH (5.8 to 6.3), using small aliquots of refrigerated (4°C) 1 N and 0.1 N KOH, respectively. After 30 min-standing in ice, potassium perchlorate was removed by centrifugation (12,000 g, 10 min, 0-4°C). Aliquots of the slightly acidic extracts were immediately assayed for cardiac metabolites (enzymatic analysis). Reextraction (n = 8) of the homogenate pellet for the key metabolites ATP, ADP, AMP, creatine phosphate, creatine, inorganic phosphate, pyruvate, and dihydroxyacetone phosphate indicated that the extraction procedure yielded the following percentage recoveries, respectively: 92.7, 89.5, 83.7, 92.8, 95.1, 95.1, 82.7, 89.8. Data presented were not corrected for measured recoveries. Most of the enzymatic tests were done in the dual beam/dual wavelength mode (measuring wavelength = 340 nm, reference wavelength = 385 nm, millimolar absorption coefficient = 6.097 cm^2 · umol^{-1}, Perkin-Elmer spectrophotometer model 556). Intracellular contents of the perfusion medium metabolites (pyruvate, lactate, inorganic phosphate) were calculated as the difference between respective total myocardial contents and extracellular amounts using the [^{14}C]inulin distribution data. The mean extracellular metabolite concentration was taken as (arterial plus venous metabolite concentration)/2. The [^{14}C]inulin space in freeze-clamped hearts was 0.57 ± 0.01 ml/g wet wt. (n = 16 mean ± S.E.M.). The intracellular solvent space was taken as the difference between total tissue water and [^{14}C]inulin space. Intracellular water was not corrected for mitochondrial water. Perfused myocardium wet weight/dry weight ratio after freeze-clamping was 9.6 ± 0.2 (n = 44).

Coronary venous purine nucleosides (adenosine, inosine) and other natural purines (hypoxanthine, xanthine, uric acid) were separated using reverse phase HPLC as described in detail previously (10,12). Purines were identified employing known retention times in combination with measured absorbance characteristics at four different wavelengths (254, 263, 273, 293 nm) using a multiwavelength detector (Waters model 490). Quantitation of the purines was accomplished by comparisons with calibrated standards. Standards and experimental samples were kept at about 4°C prior to separation by HPLC using a refrigerated automatic injector system (Waters WISP model 710B). Repetitive injections of standards showed that all purines were stable for more than 24 h under these conditions.

Data were statistically analyzed using Student's t-test for unpaired samples. For multiple comparisons Duncan's variance analysis was performed.

Results and Discussion

Effects of Excess Pyruvate vs. Lactate on Cytosolic Adenylates and Purine Nucleoside plus Xanthine Output

The oxidation-reduction state of cytosolic redox couples (e.g., [NAD$^+$]/([NADH] · [H$^+$])) was altered by perfusing the isolated working guinea-pig hearts with excess pyruvate (20 mM) or excess lactate (20 mM) in the presence of 5 mM glucose (plus insulin). Table 1 shows that, relative to lactate, pyruvate effected respective increases in total cardiac AMP content, the [CrP]/[Cr] ratio, the phosphorylation states of creatine phosphate and ATP, and the cytosolic [ATP]/[ADP] ratio; intracellular [H$^+$] and the total myocardial contents of creatine phosphate and ATP were not appreciably different between the two conditions. Similarly, the heart rate-pressure product was virtually identical with either pyruvate or lactate, but myocardial oxygen uptake tended to increase slightly in presence of lactate; this was expected from the increase in [ADP] from about 25 to 41 uM under the same conditions (6,10). In accordance with the small increase in the [CrP]/[Cr] ratio, pyruvate also ef-fected a small but definitive decrease in the cellular inorganic phosphate concentration.

The AMP content was increased almost 8-fold by excess pyruvate relative to lactate. On the other hand, calculated [AMP] fell 51% (227 vs. 108 nM, Table 1). The coronary venous concentration of adenosine was below the detection limit of 2 picomoles per 200 uL (10 nM) in the presence of pyruvate or lactate. Nevertheless, in agreement with our concept (6,10,11) that it is the cytosolic [AMP], not intramitochondrial [AMP], which functions as the common precursor of released adenosine plus inosine, coronary venous adenosine plus inosine release was minimal in the presence of pyruvate (≤ 2 nmol/min per g dry wt.), but was increased to 8.35 nmol/min per g dry wt. in the presence of lactate (Table 1).

The lactate-perfused hearts released additionally very large quantities of xanthine (44.6 ± 8.4 nmol/min per g dry wt.), the immediate precursor of uric acid; released hypoxanthine and uric acid, however, remained below the detection limit. Thus, inosine and xanthine were the major purines released from the redox-manipulated working guinea-pig hearts. Myocardial xanthine output was apparently dependent on a high cytosolic reduction potential, indicating that the native form of cardiac xanthine oxidase functions as an NAD^+-dependent dehydrogenase (3). The xanthine dehydrogenase/oxidase reaction was presumably feed-back inhibited by NADH accumulated in the cytoplasm, resulting in the output of xanthine instead of uric acid. Since xanthine oxidase is a marker enzyme for coronary capillaries (24), the xanthine of the coronary outflow tract most likely originated in the coronary capillary endothelium. Furthermore, in accordance with Eqs. 2, 6 and 7, the large output of xanthine probably reflected a fall in the endothelial phosphorylation potential in association with an increase in [AMP]. In fact, myokinase activity is about 10 times higher than that of adenosine deaminase in the coronary endothelium of the guinea pig (35).

Measured concentrations of reduced intracellular metabolites (lactate, 0.45 ± 0.11 mM; alpha-glycerophosphate, 0.02 ± 0.004 mM) were quite low in the 20 mM pyruvate-perfused hearts, but greatly increased (12.3 and 5.4 mM, respectively) in presence of 20 mM lactate. Similarly, the cellular concentration of alanine, which is likely in near-equilibrium with cytosolic pyruvate via alanine aminotransferase (e.g., ref. 42), was 13-fold higher with pyruvate (7.8 ± 0.8 mM) than with lactate.

Table 1. Effects of excess pyruvate vs. excess lactate on myocardial adenylates, free ADP, AMP, and nucleoside release

Variable	Addition to 5 mM glucose + 5 U/L insulin	
	20 mM pyruvate	20 mM lactate
pH_i	7.224	7.235
ATP content, umol/g	24.6 ± 0.6	23.9 ± 0.7
AMP content, umol/g	1.14 ± 0.13	$0.15 \pm 0.05^*$
CrP content, umol/g	50 ± 3	46 ± 3
[CrP]/[Cr]	2.30 ± 0.10	$1.74 \pm 0.10^*$
$[P_i]$, mM	4.6 ± 0.3	$5.8 \pm 0.3^{\#}$
$[CrP]/([Cr][P_i])$, mM^{-1}	0.52 ± 0.04	$0.280 \pm 0.017^*$
$[ATP]/([ADP][P_i])$, M^{-1}	57,276	31,080
[ATP]/[ADP]	254	193
[ADP], uM	24	39
[AMP], nM	108	227
V(AR + INO), nmol/min · g		
measured	< 2.0	8.4 ± 0.4
predicted from Eq. 8	2.89	8.75
HR ·ΔP, 10^{-3} cmH_2O/min	20.56 ± 0.02	$19.91 \pm 0.02^*$
MVO_2, umol/min · g	45.1 ± 1.6	50 ± 4
H_2O_{ic}, ml/g	4.1 ± 0.3	$3.18 \pm 0.08\#$

Values are means ± S.E., n = 5-7. Working hearts perfused for 20 min at P_v = 16 cmH_2O and P_a = 100 cmH_2O. Where applicable, data are expressed per g dry weight; [ ] denotes intracellular thermodynamic concentration. pH_i: Intracellular pH calculated from coronary venous pCO_2 according to Eq. 5. V(AR + INO): Coronary venous output of adenosine plus inosine. MVO_2: Myocardial oxygen consumption. H_2O_{ic}: Intracellular water space. Intracellular free $[Mg^{2+}]$ was assumed near 1 mM (Eq. 4a). K_{myk} = 1.12 (Eq. 2). Total cardiac AMP was 279 and 47 uM in the presence of 20 mM pyruvate and 20 mM lactate, respectively. Consequently, free [AMP] was less than 1% of total AMP (10). # $p < 0.05$, * $p < 0.01$.

Assuming that pyruvate distributed across the cell membrane in a manner similar to that of lactate (both metabolites are 3-carbon monocarboxylates and highly dissociated at physiological pH) the calculated intracellular [Pyr]/[Lac] ratio was about 28 in presence of excess pyruvate, but only 0.007 in presence of excess lactate, i.e., about 4000-fold lower. Clearly pyruvate, relative to lactate, oxidized the cytosolic NAD^+-NADH pool, and this imposed oxidation was associated with a significant increase in the cytosolic phosphorylation potential (Table 1). The data are consistent with the previously reported association between coronary effluent [Lac]/[Pyr] ratio and purine nucleoside release in glucose-perfused heart (6) and also with the concordant changes in lactate output and purine nucleoside release during adrenergic stimulation (see Fig. 2). On the other hand, the present reciprocity between cardiac redox and energy state apparently disagrees with the stability of the phosphorylation potential in the liver of ethanol-cyanamide treated rats (15). Here the [Lac]/[Pyr] ratio increased only about 3-fold, not several orders

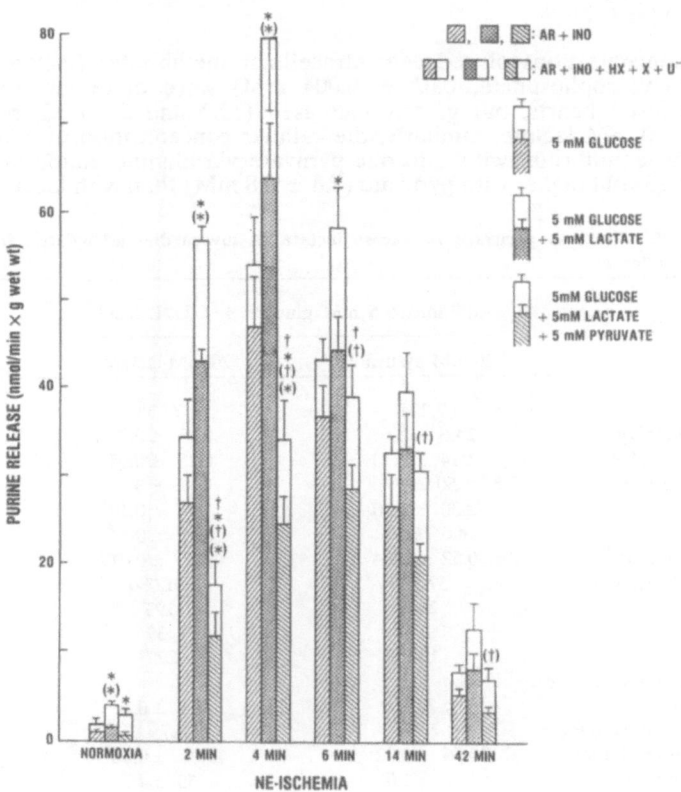

Fig. 1. Effects of external redox manipulations on coronary venous release of adenylate catabolites from isolated working guinea-pig hearts under conditions of normoxia and NE-ischemia. Data are means ± S.E., n = 6-8. NE: norepinephrine; AR: adenosine; INO: inosine; HX: hypoxanthine; X: xanthine; U^-: uric acid. HR $\cdot \Delta P$: normoxia, ~ 19,000 cmH_2O/min; during ischemia, HR $\cdot \Delta P$ fell progressively from about 4,500 cmH_2O/min at 2 min to zero at 25-45 min, and did not appreciably differ among the three conditions tested. Variance analysis was done after square-root transformation of raw data. Total purine release: [*]p<0.05 vs. 5 mM glucose. [+]p<0.05 vs. 5 mM glucose + 5 mM lactate. Adenosine plus inosine release: [(*)]p<0.05 vs. 5 mM glucose; [(+)]p<0.05 vs. 5 mM glucose + 5 mM lactate.

of magnitude; moreover, the GAP/3-PG couple can possibly compensate changes in [Lac]/[Pyr] more effectively in rat liver in vivo than in isolated working guinea-pig heart.

Of special interest was the fact that pyruvate induced a large increase in AMP content, but a decrease rather than increase in inosine release. Even if one accepts the widely held view that most of cardiac AMP is located in the extramitochondrial compartment, the pyruvate-dependent decrease in inosine release in presence of increased AMP content is difficult to explain. According to Table 1 pyruvate did not appreciably increase the already very high concentration of the putative 5'-nucleotidase inhibitor creatine phosphate (45). Pyruvate decreased (19%) the concentration of ATP slightly from about 7.5 to 6.1 mM. Allosteric ATP inhibition, not activation (13), of cytosol 5'-nucleotidase has been discussed for guinea-pig cardiac muscle (16). The minor decrease in myocardial ATP level would be expected to de-inhibit cytosolic 5'-nucleotidase in vivo, but measured nucleoside release was decreased instead of increased (Table 1).

Alternatively, the present AMP concentration and purine nucleoside release data can be discussed in terms of the recently proposed AMP-availability model (6,10,39); accordingly, the free cytosolic ATP-ADP/myokinase system can "downregulate" [AMP] into the submicromolar concentration range, effectively reducing cytosolic [AMP] to levels several orders of magnitude lower than K_{mAMP} values of 5'-nucleotidase and adenylate deaminase (13,16,47). Under such conditions cytosolic 5'-nucleotidase and adenylate deaminase fluxes would be controlled, at least in part, by [AMP]. Independent evidence for normally extremely low cytosolic AMP levels comes from digitonin fractionation of isolated rat cardiomyocytes (20); here the cytosolic fraction did not contain measurable amounts of AMP under the conditions of nor-

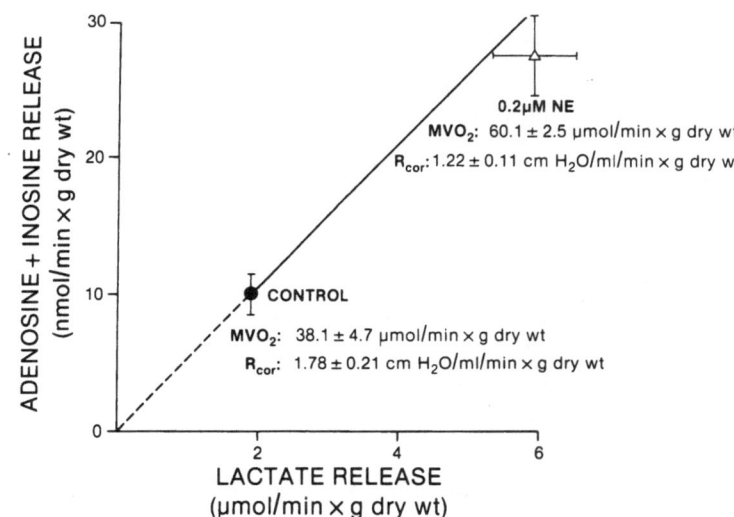

Fig. 2. Correlation between purine nucleoside release and lactate release in isolated working guinea-pig heart stimulated by norepinephrine (NE) in the presence of 1.5 mEq/L Ca^{2+} under conditions of normoxia. P_v = 12 cmH$_2$O; P_a = 90 cmH$_2$O. Data are means ± S.E., n = 6. MVO_2: myocardial oxygen consumption; R_{cor}: coronary resistance. An intraventricular needle was introduced through the cardiac apex to measure ventricular dP/dt_{max}: control = 1250 ± 85 mmHg/s; NE = 2833 ± 228 mmHg/s.

moxia; cytosolic AMP became measurable only during anoxia. Furthermore, results from two independent series of experiments in working guinea-pig hearts, one used to determine the relationship between MVO_2 and calculated [AMP] (6), the other to determine the relationship between MVO_2 and adenosine plus inosine release (6,10), allowed us to derive an operational formula that characterized the steady-state correlation between [AMP] and measured adenosine plus inosine release:

(Eq. 8) $V(AR + INO) = 0.0027 [AMP]^{1.49}$ (ref. 6)

where V(AR + INO) = adenosine plus inosine release (nmol/min per g dry wt.) and [AMP] designates free cytosolic AMP concentrations (nM). Measured and predicted V(AR + INO), using the [AMP] data from Table 1, show reasonable agreements (Table 1), supporting the concept that cytosolic AMP availability may normally be an important determinant of the rate of purine nucleoside formation and release in working guinea-pig heart. It should be emphasized that Eq. 8 was derived from predominantly myocytic, not endothelial parameters (MVO_2, ATP content, [CrP]/[Cr], [ATP]/[ADP], intracellular pH), i.e., [AMP] refers to free AMP in the myocyte cytosol.

It is clear that most (although less than 100%) of cardiac LDH, GAPDH and PGK is located in the cytoplasm of the myocytes, which make up about 85% of the myocardial wet weight (31). It thus seems reasonable to assume that most of the present effect of pyruvate/lactate infusion on inosine release originated in the free cytosolic AMP of the myocytes. The possible inosine release from endothelial cells, which comprise only about 2% of myocardial wet weight (51), is difficult to assess under the present conditions. However, dipyridamole blocked coronary venous uric acid output without inhibiting inosine release (4), and isolated heart cells release purine nucleosides in appreciable amounts (17,20,33). Taken with these reports, the present response of inosine release to cytosolic redox alterations further supports the AMP-availability model for the working heart. It is then not surprising that adenosine plus inosine release predicted from Eq. 8 and the measured adenosine plus inosine release were in reasonable agreement, both during the present redox interventions (Table 1) and during steady state low-flow ischemia (6). However, a potential complication is the fact that pyruvate can accelerate while lactate appears to decelerate hypoxanthine salvage in cardiomyocytes in culture (44). Such pyruvate stimulation of the salvage pathway via direct oxidation of cytosolic NADH cannot be ruled out with certainty under the present conditions.

In theory, the present coronary venous inosine fraction could also have been derived from cytosolic IMP generated from AMP by adenylate deaminase. However, several observations argue against this interpretation: 1) most of cardiac IMP appears to be located in the mitochondria rather than in the cytosol under the conditions of normoxia (20); 2) the thermodynamic AMP concentration is several orders of magnitude below the K_{mAMP} of adenylate deaminase; 3) guinea-pig myocardium does not appear to contain appreciable quantities of adenylate deaminase (16); and 4) the soluble 5'-nucleotidase has a higher affinity towards AMP than IMP (13,16).

Cytoplasmic 5'-nucleotidase accounts for only 3-17% of total 5'-nucleotidase activity in guinea-pig myocardium (16,47). The enzyme appears to be subcompartmented between the free cytosol and lysosomes (2,13). Nevertheless, several reports strongly support intracellular formation of adenosine (17,33,47,56). Uncertainty prevails regarding the subparenchymal site(s) of adenosine deaminase (16,47). Up to 30% of total cardiac adenosine deaminase may be localized in the endothelial compartment with only a small fraction (about 5%) in the myocytes (16). However, dipyridamole as an inhibitor of the endothelial nucleoside transporter did not appreciably block coronary venous inosine output (4); this seems to imply that inosine export from extramyocytic elements is of only minor importance in the beating heart.

Consequently, the presently released inosine fraction could actually reflect, at least in part, myocytic adenosine deaminase activity. Although the possibility has been raised that adenosine is exclusively derived from the endothelium in the resting normoxic heart (35), dipyridamole greatly increased (not decreased) cardiac adenosine output (about 20-fold), whereas uric acid release was nearly completely blocked (4). These data obviously imply that the beating myocytes can generate free interstitial adenosine which is readily taken up by and degraded within endothelial structures. However, this adenosine is also available for washout via the coronary outflow tract (1,4). Alternatively, based on studies with special isolated myocyte and endothelial preparations, it was recently emphasized (16) that guinea-pig myocytes normally release only minimum, if any, amounts of free adenosine and inosine. Instead myocytic AMP and other adenylates are released to undergo subsequent dephosphorylation by ecto-nucleotidases of various subparenchymal origins (18,53).

Effects of Excess Pyruvate vs. Lactate on Coronary Venous Purine Release during Global Low-Flow Ischemia

During states of acute myocardial oxygen- or flow-deficiency, ATP breakdown is most pronounced in the working myocytes; it is this cell type which by far predominates in cardiac parenchyma and which, due to its hydraulic workload, necessarily incurs the greatest immediate energy deficit during ischemia conditions. It is not known whether other formed elements of the myocardium (endothelial and epicardial cells, interstitial fibroblasts, leukocytes, pericytes, smooth muscle cells) develop energy deficiencies during ischemic or hypoxic perfusion. However, it appears reasonable to assume that, relative to the working myocytes, the extent and time course of any extramyocytic energy imbalance is likely to be minor and delayed. Obviously, cardiac ischemia causes ATP degradation, accumulations of lactate, NADH, and free AMP, and intracellular acidification mainly in the myocyte.

If released adenosine plus inosine are primarily derived from free cytosolic AMP of the myocytes, especially during states of acute myocardial ischemia, and if [AMP] is actually under the control, at least in part, of the cytosolic phosphorylation potential and hence linked to the NAD^+-NADH oxidation-reduction state (Table 1), ischemic purine nucleoside release would be expected to respond accordingly to an imposed (i.e., "clamped") redox state in the form of a constant arterial [Pyr]/[Lac] ratio.

We tested this rationale in a special low-flow, globally ischemic, isolated working guinea-pig heart in which coronary hypoperfusion was combined with maximum norepinephrine stimulation ("NE-ischemia") to produce a substantial rather than mild energy deficit of the myocytes. Hearts received as substrate 5 mM glucose or 5 mM glucose plus 5 mM lactate or 5 mM glucose plus 5 mM lactate plus 5 mM pyruvate. Such excess concentrations of pyruvate or lactate substantially elevate the intracellular [Pyr] and/or [Lac]. We also showed that pyruvate, when applied in a concentration of 2 mM in the presence of 5 mM lactate, is actively metabolized by ischemic or hypoxic isolated working heart, whereas lactate is oxidized only during normoxia, but released during states of oxygen deficiency (12). In addition, under low-flow ischemic conditions, external 2 mM pyruvate produces a large increase (about 12-fold) in cellular [Pyr] in association with a relative oxidation of the cytosolic NADH-NAD^+ system as evidenced by increased cellular [Pyr]/[Lac] or [dihydroxyacetone phosphate]/[alpha-glycerophosphate] ratios (data not shown). Consequently, experimental clamping of the external pyruvate-lactate oxidation-reduction potential can actually shift the cytosolic oxidation-reduction state in the globally ischemic working heart.

Figure 1 shows that those ischemic working hearts which received glucose alone or in combination with lactate released larger amounts of purines during the first 14 min of the NE-ischemia, as compared to those hearts which received excess pyruvate (arterial [Pyr]/[Lac] = 1.0). The data also show that the major portion (75-80%) of the ischemically released purines was in the form of nucleosides (adenosine plus inosine) and the remainder in the form of purine bases (xanthine compounds) plus uric acid. The decrease in purine release in the presence of the externally fixed [Pyr]/[Lac] ratio was not due to differences in residual ischemic cardiac pressure work, since the heart rate - pressure products were comparable in the absence and presence of pyruvate and/or lactate (see legend to Fig. 1). In the steady state after about 45 min NE-ischemia, purine release from the pyruvate- plus lactate-perfused hearts was only marginally smaller than that from hearts receiving glucose plus lactate. These findings thus indicated that external "clamping" or buffering of the cytosolic [NADH]/[NAD$^+$] ratio can significantly attenuate purine release, especially during short-term non-steady state ischemia conditions. Figure 1 also shows that ischemic stress, if extended for 45 min or more, is associated with an only transitory excess release in adenylate catabolites, i.e., only during early ischemia. A similar time-dependence of adenosine plus inosine release was recently reported in high-flow hypoxic working hearts (12).

The present data show unequivocally that the purine nucleosides, not the free purine bases plus uric acid, are the major adenylate catabolites during severe acute ischemic stress of the perfused working heart. Uric acid accounted for only 4% to 10% of net purine release during the first 14 min of the NE-ischemia (Fig. 1). However, uric acid output accounted for 41% to 52% of total purine release during normoxia, although after 45 min NE-ischemia only 9% to 19% of total purine release was in the form of uric acid. The situation is different in some Langendorff hearts (1,4); here uric acid release comprises more than 70% of total purine output during normoxia and also during 1 min reperfusion after a brief period (30 s) of zero-flow ischemia (1). On the other hand, the Langendorff heart, when tested under high-flow anoxia, releases adenosine plus inosine, not the free purine bases plus uric acid, as the major adenylate catabolites (1,4). Apparently, under conditions with minimum myocyte energy deficit, much of the relatively small amount of adenylate catabolite(s) can still be fully converted to uric acid due to the active purine metabolism within the endothelium. However, when the myocyte energy deficit becomes severe, cardiac adenosine plus inosine formation appears to overwhelm the metabolic capacity of the endothelium resulting in an increased appearance of nucleosides in the coronary effluent. Ultimately the interstitial nucleoside concentration may rise to nearly saturate binding, transport, and receptor sites, which would lead to an increase in the intracoronary ratio of [free nucleosides]/[free bases plus uric acid] with the uric acid output reaching a maximum (e.g., Fig. 1 and ref. 1). In fact, myocardial output of free adenosine can experimentally be increased at the expense of uric acid release when dipyridamole is applied as an adenosine transport blocker of the endothelium (4). A finite purine transport and degradation capacity of the endothelium is also consistent with Eq. 8, according to which adenosine plus inosine release increases more than proportionally with free cytosolic [AMP].

Figure 2 shows that adenosine plus inosine output and lactate release change concordantly in working hearts stimulated by a high dose of norepinephrine under the conditions of normoxia. Since these hearts received 15 mM glucose as the sole substrate, lactate release was a qualitative index of the cytosolic [NADH]/[NAD$^+$] ratio (5,6). In other words, cytosolic reduction potential of the working myocytes and coronary venous purine nucleoside output changed in parallel (6) when the phosphorylation potential decreased due to a physiologic adrenergic stimulus (10). In fact, according to Eqs. 6 and 7, the increased lactate output reflected an increase in the cytosolic [NADH]/[NAD$^+$] ratio in association with a fall in the phosphorylation potential. Consequently, [AMP] was probably increased during adrenergic stimulation due to the fall in the phosphorylation potential (6,10), most likely stimulating 5'-nucleotidase flux according to the AMP-availability model. Since adenosine and inosine appeared in the coronary outflow tract in increasing amounts, a significant fraction of myocytic adenosine plus inosine probably bypassed endothelial nucleoside transport sites. This argument is valid if one assumes that adrenergic stimulation does not appreciably alter the energy balance of the extramyocytic elements of the working heart. Indeed, increased oxygen consumption (a predominantly myocytic parameter) is associated with increased adenosine plus inosine release, regardless of whether oxidative metabolism is stimulated by, e.g., acetate (1,57) or by infusion of adrenergic agonists (6,10). Since increased oxygen consumption of perfused heart is normally associated with a fall in cytosolic phosphorylation potential (see ref. 10 and refs. therein), it appears likely that [AMP] was increased under the same conditions (6,10).

It should finally be pointed out that such energy-linked adaptations in the free concentrations of cytosolic adenylates and coronary purine nucleoside output do not necessarily identify the site(s) of primary AMP degradation. A functional coupling between cytosolic AMP degradation at the inner surface of the sarcolemma followed by vectorial adenosine export (18) is also not necessarily excluded. It appears to be important to develop an experimentally testable hypothesis to differentiate between intra- and extracellular sites of endogenous AMP metabolism in the physiologically performing heart.

Acknowledgements

This work was supported by grants from the National Institutes of Health (Heart, Lung, and Blood Institute: HL-29060, HL-36067) and from the Uniformed Services University of the Health Sciences (RO7638).

References

1. Achterberg PW, Stroeve RJ, De Jong JW: Myocardial adenosine cycling rates during normoxia and under conditions of stimulated purine release. Biochem J 235:13-17, 1986
2. Arsenis C, Touster O: Purification and properties of an acid nucleotidase from rat liver lysosomes. J Biol Chem 243:5702-5708, 1968
3. Battelli MG, Della Corte E, Stirpe F: Xanthine oxidase type D (dehydrogenase) in the intestine and other organs of the rat. Biochem J 126:747-749, 1972
4. Becker BF, Gerlach E: Uric acid, the major catabolite of cardiac adenine nucleotides and adenosine, originates in the coronary endothelium. In: Gerlach E, Becker BF, eds: Topics and perspectives in adenosine research. Berlin: Springer Verlag, 1987:209-221
5. Bünger R: Compartmented pyruvate in perfused working heart. Am J Physiol 249:H439-H449, 1985
6. Bünger R: Thermodynamic state of cytosolic adenylates in guinea pig myocardium. Energy-linked adaptive changes in free adenylates and purine nucleoside release. In: Gerlach E, Becker BF, eds: Topics and perspectives in adenosine research. Berlin: Springer Verlag, 1987:223-234
7. Bünger R, Haddy FJ, Querengässer A, Gerlach E: An isolated guinea pig heart with in-vivo-like features. Pflügers Arch 353:317-326, 1975
8. Bünger R, Permanetter B, Sommer O, Yaffe S: Adaptive changes of pyruvate oxidation in perfused heart during adrenergic stimulation. Am J Physiol 242:H30-H36, 1982
9. Bünger R, Sommer O, Walter G, Stiegler H, Gerlach E: Functional and metabolic features of an isolated perfused guinea pig heart performing pressure-volume work. Pflügers Arch 380:259-276, 1979
10. Bünger R, Soboll S: Cytosolic adenylates and adenosine release in perfused working heart. Eur J Biochem 159:203-213, 1986
11. Bünger R, Soboll S, Permanetter B: Effects of norepinephrine on coronary flow, myocardial substrate utilization, and subcellular adenylates. In: Merrill GF, Weiss HR, eds: Ca^{2+} entry blockers, adenosine, neurohumors. Baltimore: Urban and Schwarzenberg, 1983:267-279
12. Bünger R, Swindall B, Brodie D, Zdunek D, Stiegler H, Walter G: Pyruvate attenuation of hypoxia damage in isolated working guinea-pig heart. J Mol Cell Cardiol 18:423-438, 1986

13. Collinson AR, Peuhkurinen KJ, Lowenstein JM: Regulation and function of 5'-nucleotidase. In: Gerlach E, Becker BF, eds: Topics and perspectives in adenosine research. Berlin: Springer Verlag, 1987:133-144
14. Cornell N, Leadbetter M, Veech RL: Effects of free magnesium concentration and ionic strength on equilibrium constants for the glyceraldehyde phosphate dehydrogenase and phosphoglycerate kinase reactions. J Biol Chem 254:6522-6527, 1979
15. Cunningham CC, Malloy CR, Radda GK: Effect of fasting and acute ethanol administration on the energy state of in vivo liver as measured by ^{31}P-NMR spectroscopy. Biochim Biophys Acta 885:12-22, 1986
16. Dendorfer A, Lauk S, Schaff A, Nees S: New insights into the mechanism of myocardial adenosine formation. In: Gerlach E, Becker BF, eds: Topics and perspectives of adenosine research. Berlin: Springer Verlag, 1987:170-184
17. Ford DA, Rovetto MJ: Rat cardiac myocyte adenosine transport and metabolism. Am J Physiol 252:H54-H63, 1987
18. Frick GP, Lowenstein JM: Vectorial production of adenosine by 5'-nucleotidase in the perfused rat heart. J Biol Chem 253:1240-1244, 1978
19. Garfinkel L, Garfinkel D: Calculation of free Mg^{2+} in adenosine 5'-triphosphate containing solutions in vitro and in vivo. Biochemistry 23:3547-3552, 1984
20. Geisbuhler T, Altschuld RA, Trewyn RW, Ansel AZ, Lamka K, Brierley G: Adenine nucleotide metabolism and compartmentalization in isolated adult rat heart cells. Circ Res 54:536-546, 1984
21. Hebisch S, Soboll S, Schwenen M, Sies H: Compartmentation of high-energy phosphates in resting and working rat skeletal muscle. Biochim Biophys Acta 764:117-124, 1984
22. Heldt HW, Klingenberg M, Milovancev M: Differences between the ATP/ADP ratios in the mitochondrial matrix and in the extramitochondrial space. Eur J Biochem 30:434-440, 1972
23. Jacobs H, Heldt HW, Klingenberg M: High activity of creatine kinase in mitochondria from muscle and brain and evidence for a separate mitochondrial isoenzyme of creatine kinase. Biochem Biophys Res Commun 16:516-521, 1964
24. Jarasch ED, Grund C, Bruder G, Heid HW, Keenan TW, Franke WW: Localization of xanthine oxidase in mammary-gland epithelium and capillary endothelium. Cell 25:67-82, 1981
25. Klingenberg M: Energetic aspects of transport of ADP and ATP through the mitochondrial membrane. Energy transformation in biological systems. Ciba Found Symp 31 (New series), Amsterdam: Elsevier, 1975:105-121
26. Klingenberg M, Grebe K, Appel M: Temperature dependence of ADP/ATP translocation in mitochondria. Eur J Biochem 126:263-269, 1982
27. Krebs HA: Pyridine nucleotides and rate control. Symp Soc Exp Biol (Cambridge) 27:299-318, 1973
28. Krebs HA, Henseleit K: Untersuchungen über die Harnstoffbildung im Tierkörper. Hoppe-Seyler's Z Physiol Chem 210:33-66, 1932
29. Lawson JWR, Veech RL: Effects of pH and free Mg^{2+} on the Keq of the creatine kinase reaction and other phosphate hydrolyses and phosphate transfer reactions. J Biol Chem 254:6528-6537, 1979
30. Lehninger AL: Biochemistry. The molecular basis of cell structure and function. 2nd edn. New York: Worth Publ, 1975:516
31. Mall G, Mattfeldt T, Rieger P, Volk B, Frolov VA: Morphometric analysis of the rabbit myocardium after chronic ethanol feeding: Early capillary changes. Basic Res Cardiol 77:57-67, 1982
32. Malloy CR, Cunningham CC, Radda GK: The metabolic state of the rat liver in vivo measured by ^{31}P-NMR spectroscopy. Biochim Biophys Acta 885:1-11, 1986
33. Meghji P, Holmquist CA, Newby AC: Adenosine formation and release from neonatal rat heart cells in culture. Biochem J 229:799-805, 1985
34. Moreadith RW, Jacobus WE: Creatine kinase of heart mitochondria. Functional coupling of ADP transfer to the adenine nucleotide translocase. J Biol Chem 257:899-905, 1982
35. Nees S, Gerlach E: Adenine nucleotide and adenosine metabolism in cultured coronary endothelial cells: Formation and release of adenine compounds and possible functional implications. In: Berne RM, Rall TW, Rubio R, eds: Regulatory function of adenosine. The Hague: Nijhoff Publ, 1983:347-360
36. Noda L: Adenylate kinase. In: Boyer PD, ed: The enzymes. Vol 8, 3rd edn. New York: Acad Press, 1973:279-306

37. Ogunro EA, Peters TJ, Hearse DJ: Subcellular compartmentation of creatine kinase isoenzymes in guinea pig heart. Cardiovasc Res 11:250-259, 1977
38. Oliver T: A spectrophotometric method for the determination of creatine phosphokinase and myokinase. J Biol Chem 61:116-122, 1955
39. Olsson RA, Bünger R: Metabolic control of coronary blood flow. Progr Cardiovasc Dis 29:369-387, 1987
40. Perry SV: The bound nucleotide of the isolated myofibril. Biochem J 51:495-499, 1952
41. Pette D, Dölken G: Some aspects of regulation of enzyme levels in muscle energy-supplying metabolism. Adv Enzyme Regul 13:355-377, 1975
42. Peuhkurinen KJ, Nuutinen EM, Pietilainen EP, Hiltunen JK, Hassinen IE: Role of pyruvate carboxylation in the energy-linked regulation of pool sizes of tricarboxylic acid-cycle intermediates in the myocardium. Biochem J 208:577-581, 1982
43. Polimeni PI, Page E: Magnesium in heart muscle. Circ Res. 33:367-374, 1973
44. Ravid K, Diamant P, Avi-Dor Y: Regulation of the salvage pathway of purine nucleotide synthesis by the oxidation state of NAD^+ in rat heart cells. Arch Biochem Biophys 229:632-639, 1984
45. Rubio R, Bellardinelli L, Thompson CI, Berne RM: Cardiac adenosine. Electrophysiological effects, possible significance in cell function and mechanisms controlling its release. In: Baer HP, Drummond GI, eds: Physiological and regulatory functions of adenosine and adenine nucleotides. New York: Raven Press, 1979:167-182
46. Schnaitman C, Greenawalt JW: Enzymatic properties of the inner and outer membranes of rat liver mitochondria. J Cell Biol 38:158-175, 1968
47. Schütz W, Schrader J, Gerlach E: Different sites of adenosine formation in the heart. Am J Physiol 240:H963-H970, 1981
48. Seraydarian K, Mommaerts WFHM, Wallner A: The amount and compartmentalization of adenosine diphosphate in muscle. Biochim Biophys Acta 65:443-460, 1962
49. Shoubridge EA, Briggs RW, Radda GK: ^{31}P NMR saturation transfer measurements of the steady state rates of creatine kinase and ATP synthetase in the rat brain. FEBS Lett 140: 288-292, 1982
50. Siebert G, Humphrey GB: Enzymology of the nucleus. Adv Enzymol 27:239-288, 1965
51. Simionescu M, Simionescu N: Isolation and characterization of endothelial cells from the heart microvasculature. Microvasc Res 16:426-452, 1978
52. Soboll S, Bünger R: Compartmentation of adenine nucleotides in the isolated working guinea pig heart stimulated by noradrenaline. Hoppe-Seyler's Z Physiol Chem 362:125-132, 1981
53. Van Belle H, Goossens F, Wynants J: Formation and release of purine catabolites during hypoperfusion, anoxia, and ischemia. Am J Physiol 252:H886-H893, 1987
54. Veech RL, Lawson JWR, Cornell NW, Krebs HA: Cytosolic phosphorylation potential. J Biol Chem 254:6538-6547, 1979
55. Veloso D, Guynn RW, Oskarsson M, Veech RL: The concentrations of free and bound magnesium in rat tissues. Relative constancy of free Mg^{2+} concentrations. J Biol Chem 248:4811-4819, 1973
56. Worku Y, Newby AC: The mechanism of adenosine production in rat polymorphonuclear leucocytes. Biochem J 214:325-330, 1983
57. Yamada N, Bünger R, Steinhart CR, Olsson RA: Coronary vasoactivity of acetate in dog and guinea pig. Basic Res Cardiol 81:342-349, 1986

Chapter 8

Regulation of Myocardial Perfusion by Purine Metabolites

R.A. Olsson, Suncoast Cardiovascular Research Laboratory, Department of
Internal Medicine and Biochemistry, University of South Florida
College of Medicine, Tampa, Florida, U.S.A.

*This review of the roles of cardiac purines in the regulation of myocardial perfusion
begins with brief descriptions of the general characteristics of models of the
metabolic control of tissue blood flow and of cellular energy potential. The body of
the review, an analysis of the evidence for the participation of adenosine in
coronary regulation, is divided into four sections: 1) the enzymatic synthesis of
adenosine and the coupling of production to cardiac energy state; 2) the regulation
of adenosine concentration at the coronary resistance vessels; 3) the mechanism of
adenosine coronary vasodilation; and 4) the relationship of adenosine to other
vasodilatory metabolites and mechanisms that might participate in coronary flow
regulation. Two recent reviews provide additional information about the role of
adenosine in the metabolic control of coronary flow (43,57).*

The Metabolic Control of Coronary Perfusion

The coronary circulation is a vascular bed primarily under metabolic control, that is,
one in which tissue perfusion varies according to changes in tissue energy state.
The notion that energy production depends on the supply of some rate-limiting
metabolite is implicit in such a definition, as is the concept of a cellular energy
potential. The tight coupling of coronary flow to cardiac metabolism (4) stems from
attributes of both the cardiac striated and coronary smooth muscle cells. Working
cardiocytes rely almost entirely on aerobic metabolism for the large amounts of
energy needed to support their continuous activity. Because the blood concentrations
of energy-yielding substrates such as glucose or organic acids are relatively high
and the heart is unselective as to the source of reducing equivalents, oxygen is the
limiting substrate; cardiac performance begins to deteriorate within a few heartbeats
after the interruption of oxygen delivery by coronary occlusion (44). Coronary
vascular resistance is high relative to the rate of substrate usage, the results of
some poorly understood influence on - or perhaps an intrinsic property of -
coronary smooth muscle cells. Consequently, the transcoronary extraction of oxygen
is high, typically 60-80 percent under basal conditions. Since the potential for a
further increase in oxygen extraction is limited, a change in coronary flow is the
primary response to the change in the rate of cardiac energy usage.

Models of metabolic coronary vasoregulation have three fundamental characteristics.
The first is a vasodilatory chemical whose concentration is coupled to cardiac
energy state. The same working cardiocytes that generate the demand for energy-
yielding substrates also produce the vasoregulatory metabolite; thus, flow is said to
be under local control. The second characteristic is a mechanism whereby the
endogenous vasoregulator influences the contractile state of the smooth muscle cells
in the coronary resistance vessels. A third characteristic of such models is that
control lies in the direction of vasodilation. Although vasoconstriction may
determine perfusion rate under pathological conditions (35), available evidence
supports the notion that the physiological control of coronary flow consists of

dynamic metabolic vasodilation superimposed on a tonic constriction of unknown origin.

Cardiac Energy State

The energy state of a cell is a chemical potential that reflects the balance between energy generated through cellular metabolism and the consumption of that energy by cellular functions such as contraction, maintenance of a membrane potential, secretion and so forth. In "aerobic" organs such as the heart, energy metabolism consists of two sequential steps. The first is the conversion into reducing equivalents, mainly NADH, of the energy in the electrons liberated by oxidation of substrate. The second step is the orderly transfer of these electrons into mitochondria and conservation of the energy as the ATP generated by oxidative phosphorylation. Either of these steps can be the basis for a formal, thermodynamically correct statement of the cellular energy state. One is the redox potential, $[NADH][H^+]/[NAD^+]$, which is based on the concentrations of reducing equivalents. This potential is not uniform throughout the cell, mitochondria being more "reduced" than the cytoplasm (61). However, the cytosolic redox potential is still a useful indicator of energy state and, because it is coupled to the reaction catalyzed by lactic dehydrogenase, can be judged by the lactate/pyruvate ratio. The phosphorylation state of ATP in the cytosol is another indicator of cellular energy state (60). The two steps in glycolysis that are catalyzed by glyceraldehyde phosphate dehydrogenase and phosphoglycerate kinase couple the ATP potential, defined as $[ATP]_{free}/[ADP]_{free}[Pi]_{free}$, to the redox potential (Fig. 1):

$$GAP + NAD^+ + Pi \xrightleftharpoons{\text{GAPDH}} NADH + H^+ + 1,3\text{-}DPG$$

$$1,3\text{-}DPG + ADP \xrightleftharpoons{\text{PGK}} 3\text{-}PG + ATP$$

$$\frac{[ATP]_{free}}{[ADP]_{free}[Pi]} = \frac{[GAP]}{[3\text{-}PG]} \cdot \frac{[NAD^+]}{[NADH][H^+]} \cdot K_{GADPH} \cdot K_{PGK}$$

Fig. 1. Coupling of ATP potential to redox potential. 1,3-DPG = 1,3-diphosphoglycerate; GAP = glyceraldehyde-3-phosphate; GAPDH = glyceraldehyde phosphate dehydrogenase; 3-PG = 3-phosphoglycerate; PGK = phosphoglycerate kinase.

The subscript "free" in each term of the ATP potential emphasizes that the concentrations of these reactants refer to those in solution in cytosolic water rather than the total content in tissue. The ADP pool of heart muscle is highly compartmentalized, mainly through binding to actomyosin (49); ADP thus sequestered cannot be in equilibrium with ATP and Pi, which exist mainly in free solution, so calculations based on total tissue ADP are meaningless.

Although the redox and the ATP potentials are equivalent as indices of energy state, the ATP potential is perhaps preferable because ATP is the immediate source of energy for cellular function. At present it is not possible to measure $[ADP]_{free}$ directly; however, the creatine phosphate potential, $[CrP]/[Cr][Pi]$, serves as an index of the ATP potential in much the same way as the lactate/pyruvate ratio serves as an index of redox potential.

Production of Adenosine and Coupling to Cardiac Energy State

The hydrolysis of either of two substrates could account for adenosine production in heart muscle, that of AMP by 5'-nucleotidase or that of S-adenosylhomocysteine by S-adenosylhomocysteine hydrolase (Fig. 2):

$$AMP + H_2O \xrightarrow{\text{5'-NP}} Pi + Ado$$

$$AdoHcy + H_2O \xrightleftharpoons{\text{SAH}} Ado + Hcy$$

Fig. 2. Two pathways of adenosine production in heart muscle. Ado = adenosine; AdoHcy = S-adenosylhomocysteine; Hcy = homocysteine; 5'-NP = 5'-nucleotidase; SAH = S-adenosylhomocysteine hydrolase.

Cardiac 5'-Nucleotidases

Studies employing enzyme histochemistry and cell fractionation show that AMP phosphohydrolase activity resides in both the sarcolemma and the cytosol of cardiac muscle cells (11,39,52). Several lines of evidence exclude the sarcolemmal enzyme as the source of adenosine for coronary vasoregulation. This 5'-nucleotidase is an *ecto* enzyme, that is, its catalytic site is located on the external surface of the cardiac cell (18). There has never been a satisfactory explanation for how its substrate, AMP in the interior of the cell, might traverse the plasma membrane to gain access to the site of hydrolysis. ADP is a powerful inhibitor of the *ecto* nucleotidase (9); accordingly, it is not possible to reconcile the accelerated rate of adenosine production during hypoxia (2,19) and ischemia (42) with the rise in ADP levels that occur under such conditions and which should inhibit the enzyme. The most convincing evidence, however, is from studies showing that the rate of adenosine production during hypoxia or ATP substrate depletion by 2-deoxyglucose is unaffected by essentially complete inhibition of the *ecto* nucleotidase by specific antibodies, by nucleoside phosphonates or by lectins such as concanavalin A (40,62).

New information about the properties of the 5'-nucleotidase in the cardiac cytosol (11,41) and about the compartmentalization of the AMP pool of heart muscle (7,8,56) support a plausible model of adenosine production coupled to cardiac energy state (43). AMP is the preferred substrate of the cytosolic nucleotidase of rabbit heart. ATP, at the low millimolar concentrations that occur within a cell, fully activates the enzyme by increasing its affinity for substrate, the K_{app} for AMP, averaging 2.5 mM (11,41). ADP also activates the nucleotidase when the AMP concentration is low. Purification of the enzyme abolishes regulation by ATP, evidence that the enzyme may consist of regulatory as well as catalytic subunits. The activity of the enzyme in unfractionated extracts of ventricular muscle from rabbit or guinea pig is about 300 umole/min per g wet wt., more than enough to account for adenosine production at the accelerated rates, up to 1 umole/min per g wet wt., that have been measured in ischemic heart muscle (41).

The model posits that ATP potential controls adenosine production by regulating the availability of substrate, rather than by modifying the catalytic properties of the cytosolic 5'-nucleotidase. As pointed out above, the cytosolic concentration of ATP is sufficient to fully activate the enzyme and, owing to the buffering effect of the creatine phosphate system, it changes only modestly even under such extreme conditions as hypoxia. Accordingly, the enzyme probably remains fully activated at all times and adenosine production varies according to substrate availability. Very nearly the entire cardiac AMP pool is intramitochondrial (8,56); the small fraction in the cytosol probably arises from the dismutation of ADP catalyzed by myokinase (Fig. 3). Cytosolic AMP concentrations are in the submicromolar range, vary as an inverse function of the ATP potential and, according to the myokinase equilibrium, in proportion to the square root of $[ADP]_{free}$, between about 50 and 750 nM (7). Because such substrate concentrations are three to four orders of magnitude below the K_{app} of the activated cytosolic nucleotidase, any change in AMP level produces a proportional change in the rate of adenosine production.

S-Adenosylhomocysteine Hydrolase (SAH)

SAH has one clearly established function, the degradation of the S-adenosylhomocysteine that results from and is a very potent competitve inhibitor of the S-adenosylmethionine-dependent regulation of such diverse substrates as biogenic amines, proteins and membrane phospholipids (10). In most eukaryotic cells the hydrolase also reversibly binds adenosine (24,58). Adenosine sequestered in this intracellular compartment accounts for very nearly the entire cardiac pool of 1-2 nmole/g wet wt. (46), but whether it has a physiological role is unknown. Heart muscle also contains a pool of S-adenosylhomocysteine, somewhat smaller than the adenosine pool, about 0.2-0.6 nmole/g wet wt. Studies of the turnover of this pool in isolated, perfused rodent hearts suggest that adenosine formation from S-adenosylhomocysteine is somewhat less than 1 nmole/min per g under normoxic conditions and less than twice that rate during hypoxia. Since adenosine formation during hypoxia is so much greater, the transmethylation pathway may be a significant source of adenosine only under basal conditions (1,33).

One encounters important problems in trying to account for the metabolic control of coronary flow in terms of the SAH model. First, the enzyme does not appear to be under the control of cardiac energy state. SAH contains NAD^+ (47), but it is very tightly bound and there is no evidence for coupling to the cytosolic pyridine nucleotides that determine redox state. Second, the poise of the equilibrium catalyzed by SAH lies strongly in the direction of synthesis, and even in the absence of homocysteine a high affinity for adenosine promotes retention of the nucleoside in the SAH · adenosine complex (58). A physiological stimulus to the dissociation of this complex is not known. Trapping adenosine by a second enzymat-

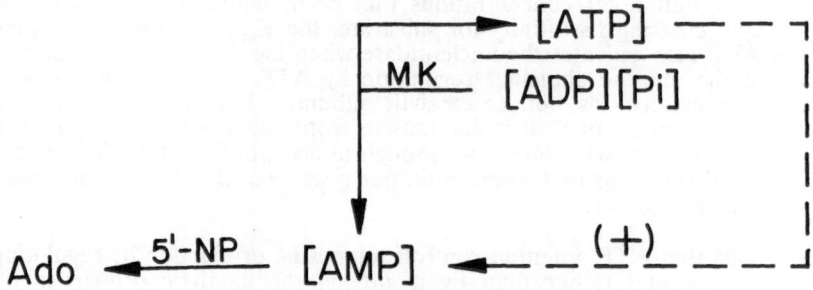

Fig. 3. Coupling of adenosine production to the ATP potential. Abbreviations are: Ado, adenosine; MK, myokinase; and 5'NP, 5'-nucleotidase. The (+) denotes ATP activation of the cytosolic 5'-nucleotidase. See text for discussion.

ic reaction is, of course, one way to effect dissociation of the SAH · adenosine complex, but the net result is not the production of adenosine but rather the product of the trapping reaction. Third, since the adenosine produced from transmethylation may have physiological significance only under basal conditions, its significance diminishes further in light of reports that basal coronary resistance is insensitive to either the destruction of endogenous adenosine by intracoronary adenosine deaminase or to the blockade of adenosine receptors by dialkylxanthines, evidence that adenosine plays no role in setting basal coronary tone (16,20,22,28,45,53,54).

Regulation of Cardiac Interstitial Adenosine Concentration

Since the coronary resistance vessels lie in the interstitial space of the heart, the concentration of adenosine in this anatomical compartment is the independent variable in those tests of the adenosine hypothesis that examine the dose-response relationship between adenosine and coronary resistance. In terms of mass balance, this concentration is the resultant of adenosine release from the cells that constitute the boundary of the interstitial space, the re-uptake of adenosine by these same cells, washout into the coronary venous drainage as well as relatively minor losses via the cardiac lymphatics and into the pericardial space. A large literature documents the movement of adenosine in either direction across cellular membranes by a combination of carrier-mediated and simple diffusion. The cardiac interstitium is not accessible to direct sampling of labile metabolites such as adenosine, a fact that greatly complicates studies of the control of adenosine concentration in this compartment. To appear in the coronary venous effluent, adenosine must first pass the endothelial barrier, the cells of which have an extremely active purine metabolism (39,48). Endothelial cells in culture not only take up adenosine avidly, but also release significant amounts of this and other purines (29,38). Perhaps as much as one-sixth of the cardiac purine efflux originates in the coronary endothelium rather than in cardiac myocytes (13). Likewise, the adenosine appearing in an epicardial transudate (17,23) or in a pericardial superfusate (21,51) will have passed through a layer of epicardial mesothelial cells. Although cardiac lymphatics drain the interstitium directly, the flow rate through these channels is so slow that the adenosine concentration in lymph probably reflects metabolism by the lymphatic endothelium rather than that in the interstitium.

Coronary Adenosine Receptor and Coupling to Effectors

The identity of the coronary artery adenosine receptor is fairly certain but that of the systems(s) through which receptor activation leads to coronary relaxation is not. Low affinity A2 (or Ra) receptors that stimulate adenylate cyclase appear to mediate coronary vasodilation by adenosine. The several lines of evidence supporting such an identification are: 1) the coronary receptor is located on the myocyte surface; 2) alkylxanthines such as theophylline competitively antagonize adenosine coronary vasodilation; 3) forskolin potentiates adenosine vasodilation; 4) the coronary vasoactivity ranking of receptor-selective analogues resembles that of A2 receptors; and 5) adenosine stimulates cyclic AMP accumulation and relaxation of coronary artery strips and in suspensions of coronary microvessels (43). The functional coupling of the coronary adenosine receptor to adenylate cyclase implies that receptor activation initiates protein phosphorylation by cyclic AMP-dependent protein kinase (A-kinase), but the identity of the substrate(s) is uncertain. A kinase can phosphorylate and thereby inactivate myosin light chain kinase (14,55); other possible substrates that could influence smooth muscle contractile state are proteins in the sarcolemma (26) or in intracellular organelles that sequester the Ca^{2+}-supporting smooth muscle contraction (59).

Two recent reports raise the possibility that adenosine causes coronary relaxation through receptors other than the A2 adenosine receptor and/or effectors other than

A-kinase. The first paper (15) describes the inhibition of the phosphatidylinositol kinase (PI-kinase) of aortic smooth muscle by adenosine and several analogues modified in the ribose, but not the purine, that compete with ATP for binding to the catalytic site. The K_i of 5'-chloro-5'-deoxyadenosine, a metabolically inert analogue, is much lower than the K_m for ATP. Cell fractionation studies show that PI-kinase purifies along with markers of sarcoplasmic reticulum, suggesting that inhibiting the formation of a precursor of phosphoinositol-4,5-diphosphate, a Ca^{2+} ionophore, could prevent the release of Ca^{2+} from intracellular stores (6). The inhibition of coronary smooth muscle PI-kinase by adenosine could well be physiologically important, but it is not a mechanism for the *metabolic* control of coronary perfusion, because adenosine production occurs inside the coronary myocyte and thus is not coupled to the cytosolic ATP potential of the working cardiocytes.

The second report describes evidence that A1 receptors coupled to guanylate cyclase mediate adenosine vasodilation, a model exactly opposite the conventional model of A2 receptors coupled to adenylate cyclase. In aortic smooth muscle cells in culture, adenosine and two analogues, N^6-cyclohexyladenosine (CHA) and N-ethyl adenosine-5'-uronamide (NECA), at concentrations between 10^{-9} and 10^{-4} M, promote the concentration-dependent 2- to 2.5-fold accumulation of cyclic GMP. Activity in the nanomolar range, an agonist potency ranking of CHA > NECA > adenosine, and antagonism by theophylline suggest that A1 adenosine receptors mediate this response. Parallel studies of the guanylate cyclases of cultured aortic smooth muscle cells and of freshly isolated media of rat aorta and also cow coronary artery show that the three adenosines stimulate the particulate, but not the soluble, guanylate cyclases of these vascular muscles (30). This report is intriguing, because it suggest that adenosine might act through the same cyclic GMP-activated protein phosphorylation cascade as, for example, the endothelium-dependent vasodilators (50). Interpretation is not straightforward, however, for these observations do not correspond to the intact heart, where the coronary vasoactivity of NECA is two orders of magnitude greater than that of CHA (31).

Relationship of Adenosine to other Endogenous Vasodilators

It is doubtful that adenosine is the sole mediator in the metabolic control of coronary blood flow. Models in which there is a single mediator of flow posit an unique quantitative relationship between agonist concentration and coronary flow, that is, one that is identical under widely differing experimental conditions (3). A number of studies purport to show a correlation between coronary flow and adenosine concentration, as reflected by one or another of the indirect indices of adenosine concentration in the interstitial space of the heart (5). While this sort of evidence is certainly consistent with the participation of adenosine in flow regulation, it is not robust enough to establish adenosine as the sole mediator at the exclusion of other endogenous regulatory substances. For want of a technique for d·-ect measurements of adenosine in the interstitial space to serve as a Gold Standard, the validity of all the indirect methods is uncertain and the inferences drawn from such data are necessarily conditional. Further, random errors in the estimates of adenosine concentration are large enough to obscure significant differences in the relationship between flow and concentration under different experimental conditions.

Two additional lines of evidence argue more directly against an unique relationship between coronary flow and interstitial adenosine concentration. The first line consists of two studies of adenosine release into the coronary effluent. Owing to adenosine release from coronary endothelial cells (38), the adenosine concentration in the coronary veins only reflects directional changes in interstitial concentration. Even so, the experiments suggest, respectively, that adenosine production may depend on the nature of the stimulus and that adenosine may only initiate hyperemic responses, which are then sustained by other agents or mechanisms. In open-chest dog preparations, raising coronary flow by means of beta-adrenergic stimulation elicits adenosine release, whereas electrical pacing to the same level of cardiac performance and coronary flow does not cause adenosine release (34). In isolated guinea-pig hearts, the infusion of norepinephrine or perfusion with a high-calcium buffer raises oxygen consumption and coronary flow to new steady-state levels that are stable over many minutes. These interventions also evoke adenosine release, but this is only transitory, reaching a peak within a minute and declining steadily thereafter toward control (12). To the extent that coronary venous adenosine reflects the interstitial concentration, the changing dose-response relationship between adenosine and coronary flow suggests the participation of other vasodilators.

The second line of evidence against an unique dependance of coronary flow on adenosine concentration comes from experiments employing specific adenosine receptor blockade by dialkylxanthines such as theophylline or the destruction of endogenous adenosine by intracoronary adenosine deaminase. These interventions reduce but do not abolish the reactive hyperemia response to ischemia (53), to beta-adrenergic stimulation (45,27), or to hypoxia (36), but have no effect on basal coronary flow or on the autoregulatory response to a change in coronary perfusion pressure (16,20,22).

One can trace a causal relationship between the release of two endogenous vasodilators, carbon dioxide and potassium, and the cytosolic ATP potential. The ATP potential controls the sarcolemmal ATPase that mediates the exchange of Na^+ for K^+ and, consequently, of K^+ release. So, too, the ATP potential regulates the rate of the oxidative substrate metabolism that generates carbon dioxide. Like adenosine, however, neither potassium nor carbon dioxide alone seems able to account for metabolic coronary flow control. Potassium lacks the potency to account for more than a fraction of coronary flow responses (37). The relationship between carbon dioxide production and oxidative metabolism, as defined by the respiratory quotient, varies according to the type of substrate. In other words, there is not an unique relationship between vasodilator release and cardiac energy state. It might well be that adenosine, potassium and carbon dioxide act in concert to control coronary flow, just as these three metabolites regulate cerebral blood flow (32). To complicate matters still further, the discovery that a high coronary flow rate by itself can cause endothelium-dependent vasodilation (35) suggests that even hyperemic responses that begin as metabolically-linked vasodilation might be sustained, in part, by non-metabolic, flow-linked vasodilation (25).

References

1. Achterberg PW, De Tombe PP, Harmsen E, De Jong JW: Myocardial S-adenosylhomocysteine hydrolase is important for adenosine production during normoxia. Biochim Biophys Acta 840:393-400, 1985
2. Berne RM: Cardiac nucleotides in hypoxia: Possible role in regulation of coronary blood flow. Am J Physiol 204:317-322, 1963
3. Berne RM: The role of adenosine in the control of coronary blood flow. Circ Res 47:807-813, 1980
4. Berne RM, Rubio R: Coronary circulation. In: Berne RM, Sperelakis N, eds: Handbook of physiology: The cardiovascular system. Section 2, Vol I. Washington DC: American Physiological Society, 1979:873-952
5. Berne RM, Winn HE, Knabb RM, Ely SW, Rubio R: Blood flow regulation by adenosine in heart, brain and skeletal muscle. In: Berne RM, Rall TW, Rubio R, eds: Regulatory function of adenosine. The Hague: Nijhoff Publ, 1983:293-317
6. Berridge MJ, Irvine RF: Inositol triphosphate, a novel second messenger in cellular signal transduction. Nature 312:315-321, 1984

7. Bünger R: Thermodynamic state of cytosolic adenylates in guinea pig myocardium. Energy-linked adaptive changes in free adenylates and purine nucleoside release. In: Gerlach E, Becker BF, eds: Topics and perspectives in adenosine research. Berlin: Springer Verlag, 1987:223-234

8. Bünger R, Soboll S: Cytosolic adenylates and adenosine release in perfused working heart. Eur J Biochem 159:203-213, 1986

9. Burger RM, Lowenstein JM: 5'-Nucleotidase from smooth muscle of small intestine and from brain. Inhibition by nucleotides. Biochemistry 14:2362-2366, 1975

10. Cantoni GL, Chiang PK: The role of S-adenosylhomocysteine and S-adenosylhomocysteine hydrolase in the control of biological methylations. In: Cavallini D, Gaull GE, Zappia V, eds: Natural sulfur compounds. New York: Plenum Press, 1980:67-80

11. Collinson AR, Peuhkurinen KJ, Lowenstein JM: Regulation and function of 5'-nucleotidases. In: Gerlach E, Becker BF, eds: Topics and perspectives in adenosine research. Berlin: Springer Verlag, 1987:133-144

12. DeWitt DF, Wangler RD, Thompson CI, Sparks HV Jr: Phasic release of adenosine during steady state metabolic stimulation in the isolated guinea pig heart. Circ Res 53:636-643, 1983

13. Deussen A, Moser G, Schrader J: Contribution of coronary endothelial cells to cardiac adenosine production. Pflügers Arch 406:608-614, 1986

14. Doctorow SR, Lowenstein JM: Adenosine and 5'-chloro-5'-deoxyadenosine inhibit the phosphorylation of phosphatidylinositol and myosin light chain in calf aorta smooth muscle. J Biol Chem 260:3469-3476, 1985

15. Doctorow SR, Lowenstein JM: Inhibition of phosphatidylinositol kinase in vascular smooth muscle membranes by adenosine and related compounds. Biochem Pharmacol 36:2255-2262, 1987

16. Dole WP, Yamada N, Bishop VS, Olsson RA: Role of adenosine in coronary blood flow regulation after reductions in coronary perfusion pressure. Circ Res 56:517-524, 1985

17. Fenton RA, Dobson JG Jr: Measurement by fluorescence of interstitial adenosine levels in normoxic, hypoxic and ischemic perfused rat hearts. Circ Res 60:177-184, 1987

18. Frick GP, Lowenstein JM: Studies of 5'-nucleotidase in perfused rat heart including measurements of the enzyme in perfused skeletal muscle and liver. J Biol Chem 251:6372-6378, 1976

19. Gerlach E, Deuticke B, Dreisbach RH: Der Nucleotid-Abbau in Herzmuskel bei Sauerstoffmangel und seine mögliche Bedeutung für die Koronardurchblutung. Naturwiss 50:228-229, 1963

20. Gerwirtz H, Brautigan DL, Olsson RA, Brown P, Most A: Role of adenosine in the maintenance of coronary vasodilation distal to a severe coronary stenosis. Circ Res 53:42-51, 1983

21. Gidday JM, Van Cleaf S, Rubio R, Berne RM: Measurement of interstitial fluid adenosine concentration by an epicardial chamber during different levels of cardiac inotropy. Physiologist 28:340, 1985 (Abstr)

22. Hanley FL, Gratton MT, Stevens MB, Hoffman JIE: Role of adenosine in coronary autoregulation. Am J Physiol 250:H558-H566, 1986

23. Heller LJ, Mohrman DE, Sunnarborg LJ: Interstitial adenosine concentration in isolated perfused rat hearts during adenosine infusions. Physiologist 28:339, 1985 (Abstr)

24. Hershfield MS, Kredich NM: S-Adenosylhomocysteine hydrolase is an adenosine-binding protein: A target for adenosine toxicity. Science 202:757-760, 1978

25. Holtz J, Forstermann U, Pohl U, Giesler M, Bassenge E: Flow-dependent endothelium-mediated dilation of epicardial coronary arteries in conscious dogs: Effects of cyclooxygenase inhibition. J Cardiovasc Pharmacol 6:1161-1169, 1984

26. Jones AW, Bylund DB, Forte LR: cAMP-dependent reduction in membrane fluxes during relaxation of arterial smooth muscle. Am J Physiol 246:H306-H311, 1984

27. Jones CE, Hurst TW, Randall Jr: Effect of aminophylline on coronary functional hyperemia and myocardial adenosine. Am J Physiol 243:H480-H487, 1982

28. Kroll K, Feigl EO: Adenosine is unimportant in controlling coronary blood flow in unstressed dog hearts. Am J Physiol 249:H1176-H1187, 1985

29. Kroll K, Schrader J, Piper HM, Henrich M: Release of adenosine and cyclic AMP from coronary endothelium in isolated guinea pig hearts: Relation to coronary flow. Circ Res 60:659-665, 1987

30. Kurtz A: Adenosine stimulates guanylate cyclase activity in vascular smooth muscle cells. J Biol Chem 262:6296-6300, 1987

31. Kusachi S, Thompson RD, Olsson RA: Ligand selectivity of dog coronary adenosine receptor

resembles that of adenylate cyclase stimulatory (Ra) receptors. J Pharmacol Exp Ther 227:316-321, 1983

32. Kuschinski W, Wahl M: Local chemical and neurogenic regulation of cerebral vascular resistance. Physiol Rev 58:656-689, 1978

33. Lloyd HGE, Schrader J: The importance of the transmethylation pathway for adenosine metabolism in the heart. In: Gerlach E, Becker BF, eds: Topics and perspectives in adenosine research. Berlin: Springer Verlag, 1987:199-207

34. Manfredi JP, Sparks HV: Adenosine's role in coronary vasodilation induced by atrial pacing and norepinephrine. Am J Physiol 243:H536-H545, 1982

35. Maseri A, Severi S, DeNes M, L'Abbate A, Chierchia S, Marzilli M, Ballestra AM, Parodi O, Biagini A, Distante A: "Variant" angina: One aspect of a continuous spectrum of vasospastic myocardial ischemia. Am J Cardiol 42:1019-1035, 1978

36. Merrill GF, Downey HF, Jones CE: Adenosine deaminase attenuates canine coronary vasodilation during hypoxia. Am J Physiol 250:H579-H583, 1986

37. Murray PA, Belloni FL, Sparks HV: The role of potassium in the metabolic control of coronary vascular resistance in the dog. Circ Res 44:767-780, 1979

38. Nees S, Gerlach E: Adenine nucleotide and adenosine metabolism in cultured coronary endothelial cells: Formation and release of adenine compounds and possible functional implications. In: Berne RM, Rall TW, Rubio R, eds: Regulatory function of adenosine. The Hague: Nijhoff Publ, 1982:347-360

39. Nees S, Herzog V, Becker BF, Bock M, Des Rosiers C, Gerlach E: The coronary endothelium: A highly active metabolic barrier for adenosine. Basic Res Cardiol 80:515-529, 1985

40. Newby AC, Holmquist CA: Adenosine production inside rat polymorphonuclear leukocytes. Biochem J 200:399-403, 1981

41. Newby AC, Worku Y, Meghji P: Critical evaluation of the role of ecto and cytosolic 5'-nucleotidase in adenosine formation. In: Gerlach E, Becker BF, eds: Topics and perspectives in adenosine research. Berlin: Springer Verlag, 1987:155-168

42. Olsson RA: Changes in content of purine nucleoside in canine myocardium during coronary occlusion, Circ Res 26:301-306, 1970

43. Olsson RA, Bünger R: Metabolic control of coronary blood flow. Progr Cardiovasc Dis 29:369-387, 1987

44. Olsson RA, Gregg DE: Myocardial reactive hyperemia in the unanesthetized dog. Am J Physiol 208:224-230, 1965

45. Olsson RA, Kusachi S: Intracoronary adenosine deaminase antagonizes beta-adrenergic stimulation of cardiac oxygen usage. Circulation 66, Suppl II:II-154, 1982 (Abstr)

46. Olsson RA, Saito D, Steinhart CR: Compartmentalization of the adenosine pool of dog and rat hearts. Circ Res 50:617-626, 1982

47. Palmer JL, Abeles RH: The mechanism of action of S-adenosylhomocysteinase. J Biol Chem 254:1217-1226, 1979

48. Pearson JD, Gordon JL: Nucleotide metabolism by endothelium. Annu Rev Physiol 47:617-627, 1985

49. Perry SV: The bound nucleotide of the isolated myofibril. Biochem J 51:495-499, 1952

50. Rapoport RM, Draznin MB, Murad F: Endothelium-dependent vasodilator- and nitrovasodilator-induced relaxation may be mediated through cyclic GMP formation and cyclic GMP-dependent protein phosphorylation. Trans Assoc Am Physicians 96:19-30, 1983

51. Rubio R, Berne RM: Release of adenosine by the normal myocardium in dogs and its relationship to the regulation of coronary resistance. Circ Res 25:407-415, 1969

52. Rubio R, Dobson JG Jr, Berne RM: Sites of adenosine production in cardiac and skeletal muscle. Am J Physiol 225:938-953, 1973

53. Saito D, Steinhart CR, Nixon DG, Olsson RA: Intracoronary adenosine deaminase reduces myocardial reactive hyperemia. Circ Res 47:875-882, 1981

54. Schütz W, Zimpfer M, Raberger G: Effect of aminophylline on coronary reactive hyperemia following brief and long occlusion periods. Cardiovasc Res 11:507-511, 1977

55. Silver PJ, Di Salvo J: Adenosine 3':5'-monophosphate mediated inhibition of myosin light chain phosphorylation in bovine aortic actomyosin. J Biol Chem 254:9951-9954, 1979

56. Soboll S, Bünger R: Compartmentation of adenine nucleotides in the working guinea pig heart stimulated by noradrenaline. Hoppe-Seyler's Z Physiol Chem 362: 125-132, 1981

57. Sparks HV Jr, Bardenheuer H: Regulation of adenosine formation in the heart. Circ Res 58:193-201, 1986

58. Ueland PM: Pharmacological and biochemical aspects of S-adenosylhomocysteine and S-adenosylhomocysteine hydrolase. Pharmacol Rev 34:223-253, 1982
59. Van Breemen C, Siegel B: The mechanism of alpha-adrenergic activation of the dog coronary artery. Circ Res 46:426-429, 1980
60. Veach RL, Randolph-Lawson JWR, Cornell NW, Krebs HA: Cytosolic phosphorylation potential. J Biol Chem 254:6538-6547, 1979
61. Williamson DH, Lund P, Krebs HA: The redox state of free nicotinamide-adenine dinucleotide in the cytoplasm and mitochondria of rat liver. Biochem J 103:514-527, 1967
62. Worku Y, Newby AC: The mechanism of adenosine production in rat polymorphonuclear leukocytes. Biochem J 214:325-330, 1983

Chapter 9

Cardiac Electrophysiological Effects of Adenosine

L. Belardinelli and G.A. West, Departments of Medicine
(Division of Cardiology) and Physiology,
University of Virginia, Charlottesville, Virginia, U.S.A.

Adenosine, in addition to its well-established vasodilator actions, also modulates a) pacemaker activity, b) atrioventricular (AV) transmission, c) atrial contractility, and d) the myocardial actions of catecholamines. Based on pharmacological response and mechanism of action, the above cardiac effects of adenosine can be tentatively subdivided into two categories, i.e., non-cAMP-mediated effects which appear to be due to an activation of K^+ channels, and cAMP-dependent effects characterized by a modulation of calcium inward current. In working ventricular myocardium, only the cAMP-dependent mechanism appears to be operative, whereas in the atrial myocardium and specialized tissues [e.g., sinoatrial (SA) and AV node], both the cAMP-independent and cAMP-dependent mechanisms are operative. In the SA node, adenosine causes sinus slowing and a shift in location of the primary pacemaker site. In atrial myocytes, adenosine depresses contractility and shortens the action potential; in the AV node, it depresses the nodal (N-cell) action potential which is accompanied by prolongation of AV conduction time. In ventricular myocardium, adenosine antagonizes the stimulatory effect of agents that stimulate adenylate cyclase activity but has no effect in their absence. In summary, adenosine can affect cardiac function at several levels by different mechanisms and may have greater physiological importance than previously recognized.

Introduction

In addition to its well-established vasodilator effects, the nucleoside adenosine has many other cardiac actions. Adenosine has a depressant effect on a) sinoatrial (SA) and atrioventricular (AV) node function (i.e., negative chronotropic and dromotropic action, respectively), b) atrial contractility (negative inotropic effect), and c) the cardiac stimulatory actions of catecholamines. Based on pharmacological response and mechanism of action, the above cardiac effects of adenosine can be tentatively subdivided into two categories: cAMP-independent and cAMP-dependent (Fig. 1). The cAMP-independent effects appear to be due to an increase in K^+-conductance (e.g., hyperpolarization of SA-node cells). On the other hand, the cAMP-dependent effects appear to be the result of an inhibition of adenylate cyclase. They are characterized by an attenuation of the catecholamine-enhanced calcium inward current (e.g., antagonism of the positive inotropic and arrhythmogenic effect of isoproterenol). In working ventricular myocardium, only the cAMP-dependent mechanism appears to be operative, whereas in the atrial myocardium and specialized tissues (e.g., SA and AV node), both the direct and indirect mechanisms are operative (Fig. 1).

Current evidence indicates that all of the aforementioned actions of adenosine are a) mediated by extracellular receptors - subtype A_1 (or R_1), b) potentiated by nucleoside transport blockers (dipyridamole) and adenosine deaminase inhibitors, e.g., erythro-6-amino-9-(2-hydroxy-3-nonyl)purine, and c) antagonized competitively by alkylxanthines (such as theophylline) and abolished by adenosine deaminase. In summary, the adenosine system consists of three components: a receptor, a mechanism for production of adenosine, and a mechanism for its removal.

J.W. de Jong (Ed.), Myocardial Energy Metabolism, Martinus Nijhoff Publishers, Dordrecht/Boston/Lancaster, 1988

The intent of this review is to describe the more general features of the effects of adenosine on the cardiac functions listed above. In addition, the present state of knowledge regarding the mechanism of action of adenosine in various cardiac cell types, as well as the possible role of adenosine as a modulator of certain rhythm disturbances associated with myocardial hypoxia/ischemia will be discussed.

I. Chronotropic Effect (SA Node)

Adenosine, ATP and adenine derivatives (e.g., ADP, AMP) exert a negative chronotropic effect on the sinoatrial node. This effect of the nucleoside has been demonstrated in a variety of isolated preparations from mammalian (7,13,51,56,57) and non-mammalian species (25) as well as in intact anesthetized (7,22,34,45,54) and awake animals (29). Recently, sinus slowing and arrest was shown in humans after intravenous injection of adenosine (19). As illustrated in Fig. 2, adenosine, in addition to slowing sinus rate, causes pacemaker shift, i.e., in the presence of adenosine the earliest site of activation shifts from the SA node to the right atria (56). At high concentrations adenosine also causes sinoatrial exit block (56). Similar to its many other actions, the depressant effect of adenosine in pacemakers (SA, atrial or ventricular) can be antagonized by methylxanthines (23,28,45) and potentiated by nucleoside transport inhibitors (13,25,37,45,56). As shown in Fig. 3, the negative chronotropic effect of adenosine in atrial and His-bundle pacemakers are antagonized by aminophylline (Fig. 3, panel A) but enhanced by dipyridamole (panel B). From this figure it can be seen, as has been reported previously (7,51), that in the guinea-pig heart the His-bundle pacemaker is more sensitive to adenosine than atrial pacemaker, i.e., IC_{50} of 1×10^{-5} and 2×10^{-4} M, respectively.

Adenosine-induced sinus slowing in frog sinus venosus (25) and isolated rabbit SA node (56) is associated with a hyperpolarization to about the K^+-equilibrium potential (E_K). Like the sinus slowing, the adenosine-induced hyperpolarization can be reversed by aminophylline (57). It is worth noting that the magnitude of hyperpolarization caused by adenosine correlates with the degree of slowing (57), i.e., the greater the hyperpolarization the greater the sinus slowing. Since hyperpolarization of pacemaker cells causes slowing of pacemaker firing (11), the above findings suggest that the mechanism of the adenosine-induced slowing is the same as the hyperpolarization. Because the hyperpolarization caused by adenosine is

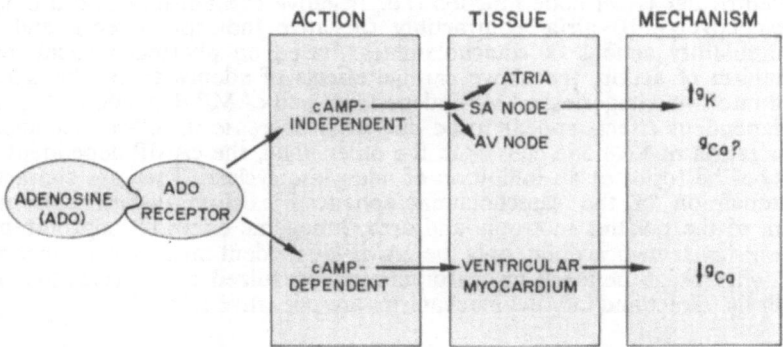

Fig. 1. Cardiac actions of adenosine. Mechanisms involved in the cAMP-independent and cAMP-dependent actions of adenosine on sinoatrial (SA), atrioventricular (AV) node, atrial and ventricular myocardium. The indirect actions of adenosine, which can be demonstrated only when cAMP levels have been elevated, seem to be mediated by an inhibitory effect on the adenylate cyclase. See text for discussion. g = conductance.

neither blocked by ouabain nor low concentrations (1-3 mM) of cesium chloride, it is unlikely that the adenosine-induced hyperpolarization is due to an effect on Na^+/K^+ pump and/or i_f, i.e., the time- and voltage-dependent inward current activated by hyperpolarization (-80 to -120 mV; ref. 57).

Recent studies (59) of the macroscopic and microscopic currents in single isolated SA-node cells have confirmed as well as extended the results obtained in the multicellular preparations. Adenosine caused a dose-dependent slowing of spontaneous rate and hyperpolarization in isolated nodal cells. In whole-cell voltage clamp experiments, adenosine caused an increase in background K^+ current, which was dose-dependent. Adenosine alone had no significant effect on the control (basal) calcium current i_{Ca} nor the hyperpolarization-activated inward current i_f. However, adenosine did attenuate isoproterenol-enhanced i_{Ca} and i_f. In studies of microscopic current in SA nodal cells, adenosine-activated channels had singel-channel conductance of 8 ± 2 pA and showed significant inward rectification. The activation of the K^+ channel by adenosine, as has been shown in atrial cells (33,38), involves a GTP-dependent coupling protein, e.g., G_o, mediating the coupling of the receptor and K^+ channel. The K^+ channel is thought to be independent of cAMP. Thus, in the modulation of pacemaker rate, adenosine activates K^+ channels and, in addition, during increased adrenergic tone, affects i_{Ca} and i_f through a cAMP-dependent mechanism.

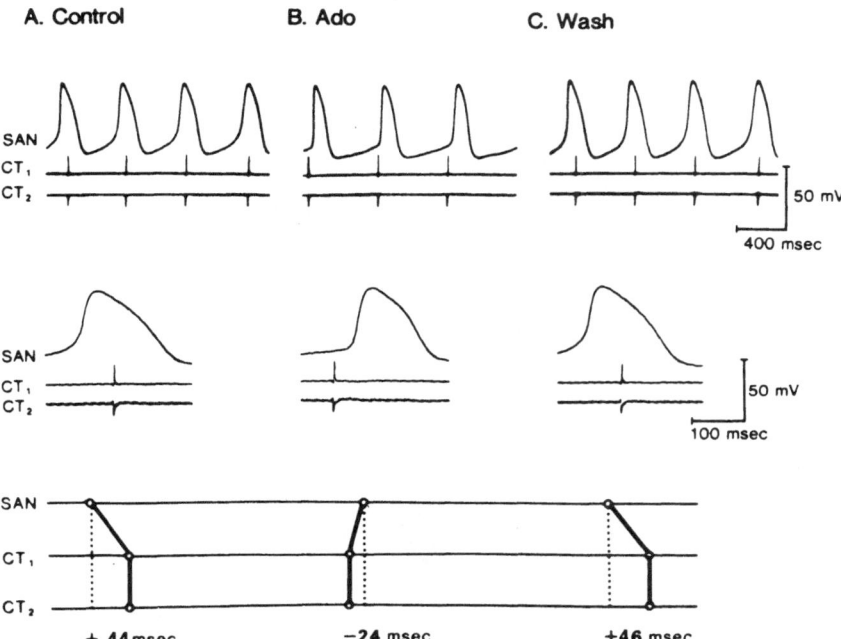

Fig. 2. Adenosine (Ado)-induced sinus slowing and shift in pacemaker site in SA-node preparation. First two rows show same recording at two different sweep speeds. Ladder diagram depicts relationship between activation of SA node with respects to activation of crista terminalis during each condition. Impalement was maintained throughout control, adenosine and washout. A: Control. Cycle length = 460 milliseconds, activation of SAN precedes that of CT_1 and CT_2 by 44 ms. B: Adenosine, 1 x 10^{-4} M. Cycle length = 540 ms. Note change in configuration of SA-node action potential and that activation of CT_1 and CT_2 precedes that of SAN by 24 ms. C: Wash (3 min). Cycle length = 465 ms, SAN again leads CT_1 and CT_2 by 46 ms. SAN = intracellular recording from leading pacemaker cell. CT_1 and CT_2 = extracellular recording from upper and lower crista terminalis, respectively. Reproduced from ref. 56, with permission.

II. Dromotropic Effect (AV Node)

Since 1929, adenosine and related compounds have been known to cause heart block (22). In the 1970s and through the 1980s, numerous studies confirmed the observations that adenosine and adenine nucleotides (e.g., ATP) depress impulse conduction through the AV node in various preparations and species including humans (4,5,14,19,50,53). Although the depressant effect of adenosine in the AV node is a reproducible finding, the magnitude of the slowing of AV-node conduction varies greatly with the species and preparation (7,14,53). This negative dromotropic effect of adenosine is due to activation of a cell surface receptor that can be competitively blocked by alkylxanthines (2). Three key observations support the hypothesis that adenosine action in the AV node is mediated by a specific extracellular receptor, i.e., a) the effect of adenosine is enhanced by nucleoside transport inhibitors such as dipyridamole (2,50), b) adenosine analogs, too large to permeate the cell membrane, cause a similar or greater effect than does free adenosine (2), and c) the alkylxanthine 8-(p-sulfophenyl)theophylline, which has a polar subgroup that limits its cell permeability (28), is more potent than theophylline and only slightly less potent than its non-sulfonated congener (8-phenyltheophylline) to antagonize adenosine. Furthermore, the adenosine P-site analog 2'-deoxyadenosine has no dromotropic effect (2). Despite a large number of

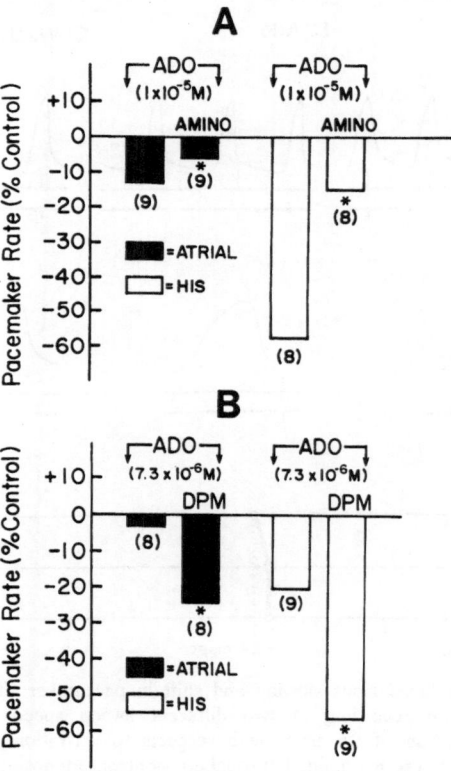

Fig. 3. Attenuation (Panel A) and potentiation (Panel B) of the negative chronotropic effect of adenosine (ADO) on atrial and His-bundle rate by 30 uM aminophylline (AMINO) and 5 uM dipyridamole (DPM), respectively. The control (i.e., absence of adenosine) atrial and His-bundle rates were 190 ± 12 and 90 ± 5 beats/min, respectively. Numbers in parentheses indicate number of guinea-pig hearts studied. *Significantly different from ADO alone (P < 0.05).

reports demonstrating the negative dromotropic effect of adenosine, few or no attempts have been made to elucidate the mechanism and specific site of adenosine action responsible for its negative dromotropic effect. Previous studies (2,5), including those in human subjects (8,19), have shown that the AV block caused by adenosine is due to an increase in conduction time between the atrium and His bundle (i.e., an increase of the AH interval), whereas the His-bundle to ventricle (H-V interval) conduction time is not affected (2,5,53). In a more recent study, it was found that the electrophysiological effects of adenosine on the cells of the AV-node region are concentration-dependent and varied according to the cell type (15). As shown in Fig. 4, adenosine causes an overall depression (i.e., amplitude, duration and rate of rise) of the nodal (N) cell action potential. In comparison, in atrial and atrial-nodal (AN) cells the only significant effect of adenosine is a shortening of the action potential. Thus, the depression and eventual abolition of the action potential of N-cells identifies the N-zone of the AV node as the site of adenosine-induced AV block (15). In contrast, the nucleoside had no effect on the action potentials recorded from nodal-His (NH), His-bundle (HB) and ventricular (V) cells (15), as expected from results which showed a lack of effect on H-V interval.

Little is known regarding the cellular mechanism by which adenosine depresses the N-cell action potential and hence impairs AV conduction. The depression of the N-cell action potential could be explained by a) an increase in K^+-conductance as is the case of SA-node cells and atrial myocytes, b) a decrease in calcium inward current (i_{Ca}), or c) both. As mentioned previously, it should be noted that adenosine does not decrease i_{Ca} in sinoatrial-node cells in the absence of catecholamine stimulation. Furthermore, adenosine has been shown to increase K^+-conductance not only in sinoatrial-node cells (59) and atrial myocytes (4), but also in cultured striatal neurons (52), Xenopus oocytes (42), and smooth muscle from urinary bladder (G. Isenberg, personal communication). Thus, it would not be surprising if the mechanism by which adenosine causes AV block is due to an increase in K^+-conductance. Additional support comes from data showing that acetylcholine (Ach), which in the heart has similar effects as adenosine, increases single channel K^+-currents in both isolated SA- and AV-node cells (48,59). Furthermore, studies in isolated hearts from animals treated with pertussis toxin show an attenuation of the adenosine-induced AH prolongation (60), similar to the effects in SA-node and atrial tissue. This implies a coupling of adenosine receptor and G-protein to the effector. Although indirect, this latter finding further suggests that the action of adenosine in the N-cells may also be due to an increase in K^+-conductance.

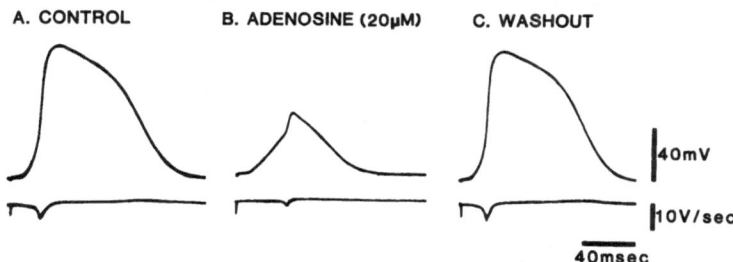

Fig. 4. Effect of adenosine on the action potential of a nodal (N) cell from a small (1 mm²) AV-node preparation of a guinea-pig heart. Traces under each action potential are dV/dt. Panel A: Control; Panel B: Overall depression 2 min after addition of adenosine; and C: Five min after washout of adenosine the action potential returned to control. Preparation was paced at a rate of 3 Hz. All records were obtained from a single impalement. Time and voltage calibrations in panel C apply to all panels.

III. Electrophysiological and Inotropic Effect (Atrium and Ventricle)

In this section, the electrophysiological and inotropic effects of adenosine on atrial and ventricular myocardium are described. Also discussed are the differences in response and mechanism of action of adenosine in atrial and ventricular tissue.

A. Atrial Myocardium

Since the early study of Drury and Szent-Györgyi (22), the negative inotropic effect of adenosine in atrial tissue has been confirmed in various preparations and species (12,18,20,24,43,47). In contrast to the effect on ventricular tissue, in atrial myocardium adenosine and its analogs exert a concentration-dependent negative inotropic effect that is independent of prior catecholamine stimulation, i.e., adenosine depresses basal contractility. Furthermore, this direct negative inotropic effect of adenosine seems to be independent of a decrease in cellular cAMP levels

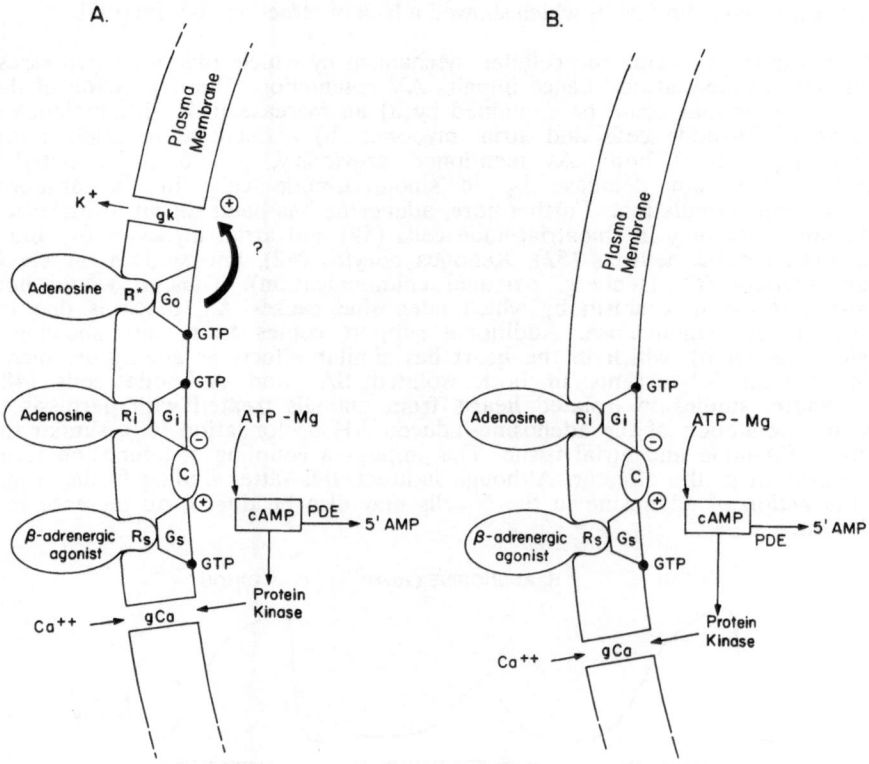

Fig. 5. Schematic of a model to describe the cAMP-independent and cAMP-dependent actions of adenosine. In A, both the cAMP-dependent and -independent actions are shown, both being present in atria, SA node and AV node. The cAMP-dependent action of adenosine is thought to result in an increase in K^+-conductance (g_K), whereas the cAMP-independent action involves the inhibition of adenylate cyclase and hence a decrease in cAMP levels. In B, as occurs in ventricular tissue, only the cAMP-dependent action is shown. The adenosine receptors mediating both the cAMP-dependent and -independent effects are of the subtype A_1, i.e., both the R_i (inhibitory on adenylate cyclase) and R^* (increasing K^+-conductance) are the same receptor types termed A_1. G_s, G_i and G_o represent guanine-nucleotide coupling proteins, and C is the catalytic subunit of adenylate cyclase. See text for discussion. PDE = phosphodiesterase.

(12,23). Thus, like in the SA and AV node, in atrial tissue adenosine has a direct depressant effect (Fig. 1). The negative inotropic effect of adenosine in atrial myocardium is accompanied by a shortening of the action potential. It is conceivable that this shortening of the atrial action potential (4,12,30,35,36) indirectly limits calcium entry and, hence, accounts for the decrease in contractility (12,35). The marked abbreviation of the action potential and attenuation of plateau can be largely accounted for by an adenosine-induced increase in steady-state outward potassium currents (4,35). In support of this hypothesis, adenosine has been also shown to increase ^{42}K efflux in atrial trabeculae (35) and sinus venosus (31). It is worth noting that these actions of adenosine in atrial tissue are similar to those of Ach, but due to activation of specific adenosine receptors that are blocked by theophylline and not atropine. Furthermore, it appears that both Ach and adenosine have a common mechanism of action, i.e., activation of a potassium conductance (4). Recent patch-clamp studies in atrial myocytes indicate that adenosine activates the same population of K^+-channels as does Ach (38), but the activation of the channels is mediated by different receptors and both receptors are coupled to the K^+-channel by a GTP-dependent protein.

Pharmacological and ligand binding studies indicate that the above effects of adenosine are mediated by the extracellular A_1-receptor (12,24,41). As discussed below and depicted in Fig. 5, the adenosine antagonism of the effects of isoproterenol in ventricular muscle is mediated by A_1-receptors coupled to adenylate cyclase, whereas in atrial tissue this receptor seems to be coupled to a K^+-channel. Whether this atrial A_1-receptor is coupled to K^+-channels by a guanine regulatory protein (G), as has been suggested for Ach receptors in atrial tissue (46), is not known. It has also been suggested that the A_1-receptors that mediate the direct action of adenosine in atrial tissue, are also coupled to adenylate cyclase and the negative inotropic effect of adenosine is mediated by cAMP (40). However, much of the current evidence does not support such a hypothesis (19,12,23). It has yet to be determined whether the adenosine receptors that mediate the cAMP-independent and cAMP-dependent actions of adenosine in the atria are the same, i.e., the receptor coupled to a K^+-channel and the receptor linked to adenylate cyclase are the same. However, that both induce K^+-channel activation and adenylate cyclase inhibition can be clearly demonstrated. Regardless of the molecular basis of the adenosine action, the adenosine-induced increase in i_K is a major mechanism in the action of this nucleoside in atrial myocardium.

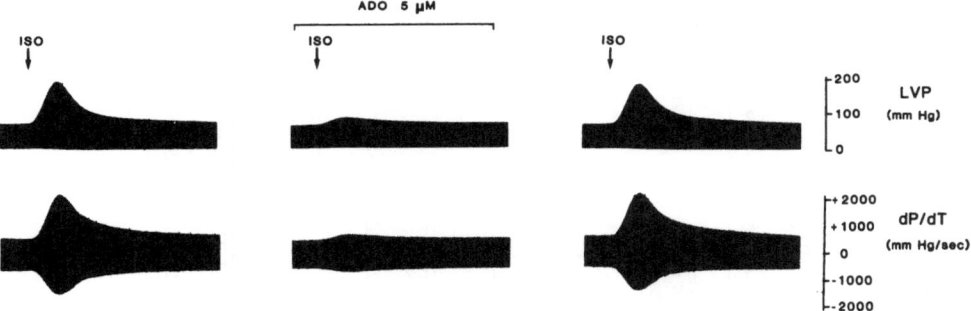

Fig. 6. Example of the anti-adrenergic effect of adenosine in an isolated guinea-pig heart. In the first panel, a bolus (pressure injected, 0.6 s pulse) of 1 uM isoproterenol (ISO) caused a 175% and 285% increase in left ventricular pressure (LVP) and dP/dt, respectively. As shown in the middle panel, in the presence of 5 uM adenosine (ADO), ISO caused only a 32% and 45% increase in LVP and dP/dt, respectively. Note that ADO alone had no direct effect on LVP or dP/dt and only attenuated the ISO-induced increase in LVP and dP/dt. After washout of adenosine, third panel, the response to ISO was the same as during control.

B. Ventricular Myocardium ("Anti-Adrenergic" Effect)

Evidence has accumulated that adenosine modulates some of the myocardial actions of catecholamines (e.g., inotropic and arrhythmogenic; refs. 1,6,21,49). This modulatory role of adenosine on the cardiac effects of catecholamines occurs at two levels: a) pre-junctionally at the level of adrenergic nerve terminals (27,55), and b) post-junctionally at the level of cardiac myocytes (3,32).

In contrast to atrial myocardium, the negative inotropic action of adenosine in ventricular muscle can only be demonstrated when the adenylate cyclase activity has been previously stimulated above basal levels. As illustrated in Fig. 6, adenosine alone does not exert a negative inotropic effect, but it significantly attenuates the positive inotropic effect of isoproterenol. In fact, adenosine attenuates the positive inotropic action of forskolin, amrinone or histamine, agents known to cause an increase in cellular cyclic AMP levels (1,10,58). Similarly, adenosine and its non-metabolizable analogs have also been shown to antagonize the electrophysiologic effects of catecholamines and forskolin (3,32,58). For instance, in isolated guinea-pig and bovine-ventricular myocytes, adenosine antagonizes the isoproterenol- and forskolin-induced a) prolongation of the action potential, b) displacement of the plateau to more positive potentials, c) depolarizing afterpotentials, and d) triggered activity (3,58). These actions of adenosine, i.e., antagonism of the inotropic and electrophysiological effects of isoproterenol and forskolin, can be explained by the ability of the nucleoside to attenuate the isoproterenol- and forskolin-induced increase in calcium inward current (i_{Ca}; ref. 32).

There is evidence to indicate that the adenosine antagonism of the aforementioned actions of catecholamines, as well as forskolin and histamine, is due to the ability of the nucleoside to antagonize the increased accumulation of cAMP caused by these substances (Fig. 7). In keeping with these findings is the observation that adenosine does not attenuate either the positive inotropic or electrophysiological effects of dibutyryl-cAMP, i.e., the site of action of adenosine is likely to be due to attenuation of adenylate cyclase activity (47,58). Based on biochemical and pharmacological studies, it appears that the adenosine receptor responsible for this inhibitory action is the A_1-receptor as identified in other preparations (40,41). As

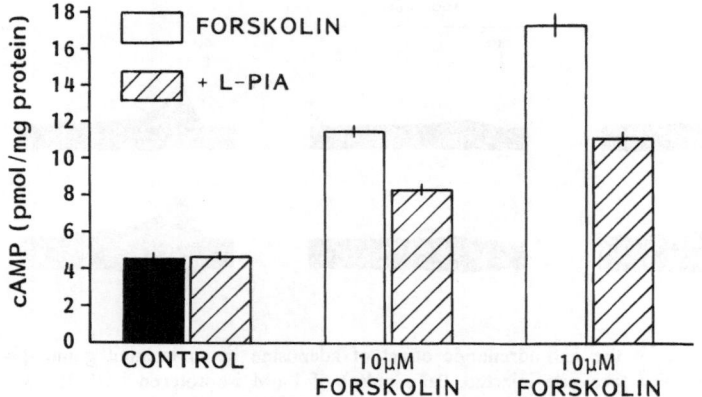

Fig. 7. Inhibition by 0.5 uM N^6-R-phenylisopropyladenosine (L-PIA) of forskolin-induced cyclic AMP accumulation in dispersed guinea-pig ventricular myocytes. Basal levels of cAMP were 4.53 ± 0.64 pmol/mg protein. Note that L-PIA alone had no effect on basal cAMP levels, but it significantly (P <0.05) attenuated the forskolin-induced increase in cAMP. Reproduced from ref. 59, with permission.

illustrated in **Fig. 5**, in ventricular cells this receptor is coupled to the catalytic subunit of adenylate cyclase via the guanine regulatory protein (G_i). Likewise, in the atria the receptor that mediates the indirect action ("anti-adrenergic") seems also to be linked to the adenylate cyclase via G_i (40). In support of the working model depicted in Fig. 5 is the observation that pertussis toxin, known to cause ADP-ribosylation of G_i, and, hence, uncouple the inhibitory receptors (e.g., muscarinic and adenosine-A_1) from adenylate cyclase, significantly diminishes the ability of N^6-methyladenosine to antagonize the isoproterenol-induced cAMP accumulation (26). Thus, the susceptibility of the adenosine-mediated inhibition of adenylate cyclase activity to pertussis-toxin treatment provides further evidence that the adenosine A_1-receptors in ventricular myocytes are located on the cell-surface membrane and are linked to adenylate cyclase.

It should be also noted that there are a few reports suggesting the mechanism of action of adenosine in ventricular myocardium to be independent of adenylate cyclase (9,10,12). The reasons for the contradictory results are not clear, but in part may be explained by experimental technique.

IV. The Physiological Role of Adenosine

Adenosine is normally formed and released by myocardial cells into the intracellular space. Whether the amount of adenosine that accumulates in the interstitium under normal physiological conditions (i.e., well-oxygenated heart) is sufficient to affect any of the cardiac functions discussed in the previous sections remains uncertain. Although the components and their interactions of the adenosine system in the heart is becoming better understood, there are still many aspects to be elucidated. For example, the true interstitial adenosine concentration under various physiological conditions is still not known. The lack of potent and specific adenosine antagonists has hindered the characterization of adenosine receptors, as well as its physiological roles. Although recent discoveries of selective and very potent adenosine receptor antagonists are beginning to appear (16,17), the full applications and interpretation of the results with these substances remain to be clarified. Likewise, nucleoside transport blockers such as dipyridamole have multiple actions and a debate exists whether they block both uptake as well as release of adenosine. Thus, more potent and specific tools that modify the various components of the adenosine system are needed before one can more precisely define the physiological role of the nucleoside.

Despite some lack of understanding the details of the process, in recent years increasing evidence implicating adenosine as a biochemical mediator of sinus slowing and AV-conduction disturbances due to hypoxia and ischemia has accumulated (2,5,44). The observation that the effects of hypoxia and ischemia on sinus- and AV-node function can be modified in a more or less predictable manner by modulators of the actions of adenosine supports the hypothesis that the chronotropic and dromotropic effects of hypoxia and ischemia are at least in part mediated by adenosine released from oxygen-deprived myocardial cells (2,5,44). Furthermore, since adenosine has been shown to have negative chronotropic, dromotropic and anti-adrenergic effects in the human heart (19,39), the recognition and understanding of its actions may provide new insight into the etiology and treatment of certain heart diseases.

Acknowledgement

This work was supported in part by grants from the National Heart, Lung and Blood Institute (HL31111 and HL35272).

References

1. Baumann G, Schrader J, Gerlach E: Inhibitory action of adenosine on histamine- and dopamine-stimulated cardiac contractility and adenylate cyclase in guinea pigs. Circ Res 48:259-266, 1981
2. Belardinelli L, Fenton AR, West GA, Linden J, Althaus JS, Berne RM: Extracellular action of adenosine and the antagonism by aminophylline on the atrioventricular conduction of isolated perfused guinea pig and rat hearts. Circ Res 51:569-579, 1982
3. Belardinelli L, Isenberg G: Actions of adenosine and isoproterenol on isolated mammalian ventricular myocytes. Circ Res 53:287-297, 1983
4. Belardinelli L, Isenberg G: Isolated atrial myocytes: Adenosine and acetylcholine increase potassium conductance. Am J Physiol 224:H734-H737, 1983
5. Belardinelli L, Mattos EC, Berne RM: Evidence for adenosine mediation of atrioventricular block in ischemic canine myocardium. J Clin Invest 68:195-205, 1981
6. Belardinelli L, Vogel S, Linden J, Berne RM: Anti-adrenergic actions of adenosine on ventricular myocardium in embryonic chick hearts. J Mol Cell Cardiol 14:291-294, 1982
7. Belardinelli L, West GA, Crampton R, Berne RM: Chronotropic and dromotropic actions of adenosine. In: Berne RM, Rall TW, Rubio R, eds: The regulatory function of adenosine. The Hague: Nijhoff Publ, 1983:378-398
8. Belhassen B, Pelleg A, Shoshani D, Geva B, Laniado S: Electrophysiologic effects on adenosine-5'-triphosphate on atrioventricular reentrant tachycardia. Circulation 68:827-833, 1983
9. Böhm M, Brückner R, Meyer W, Nose M, Schmitz W, Scholz H, Starbatty J: Evidence for adenosine receptor-mediated isoprenaline-antagonistic effects of adenosine analogs PIA and NECA on force of contraction in guinea pig atrial and ventricular cardiac preparations. Naunyn-Schmiedeberg's Arch Pharmacol 331:131-139, 1985
10. Böhm M, Bürmann H, Meyer W, Nose M, Schmitz W, Scholz H: Positive inotropic effect of Bay K 8644: cAMP-independence and lack of inhibitory effect of adenosine. Naunyn-Schmiedeberg's Arch Pharmacol 329:447-450, 1985
11. Brooks CM, Lu H-H: The sinoatrial pacemaker of the heart. Springfield, Ill: Thomas Publ, 1972
12. Brückner R, Fenner L, Meyer W, Nobis T-M, Schmitz W, Scholz H: Cardiac effects of adenosine and adenosine analogs in guinea pig atrial and ventricular preparations: Evidence against a role of cyclic AMP and cyclic GMP. J Pharmacol Exp Ther 234:766-774, 1985
13. Chiba S: Potentiation of the negative chronotropic and inotropic effects of adenosine by dipyridamole. Tohoku J Exp Med 114:45-48, 1974
14. Chiba S, Hashimoto K: Differences in chronotropic and dromotropic responses of the SA and AV nodes to adenosine and acetylcholine. Jpn J Pharmacol 22:273-274, 1972
15. Clemo HF, Belardinelli L: Effect of adenosine on atrioventricular conduction. Site and characterization of adenosine action in the guinea pig atrioventricular node. Circ Res 59:427-436, 1986
16. Clemo HF, Bourassa A, Linden J, Belardinelli L: Antagonism of the effects of adenosine and hypoxia on atrioventricular conduction time by two novel alkylxanthines: correlation with binding of adenosine to A_1 receptors. J Pharmacol Exp Ther 242:478-484, 1987
17. Daly JW, Ukena D, Jacobson KA: Analogues of adenosine, theophylline, and caffeine: selective interactions with A_1 and A_2 adenosine receptors. In: Gerlach E, Becker B, eds: Topics and perspectives in adenosine research. Berlin: Springer Verlag, 1987:23-35
18. De Gubareff T, Sleator W: Effects of caffeine on mammalian atrial muscle and its interaction with adenosine and calcium. J Pharmacol Exp Ther 148:202-214, 1965
19. DiMarco JP, Sellers TD, Lerman BB, Greenberg ML, Berne RM, Belardinelli L: Diagnostic and therapeutic use of adenosine in patients with supraventricular tachycardia. J Am Coll Cardiol 6:417-425, 1985
20. Dobson Jr JG: Interaction between adenosine and inotropic interventions in guinea pig atria. Am J Physiol 245:H475-H480, 1983
21. Dobson Jr JG: Mechanism of adenosine inhibition of catecholamine-induced responses in heart. Circ Res 52:151-160, 1983
22. Drury AN, Szent-Györgyi A: The physiological activity of adenine compounds with especial reference to their action upon the mammalian heart. J Physiol (Lond) 68:213-237, 1929
23. Endoh M, Maruyama M, Taira N: Modification by islet-activating protein of direct and indirect inhibitory actions of adenosine on rat atrial contraction in relation to cyclic nucleotide metabolism. J Cardiovasc Pharmacol 5:131-142, 1983

24. Evans DB, Schenda JA: Adenosine receptors mediating cardiac depression. Life Sci 31:2425-2432, 1982
25. Hartzell HC: Adenosine receptors in frop sinus venosus: slow inhibitory potentials produced by adenine compounds and acetylcholine. J Physiol (Lond) 293:23-49, 1979
26. Hazeki O, Ui M: Modification by islet-activating protein of receptor-mediated regulation of cyclic AMP accumulation in isolated rat heart cells. J Biol Chem 256:2856-2862, 1981
27. Hedqvist P, Fredholm BB: Inhibitory effect of adenosine on adrenergic neuroeffector transmission in the rabbit heart. Acta Physiol Scand 105:120-122, 1979
28. Heller LJ, Olsson RA: Inhibition of rat ventiruclar automaticity by adenosine. Am J Physiol 248:H907-H913, 1985
29. Hintze TH, Belloni FL, Harrison JE, Shapiro GC: Apparent reduction in baroreflex sensitivity to adenosine in conscious dogs. Am J Physiol 249:H554-H599, 1985
30. Hollander PB, Webb JL: Effects of adenine nucleotides on the contractility and membrane potentials of rat atrium. Circ Res 5:349-353, 1957
31. Hutter OF, Rankin AC: Ionic basis of the hyperpolarizing action of adenyl compounds on sinus venosus of the tortoise heart. J Physiol (Lond) 353:111-125, 1984
32. Isenberg G, Belardinelli L: Ionic basis for the antagonism between adenosine and isoproterenrol on isolated mammalian ventricular myocytes. Circ Res 55: 309-325, 1984
33. Isenberg G, Gerbai E, Klockner U: Ionic channels and adenosine in isolated heart cells. In: Gerlach E, Becker B, eds: Topics and perspectives in adenosine research. Berlin: Springer Verlag, 1987:323-334
34. James TN: The chronotropic action of ATP and related compounds studied by direct perfusion of the sinus node. J Pharmacol Exp Ther 149:233-247, 1965
35. Jochem G, Nawrath H: Adenosine activates a potassium conductance in guinea-pig atrial heart muscle. Experientia 39:1347-1349, 1983
36. Johnson EA, McKinnon MG: Effect of acetylcholine and adenosine on cardiac cellular potentials. Nature 178:1174-1175, 1956
37. Kolassa N, Pfleger K, Träm M: Species differences in action and elimination of adenosine after dipyridamole and hexobendine. Eur J Pharmacol 13:320-325, 1971
38. Kurachi Y, Nakajima T, Sugimoto T: On the mechanism of activation of muscarinic K^+ channels by adenosine in isolated atrial cells: involvement of GTP-binding proteins. Pflügers Arch 407:264-274, 1986
39. Lerman BB, Belardinelli L, DiMarco JP: Adenosine sensitive ventricular tachycardia: Evidence suggesting cyclic AMP mediated triggered activity. J Am Coll Cardiol 7:155A, 1986 (Abstr)
40. Linden J, Hollen CE, Patel A: The mechanism by which adenosine and cholinergic agents reduce contractility in rat myocardium. Circ Res 56:728-735, 1985
41. Lohse MJ, Ukena D, Schwabe U: Adenosine receptors on heart muscle. Lancet 2:355, 1984
42. Lotan I, Dascal N, Oron Y, Cohen S, Lass Y: Adenosine-induced K^+ current in Xenopus oocyte and the role of adenosine 3',5'-monophosphate. Mol Pharmacol 28:170-177, 1985
43. Meinertz T, Nawrath H, Scholz H: Influence of cyclization and acyl substitution on the inotropic effects of adenine nucleotides. Naunyn-Schmiedeberg's Arch Pharmacol 278:165-178, 1973
44. Motomura S, Hashimoto K: Reperfusion-induced bradycardia in the isolated, blood-perfused sino-atrial node and papillary muscle preparations of the dog. Jpn Heart J 23:112-114, 1982
45. Pelleg A, Belhassen B, Ilia R, Laniado S: Comparative electrophysiologic effects of adenosine triphosphate and adenosine in the canine heart: Influence of atropine, propranolol, vagotomy, dipyridamole and aminophylline. Am J Cardiol 55:571-576, 1985
46. Pfaffinger PJ, Martin JM, Hunter DD, Nathanson NM, Hille B: GTP-binding proteins couple cardiac muscarinic receptors to a K channel. Nature 317:536-538, 1985
47. Rockoff JB, Dobson JG: Inhibition by adenosine of catecholamine-induced increase in rat atrial contractility. Am J Physiol 239:H365-H370, 1980
48. Sackmann B, Noma A, Trautwein W: Acetylcholine activation of single muscarinic K^+ channels in isolated pacemaker cells of the mammalian heart. Nature 303:250-253, 1983
49. Schrader J, Baumann G, Gerlach E: Adenosine as inhibitor of myocardial effects of catecholamines. Pflügers Arch 372:29-35, 1977
50. Stafford A: Potentiation of adenosine and the adenine nucleotides by dipyridamole. Br J Pharmacol Chemother 28:218-227, 1966
51. Szentmiklosi AJ, Nemeth M, Szegi J, Papp JG, Szekeres L: Effect of adenosine on sinoatrial and

ventricular automaticity of the guinea pig. Naunyn-Schmiedeberg's Arch Pharmacol 311:147-149, 1980
52. Trussel LO, Jackson MB: Adenosine-activivated potassium conductance in cultured striatal neurons. Proc Natl Acad Sci (USA) 82:4857-4861, 1985
53. Urthaler F, James TN: Effects of adenosine and ATP on AV conduction and on AV junctional rhythm. J Lab Clin Med 79:96-105, 1972
54. Versprille A: The chronotropic effect of adenine- and hypoxanthine derivatives on isolated rat hearts before and after removing the sino-auricular node. Pflügers Arch 291:261-267, 1966
55. Wakade AR, Wakade TD: Inhibition of noradrenaline release by adenosine. J Physiol (Lond) 282:35-49, 1978
56. West GA, Belardinelli L: Sinus slowing and pacemaker shift caused by adenosine in rabbit SA node. Pflügers Arch 403:66-74, 1985
57. West GA, Belardinelli L: Correlation of sinus slowing and hyperpolarization caused by adenosine in sinus node. Pflügers Arch 403:75-81, 1985
58. West GA, Isenberg G, Belardinelli L: Antagonism of forskolin effects by adenosine in guinea pig isolated hearts and ventricular myocytes: Evidence that anti-adrenergic affects of adenosine are due to inhibition of adenylate cyclase. Am J Physiol 250:H769-H777, 1986
59. West GA, Giles W, Belardinelli L: The negative chronotropic effect of adenosine in sinus node cells. In: Gerlach E, Becker B, eds: Topics and perspectives in adenosine research. Berlin: Springer Verlag, 1987:336-342
60. Wilson WW, West GA, Hewlett EL, Belardinelli L: Attenuation of the inhibitory effects of adenosine by pertussis toxin in isolated guinea pig hearts. Fed Proc 46:6536, 1987 (Abstr)

Chapter 10

Acceleration of Adenine Nucleotide Biosynthesis after Ischemic Insult

H.-G. Zimmer, Department of Physiology,
University of Munich, Munich, G.F.R.

The possibility was examined that it is the slow rate of cardiac adenine nucleotide biosynthesis that may be responsible for the retarded metabolic recovery of the myocardium during the reperfusion period following a brief ischemic insult. The approach that was used is based on the fact that ribose expands the available pool of 5-phosphoribosyl-1-pyrophosphate and thus leads to the stimulation of the biosynthesis of adenine nucleotides in heart and skeletal muscle. The effect of this stimulation on the cardiac ATP pool was then studied. Two experimental models of myocardial ischemia and reperfusion were examined in rats: recovery from multiple asphyxic periods and recovery from temporary regional ischemia. The spontaneous increase in adenine nucleotide biosynthesis that occurred in the myocardium during the reperfusion period in both models was further stimulated with ribose. During the recovery from a 15-minutes' period of regional ischemia, the restoration of the ATP pool was achieved much earlier when ribose had been applied. It can therefore be concluded that the slow adenine nucleotide biosynthesis determines the speed of ATP repletion, and that ribose accelerates this metabolic recovery process. It is shown that ribose fulfills the essential criteria that should be met by a suitable cardioprotective agent.

The Reperfusion Problem

Since the introduction of nonsurgical recanalization of occluded coronary arteries by means of intracoronary thrombolysis (22) and percutaneous transluminal coronary angioplasty (10), reperfusion of the previously ischemic myocardium has received much attention (2,3). Although the restoration of normal blood flow is the adequate therapeutic intervention to prevent the further development of ischemia and ultimately infarction, there are several risks involved. Reperfusion leads to the washout of adenine nucleotide degradation products that have accumulated during ischemia (8). They can therefore not be utilized to a full extent for the resynthesis of adenine nucleotides so that the previously ischemic myocardium is dependent predominantly on the de novo synthesis process (29). A further risk is the readmission of calcium which may ultimately culminate in the calcium paradox, i.e., tissue disruption, enzyme release, development of contracture and marked reduction in high-energy phosphates (31). Concomitantly, cell volume regulation may also be disturbed resulting in swelling of myocytes and endothelial cells which may contribute to vascular compression, a phenomenon termed "no-reflow" (16). Leukocytes usually migrate into the ischemic myocardium from the adjacent tissue. During reperfusion, this migration is accelerated, particularly when the vascular endothelium has been injured. That leukocytes may be involved in aggravating myocardial necrosis has been suggested, since leukocyte depletion achieved by administration of neutrophil antiserum has been shown to result in reduction of infarct size in dogs (23). Neutrophils may also contribute to the production of oxygen free radicals. It has been reported there is a burst of oxygen radical generation during reperfusion (33), that organic spin trap agents can reduce reperfusion-induced arrhythmias (13), and that superoxide dismutase and catalase protect the jeopardized myocardium (15). However, there are also studies with negative results (18,24). Another concern about reperfusion is that hemorrhage may occur and lead to the extension of myocardial necrosis.

Of particular experimental interest are the findings that the myocardial ATP content remains low for days during reperfusion after a brief ischemic period (4,21,27), and that regional heart function is impaired for a prolonged period (14). This phenomenon has been termed "stunned myocardium" (2). Several mechanisms have been proposed to be responsible for this phenomenon. First, there may be a depressed Ca^{++}-activation of contraction and a shift in Ca^{++}-sensitivity (17). Furthermore, oxygen

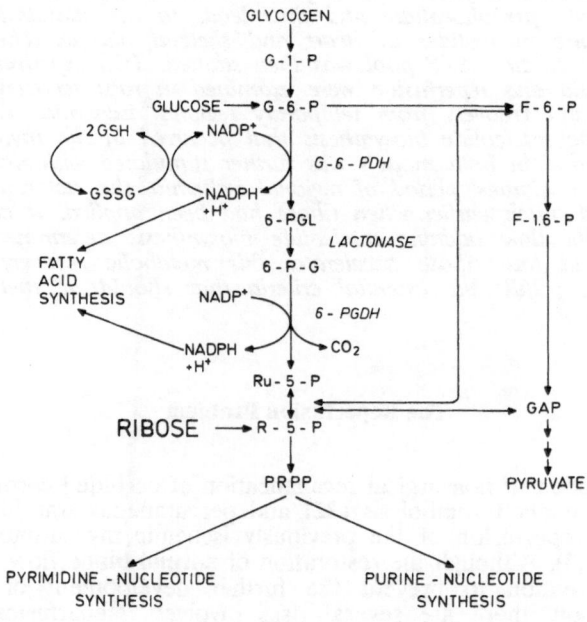

Fig. 1. Schematic presentation of the pentose phosphate pathway. The reactions of the oxidative branch are shown in detail in the center. G-1-P: Glucose-1-phosphate; G-6-P: Glucose-6-phosphate; G-6-PDH: Glucose-6-phosphate dehydrogenase; 6-PGL: 6-Phosphogluconolactone; 6-P-G: 6-Phosphogluconate; 6-PGDH: 6-Phosphogluconate dehydrogenase; Ru-5-P: Ribulose-5-phosphate; R-5-P: Ribose-5-phosphate; PRPP: 5-phosphoribosyl-1-pyrophosphate; F-6-P: Fructose-6-phosphate; F-1.6-P: Frucose-1.6-diphosphate; GAP: Glyceraldehyde-3-phosphate; $NADP^+$: Nicotinamide adenine dinucleotide phosphate; GSSG: Oxidized glutathione; GSH: Reduced glutathione.

free radicals may be responsible, since superoxide dismutase and catalase improved segment shortening of the previously ischemic myocardium (20). In addition, there seems to be a disruption of myofibrillar energy use due to a decreased creatine kinase activity associated with the myofibril and to limitation of substrate necessary for maximal creatine kinase activity (9). Finally, the recovery of ATP seems to be limited by an insufficient increase in adenine nucleotide biosynthesis (29). In this contribution, it will be examined whether the restoration of the ATP pool during reperfusion can be accelerated by a metabolic intervention directed at stimulating adenine nucleotide biosynthesis. This can be done with ribose.

Mode of Action of Ribose

To understand the ribose effects, one has to examine the oxidative pentose phosphate pathway. Glucose-6-phosphate originating from glycogenolysis or from glucose taken up by the cell is metabolized mainly via glycolysis (Fig. 1, right hand side).

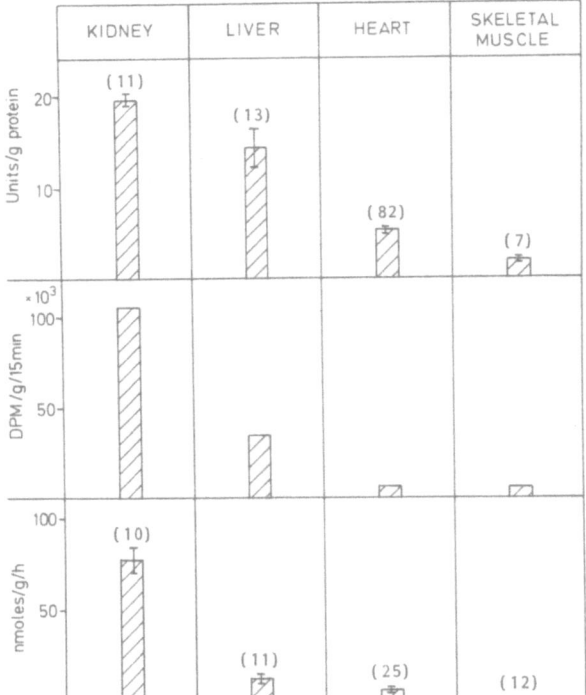

Fig. 2. Activities of glucose-6-phosphate dehydrogenase (upper panel), the available pool of 5-phosphoribosyl-1-pyrophosphate measured as the incorporation of ^{14}C-adenine into adenine nucleotides after 15 minutes of in vivo exposure (middle panel), and rates of adenine nucleotide biosynthesis (bottom panel) in various rat organs. Mean values ± SEM; number of experiments in parentheses; middle panel, n = 2.

An alternative possibility is the processing via the oxidative pentose phosphate pathway (central portion in Fig. 1) of which glucose-6-phosphate dehydrogenase (G-6-PDH) is the first and rate-limiting enzyme (6). After a further step catalyzed by lactonase, 6-phosphogluconate is converted to ribulose-5-phosphate by 6-phosphogluconate dehydrogenase. From ribulose-5-phosphate originates ribose-5-phosphate, the immediate precursor for 5-phosphoribosyl-1-pyrophosphate (PRPP). This substrate is essential both for pyrimidine and purine nucleotide biosynthesis as well as for the salvage pathway.

The oxidative pentose phosphate pathway has many important functions. It provides reducing equivalents in the form of NADPH. In this way, oxidized glutathione (GSSG) can be transformed to reduced glutathione. In the myocardium, an increase in GSSG may indicate oxidative stress such as during post-ischemic reperfusion. Moreover, NADPH is needed for the synthesis of fatty acids, particularly in liver, mammary gland and adrenal cortex. There are connections to glycolysis via the transaldolase and transketolase reactions, which have not been investigated exten-

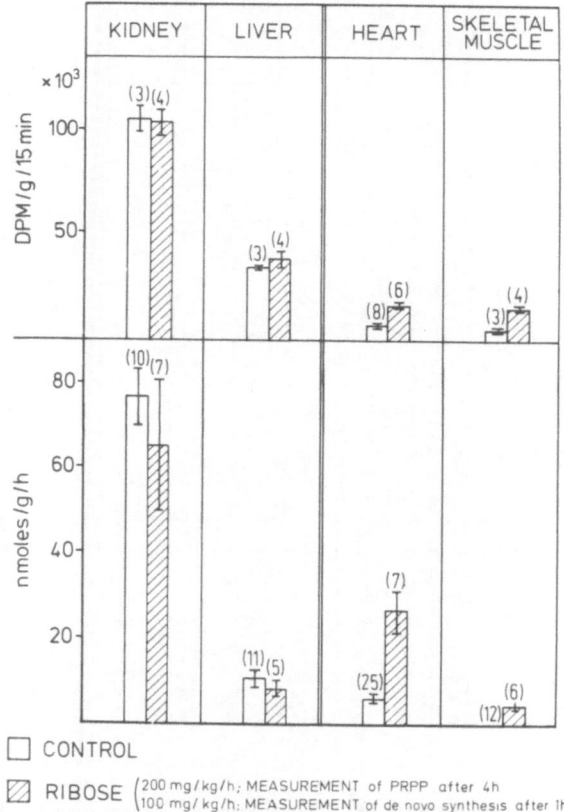

Fig. 3. Effect of ribose on the available pool of 5-phosphoribosyl-1-pyrophosphate (upper panel) and on the rate of adenine nucleotide biosynthesis (bottom panel) in various rat organs. Mean values ± SEM; number of experiments in parentheses.

sively in the myocardium. The most important function in the present context is the generation of ribose-5-phosphate for the synthesis of PRPP.

To assess the capacity of the oxidative pentose phosphate pathway in various rat organs, the activity of G-6-PDH, the available pool of PRPP and the rate of adenine nucleotide biosynthesis were determined. Fig. 2 shows that the highest activity of G-6-PDH, the greatest PRPP pool and the fastest rate of adenine nucleotide biosynthesis were found in kidney, followed by liver, heart and skeletal muscle. From these results it appears that the capacity of the oxidative pentose phosphate pathway is very small in muscular organs. The consequence of this is that the PRPP pool is minute and that adenine nucleotide biosynthesis is small in the heart and not measureable at all in the resting skeletal muscle.

The limitation of the PRPP pool can be overcome by bypassing the critical step in the oxidative pentose phosphate pathway with ribose (Fig. 1). It was therefore-examined which effect ribose has on the available pool of PRPP and on the rate of adenine nucleotide biosynsthesis in these organs (Fig. 3). Ribose had no effect in kidney and liver. In heart and skeletal muscle, however, there was a threefold increase. Thus, ribose had a stimulating effect only in muscular organs in which the activity of G-6-PDH is very low. This concept has been confirmed in studies on the isolated perfused rat heart (12) and on mature rat cardiac myocytes (5). This investigation and other studies demonstrated that the metabolic influence of ribose is specific for muscle cells, so endothelial or interstitial cells seem not to contribute to the results obtained in the heart in vivo (26). Since ribose had such a marked stimulating effect already in the normal heart, it was of interest to examine whether it might enhance cardiac adenine nucleotide biosynthesis during reperfusion following an ischemic insult, and whether this would then result in changes in the adenine nucleotide content.

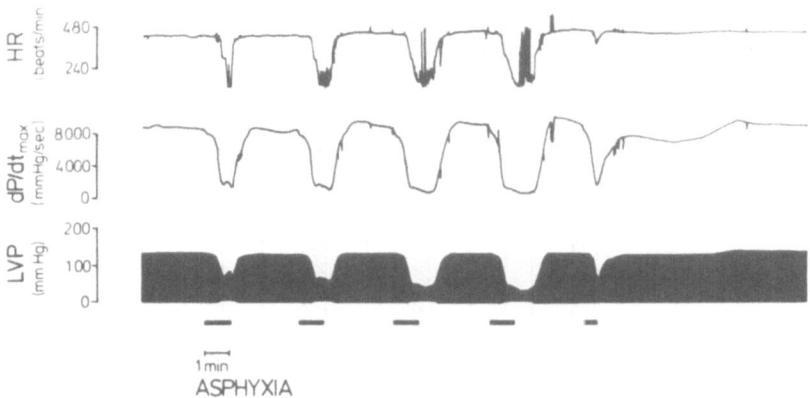

Fig. 4. Effect of successive periods of asphyxia (horizontal black bars) on heart rate (HR), on the maximal rate of rise in left ventricular pressure (dP/dt$_{max}$), and on left ventricular pressure (LVP) in rats anesthetized with ether and ventilated artificially using a Rodent respirator.

Experimental Models to Study Post-Ischemic Recovery

Two experimental models were used to induce myocardial ischemia. In the first, successive periods of total myocardial ischemia were achieved by intermittent asphyxia of 1-min duration in rats induced by interruption of the artificial respiration. Each asphyxic period was characterized by a decline in heart rate, LV dp/dt_{max} and in left ventricular systolic pressure, while diastolic pressure was elevated (Fig. 4). The decrease in left ventricular systolic pressure was more pronounced with each successive asphyxic period so coronary artery perfusion must have been severely impaired. As a result, the ATP content was reduced from 4.4 ± 0.008 umol/g wet wt. (n = 25) to 3.7 ± 0.3 umol/g (n = 6; $P < 0.05$) at the end of the last asphyxic period (25).

In the second model, regional myocardial ischemia was induced in rats by ligation of the descending branch of the left coronary artery for a brief period (1). Thereafter, reperfusion was allowed for 24 h (27). There was a progressive decline in the adenine nucleotide content depending on the duration of the ischemic period (Fig. 5). For a further detailed analysis of adenine nucleotide metabolism, an ischemic period of 15 min was chosen.

Ribose and Adenine Nucleotide Metabolism during Reperfusion

During the first hour of recovery from multiple asphyxic periods, there was an increase in cardiac adenine nucleotide biosynthesis, when no intervention was made. When ribose was administered as continuous i.v. infusion, there was a further stimulation of this biosynthetic process in a dose-dependent fashion (Fig. 6). However, in this model it could not be examined whether this increase might have an effect on

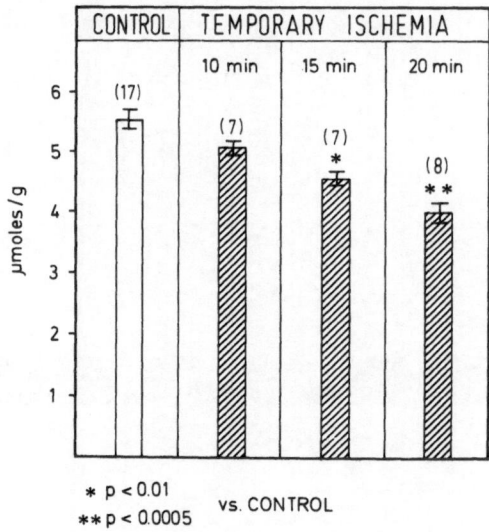

Fig. 5. Adenine nucleotide content (umol/g wet wt.) in rat hearts after 24 hours of reperfusion following a 10, 15 or 20 min period of regional myocardial ischemia. Mean values ± SEM; number of experiments in parentheses.

the actual ATP content in the myocardium, since the recovery period could not be extended for a longer period of time, because of the artifical respiration and the ether anesthesia that was used in this experimental series (25).

Essentially similar results were obtained during the reperfusion period following a 15-minutes' regional ischemia (Fig. 7). After 5 h of recovery there was a spontaneous enhancement of adenine nucleotide biosynthesis which became more pronounced when ribose had been applied. This ribose-induced additional increase was of such an extent that it should have an effect on the ATP content. Fig. 8 shows the decline of the ATP pool at the end of the 15-minutes' period of regional ischemia. Without any intervention there was a spontaneous increase after 5 h of post-ischemic recovery, and ribose had no effect at this time. However, after 12 h, the ATP pool remained unchanged compared to the 5-h value in the untreated hearts, but had become normalized when ribose had been infused. Also after 24 h, the ATP in the untreated group was reduced to the same extent as after 12 h, and was normal with ribose treatment. Thus, the time required to reach complete ATP recovery was markedly shortened by ribose. Studies on isolated perfused working rat hearts (19) and on dog hearts in vivo (7,11) have shown that also the recovery of function was improved with ribose administration. It can therefore be concluded that in these experimental studies ribose had a cardioprotective effect.

A Suitable Substrate to Improve Post-Ischemic Recovery

Since there are many competing substrates for metabolic intervention, it may be useful to have some criteria for the selection of the most suitable one. Such criteria include:
1. Its mechanism of action on cardiac metabolism must be known.
2. There should be a marked and long-term effect on myocardial metabolism in several pathophysiological situations.

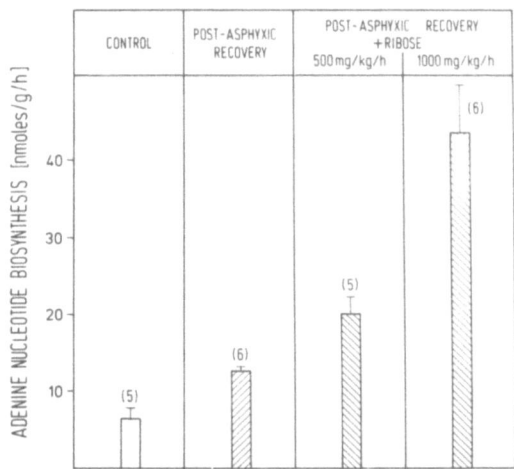

Fig. 6. Rates of adenine nucleotide biosynthesis in rat hearts at the end of 60 minutes of recovery from 5 asphyxic periods of 4.5-minutes' total duration (see Fig. 4), and the effect of ribose. Mean values ± SEM; number of experiments in parentheses.

3. The beneficial effects should outweigh the possible risks of adverse side effects.
4. It should be effective in man.
5. It must not have hemodynamic or vasoactive properties.
6. Its combination with conventional therapy must be possible.

Ribose fulfills all these criteria. As to its possible effects in man, it is worth mentioning that the activity of glucose-6-phosphate dehydrogenase, the first and regulating enzyme of the oxidative pentose phosphate pathway, is even lower in the human heart than its activity in the rat or guinea-pig heart in which the cardioprotective influence of ribose has been demonstrated (28). That ribose is effective in man is also suggested by the therapeutic effectiveness of ribose in a patient with myoadenylate deaminase deficiency in skeletal musle (32). Of great advantage is the fact that ribose has no hemodynamic influence of its own and that it does not affect the functional effects of calcium antagonists and ß-receptor blockers (30). Moreover, even when heart function is severely depressed by negative inotropic drugs, the metabolic stimulation of ribose is still preserved (30).

Acknowledgements

The results reported in this chapter were obtained in studies supported by the Deutsche Forschungsgemeinschaft (Zi 199/4-4, Zi 199/4-5). The excellent technical assistance of Ms. G. Steinkopff and Ms. B. Sender is gratefully acknowledged.

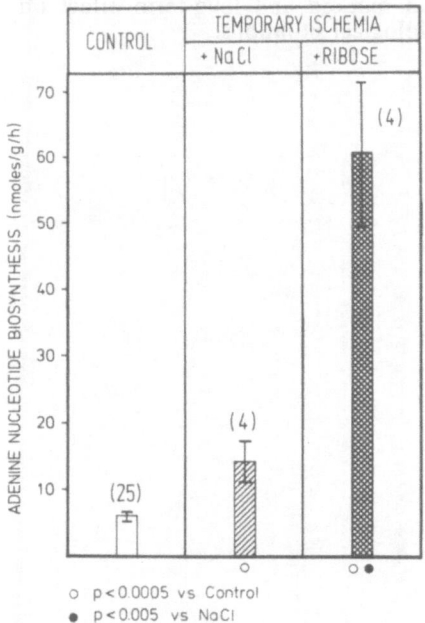

Fig. 7. Rates of adenine nucleotide biosynthesis after 5 hours of recovery from a 15-minutes' period of regional myocardial ischemia in rats with continuous i.v. infusion of 0.9% NaCl or ribose (200 mg/kg/h). Mean values ± SEM; number of experiments in parentheses.

References

1. Bechtelsheimer H, Zimmer H-G, Gedigk P: Histologische und histochemische Untersuchungen am nicht infarzierten Myokardanteil beim experimentellen Herzinfarkt der Ratte. Virch Arch Abt B Zellpath 9:115-124, 1971
2. Braunwald E, Kloner RA: The stunned myocardium: Prolonged post-ischemic ventricular dysfunction. Circulation 66:1146-1149, 1982
3. Braunwald E, Kloner RA: Myocardial reperfusion: A double-edged sword? J Clin Invest 76:1713-1719, 1985
4. DeBoer LWV, Ingwall JS, Kloner RA, Braunwald E: Prolonged derangements of canine myocardial purine metabolism after a brief coronary occlusion not associated with anatomic evidence of necrosis. Proc Natl Acad Sci USA 77:5471-5475, 1980
5. Dow JW, Nigdikar S, Bowditch J: Adenine nucleotide synthesis de novo in mature rat cardiac myocytes. Biochim Biophys Acta 847:223-227, 1985
6. Eggleston LV, Krebs HA: Regulation of the pentose phosphate cycle. Biochem J 138:425-435, 1974
7. Foker JE, Ward HB, St Cyr JA, Alyono D, Bianco RW, Kriett JM: Enhanced ATP return after global ischemia. Eur Heart J 5, Suppl 1:377, 1984 (Abstr)
8. Gerlach E, Deuticke B, Dreisbach RH: Der Nucleotid-Abbau im Herzmuskel bei Sauerstoffmangel und seine mögliche Bedeutung für die Coronardurchblutung. Naturwiss 50:228-229, 1963
9. Greenfield RA, Swain JL: Disruption of myofibrillar energy use: Dual mechanisms that may contribute to postischemic dysfunction in stunned myocardium. Circ Res 60: 283-289, 1987
10. Grüntzig AR: Transluminal dilatation of coronary-artery stenosis. Lancet 1:263, 1978 (Abstr)
11. Haas GS, DeBoer LWV, O'Keefe DD, Bodenhamer RM, Geffin GA, Drop LJ, Teplick RS,

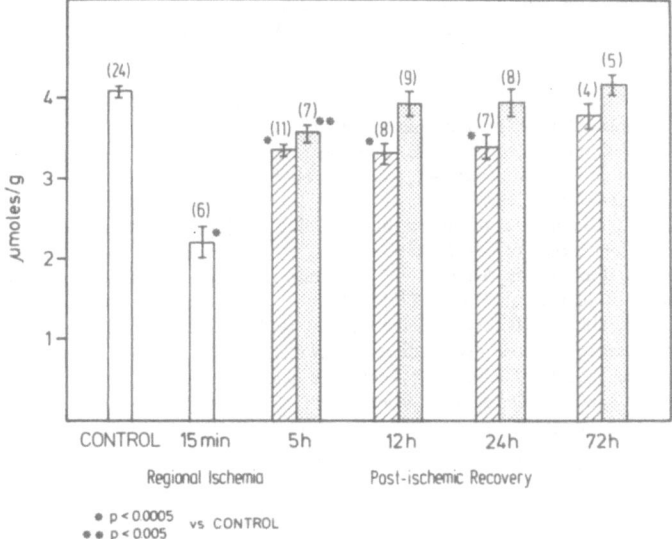

Fig. 8. Alterations in the cardiac ATP pool (umol/g wet wt.) at the end of temporary regional ischemia induced by ligation for 15 minutes of the descending branch of the left coronary artery in rats and during the subsequent post-ischemic recovery period. Infusion of 0.9% NaCl (hatched bars) or ribose (stippled bars) at a dose of 200 mg/kg/h was started at the end of the ischemic period and maintained throughout the repective recovery periods indicated on the abscissa. Infusion rate: 5 ml/kg/h.

Daggett WM: Reduction of postischemic myocardial dysfunction by substrate repletion during reperfusion. Circulation 70, Suppl I:65-74, 1984

12. Harmsen E, De Tombe PP, De Jong JW, Achterberg PW: Enhanced ATP and GTP synthesis from hypoxanthine or inosine after myocardial ischemia. Am J Physiol 246:H37-H43, 1984

13. Hearse DJ, Tosaki A: Free radicals and reperfusion-induced arrhythmias: Protection by spin trap agent PBN in the rat heart. Circ Res 60:375-383, 1987

14. Heyndrickx GR, Millard RW, McRitchie RJ, Maroko PR, Vatner SF: Regional myocardial functional and electrophysiological alterations after brief coronary artery occlusion in conscious dogs. J Clin Invest 56:978-985, 1975

15. Jolly SR, Kane WJ, Bailie MB, Abrams GD, Lucchesi BR: Canine myocardial reperfusion injury. Its reduction by the combined administration of superoxide dismutase and catalase. Circ Res 54:277-285, 1984

16. Kloner RA, Canote CE, Jennings RB: The "no-reflow" phenomenon after temporary coronary occlusion in the dog. J Clin Invest 54: 1496-1508, 1974

17. Kusuoka H, Porterfield JK, Weisman HF, Weisfeldt ML, Marban E: Pathophysiology and pathogenesis of stunned myocardium. Depressed Ca^{++} activation of contraction as consequence of reperfusion-induced cellular calcium overload in ferret hearts. J Clin Invest 79:950-961 1987

18. Parratt JR, Wainright CL: Failure of allopurinol and a spin trapping agent N-t-butyl-alpha-phenyl nitrone to modify significantly ischaemia and reperfusion-induced arrhythmias. Br J Pharmacol 91:49-59, 1987

19. Pasque MK, Spray TL, Pellom GL, Van Trigt P, Peyton RN, Currie WD, Wechsler AS: Ribose-enhanced myocardial recovery following ischemia in the isolated working rat heart. J Thorac Cardiovasc Surg 83:390-398, 1982

20. Przyklenk K, Kloner RA: Superoxide dismutase plus catalase improve contractile function in the canine model of the "stunned myocardium". Circ Res 73:148-156, 1986

21. Reimer KA, Hill ML, Jennings RB: Prolonged depletion of ATP and of the adenine nucleotide pool due to delayed resynthesis of adenine nucleotides following reversible myocardial ischemic injury in dogs. J Mol Cell Cardiol 13:229-239, 1981

22. Rentrop P, Blanke H, Karsch KR, Wiegand V, Köstering H, Rahlf G, Oster H, Leitz K: Wiedereröffnung des Infarktgefässes durch transluminale Rekanalisation und intrakoronare Streptokinase-Applikation. Dtsch Med Wschr 104:1438-1440, 1979

23. Romson J, Hook B, Kunkel S, Abrams G, Schork A, Lucchesi BR: Reduction in the extent of ischemic myocardial injury by neutrophil depletion in the dog. Circulation 67:1016-1023, 1982

24. Uraizee A, Reimer KA, Murry CE, Jennings RB: Failure of superoxide dismutase to limit size of myocardial infarction after 40 minutes of ischemia and 4 days of reperfusion in dogs. Circulation 75:1237-1248, 1978

25. Zimmer H-G: Restitution of myocardial adenine nucleotides: Acceleration by administration of ribose. J Physiol (Paris) 76:769-775, 1980

26. Zimmer H-G, Gerlach E: Stimulation of myocardial adenine nucleotide biosynthesis by pentoses and pentitols. Pflügers Arch 376:223-227, 1978

27. Zimmer H-G, Ibel H: Ribose accelerates the repletion of the ATP pool during recovery from reversible ischemia of the rat myocardium. J Mol Cell Cardiol 16:863-866, 1984

28. Zimmer H-G, Ibel H, Suchner U, Schad H: Ribose intervention in the cardiac pentose phosphate pathway is not species-specific. Science 223:712-714, 1984

29. Zimmer H-G, Trendelenburg C, Kammermeier H, Gerlach E: De novo synthesis of myocardial adenine nucleotides in the rat. Acceleration during recovery from oxygen deficiency. Circ Res 32:635-642, 1973

30. Zimmer H-G, Zierhut W, Marschner G: Combination of ribose with calcium antagonist and ß-blocker treatment in closed-chest rats. J Mol Cell Cardiol 19:635-639, 1987

31. Zimmerman ANE, Hülsmann WC: Paradoxical influence of calcium ions on the permeability of the cell membranes of the isolated rat heart. Nature 211:646-647, 1966

32. Zöller N, Reiter S, Gross M, Pongratz D, Reimers CD, Gerbitz K, Paetzke I, Deufel T, Hübner G: Myoadenylate deaminase deficiency: Successful symptomatic therapy by high dose oral administration of ribose. Klin Wschr 64:1281-1290, 1986

33. Zweier JL, Flaherty JT, Weisfeldt ML: Direct measurement of free radical generation following reperfusion of ischemic myocardium. Proc Natl Acad Sci USA 84:1404-1407, 1987

Technical Considerations/Models

Chapter 11

The Importance of the Determination
of ATP and Catabolites

E. Harmsen and A.-M.L. Seymour, Department of Biochemistry,
University of Oxford, Oxford, U.K.

*Since 30 years there has been a great deal of interest in the regulatory properties
of the high-energy phosphates (HEP) ATP and phosphocreatine (PCr), and their
breakdown products. The determination and interpretation of these compounds can
be full of pitfalls, mainly as a consequence of their high turnover rate and
compartmentation. Three different methods to determine the myocardial HEP or
breakdown products are compared:*
*1. Cardiac tissue can be rapidly frozen, and after appropriate extraction, the HEP
can be determined either enzymatically or chromatographically. These methods are
sensitive and selective, but evidently destructive.*
2. ATP and PCr levels can be determined by ^{31}P-NMR in vivo and in situ.
*Furthermore, as a result of the chemical shift of the inorganic phosphate, the
intracellular pH can be estimated. This method is less selective and sensitive than
the destructive measurements, but the only method available to directly correlate
intracellular biochemical with physiological parameters.*
*3. The apolar products of ATP breakdown such as adenosine and its catabolites
inosine, (hypo)xanthine and uric acid can be determined in the coronary effluent.
This method is sensitive. However, it is an indirect assessment of ATP breakdown.*
*A complete understanding of myocardial HEP turnover and breakdown will require
complementary measurements of these compounds with each of the three measured
paradigms.*

Introduction

In the normoxic heart a delicate balance exists between ATP production in the
mitochondria and ATP utilization by the contractile apparatus. In response to an
increase in work, the heart increases its oxygen consumption and rate of ATP
synthesis instantaneously. Throughout ischemia, when flow is inadequate and hence
oxygen delivery is limited, the normal equilibrium is perturbed. Despite a decline in
contractility, ATP utilization is greater than its production resulting in a decrease
in ATP concentration. During reperfusion, the oxygen supply is restored, but neither
ATP nor contractility are restored to their pre-ischemic values.

ATP and its catabolites are central to the regulation of energy metabolism. For this
reason, it is essential to measure the concentration of these metabolites in the cell
and be aware of the limitations of the techniques presently available.

Initially, we will give a brief overview of myocardial ATP metabolism, then discuss
the techniques available to determine ATP and its catabolites and finally evaluate
the results of these measurements.

J.W. de Jong (Ed.), Myocardial Energy Metabolism, Martinus Nijhoff Publishers, Dordrecht/Boston/Lancaster, 1988   117

Normoxia

The normoxic myocyte maintains remarkably constant ATP concentrations under different conditions of work. This observation suggests a tight coupling between the production of ATP by oxidative phosphorylation in the mitochondria, and its consumption by the contractile apparatus (Fig. 1, top panel). It is thought that ATP or its products of hydrolysis (ADP and inorganic phosphate) play a major role in mitochondrial regulation, either directly as ADP (17), or through cytosolic ratios such as the phosphate potential ($[ATP]/[ADP][P_i]$; (see refs. 31,46,69), the $[ATP]/[ADP]$ ratio (65) or the energy charge, $([ATP] + 0.5 [ADP])/([ATP] + [ADP] + [AMP])$ (see ref. 5). Other important candidates involved in mitochondrial regulation are the NAD/NADH ratio (23) and Ca^{2+} ions (22,63).

Normoxia

Ischemia

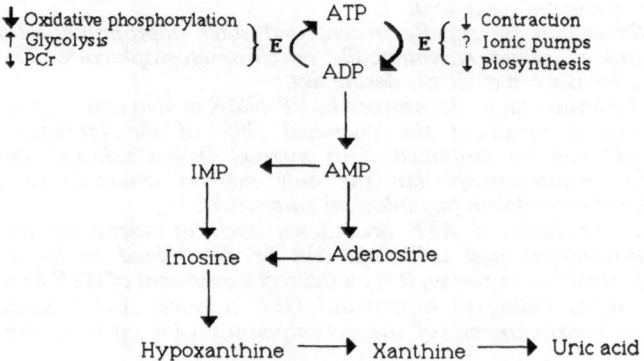

Reperfusion

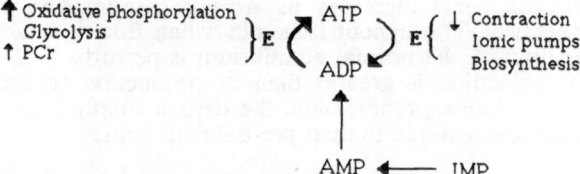

Fig. 1. Schematic view of ATP breakdown during normoxia, ischemia and reperfusion. PCr = phosphocreatine; E = energy.

Phosphocreatine (PCr) is in equilibrium with ATP and ADP via the creatine kinase reaction (64). PCr is thought to play a role either as an energy buffer, which keeps ATP levels constant during a sudden increase in energy demand (37), or as an energy shuttle (11,58). In the latter model, it is proposed that the energy produced by the mitochondrion is predominantly transported to the myofibrils as PCr. Recently these theories have been refined by the suggestion that PCr acts either as an ADP buffer or ADP transporter (39,40). However, some investigators report that, in a normoxic heart, ATP, PCr, and in turn, ADP, AMP and inorganic phosphate (P_i) levels are remarkably constant, giving a constant ATP/ADP ratio, energy charge and phosphate potential at various workloads (8,25,63,73). Thus, the signal which enhances mitochondrial function as a result of increased cardiac work still remains elusive.

Ischemia

During flow restriction and hence diminished oxygen delivery to the heart, the balance between ATP production and consumption is disturbed. Despite a rise in anaerobic glycolysis, and a decrease in contractility, ATP concentrations decline, with a concomitant rise in ADP and P_i levels. In turn, ADP is metabolized to AMP (by adenylate kinase), which subsequently is dephosphorylated by 5'-nucleotidase to adenosine. This apolar compound leaves the cell and enters the bloodstream, where it is further converted to inosine, hypoxanthine and uric acid (refs. 1,16,21,34,35; see Fig. 1, middle panel). The rapid decline in contractility has been attributed to a number of different process - inadequate mitochondrial ATP production, insufficient delivery of ATP (37), an increase of ADP (19), an increase of P_i (3,44), a decrease in the phosphate potential (18,31,41), developing acidosis (43), low oxygen tension (53), or a collapse of the coronary system.

It should be stressed that ischemia is only a relative term, indicating a restriction of coronary flow. However, the severity of ischemia depends primarily on the extent of imbalance between ATP production and ATP consumption. This is reflected by the rise in ADP concentration and the parallel fall in PCr concentration (35). The concentrations of ADP and AMP are determined in turn by the adenylate kinase reaction, an equilibrium reaction. As a result, a high concentration of AMP induces a high level of release of adenosine and its catabolites with a consequently steeper drop in ATP levels (see Fig. 2). According to this model, the fall in ATP, the decline in PCr levels and the purine release are a true reflection of the severity of ischemia (16,35).

Reperfusion

During reperfusion (when both circulation of blood and adeqate supply of oxygen are re-established), the mono- and diphosphates are rapidly reconverted to ATP (Fig. 1, bottom panel). The ATP levels remain depressed, however, as a result of washout of adenosine and its catabolites (1,6,20,21). Unless the cell is irreversibly damaged, the PCr concentration recovers to greater than its pre-ischemic levels, reflecting low ADP levels (6,59). This could indicate an adequate restoration of mitochondrial activity accompanying depressed myocardial function. Evidence from the literature suggests that there is a correlation between the concentration of ATP at the end of ischemia or at the time of reperfusion and subsequent restoration of function (56,62,68). However, there are many other factors resulting from the severity of the ischemic insult such as accumulation of H^+ and lactate which may cause failure of functional recovery (55). Indeed, it has been demonstrated that restoration of ATP levels in itself does not necessarily increase function (57). Changes in the contractile apparatus or Ca^{2+}-handling by the sarcoplasmic reticulum may also be involved.

If the mitochondria are not able to generate sufficient ATP, as reflected by persistently elevated ADP and low PCr concentrations, then reperfusion may be incomplete (59). In this situation the mitochondria are irreversible damaged and the cells ultimately die.

Compartmentalization

There are several distinct domains in the heart, in which one can expect to find different concentrations of high-energy phosphates. These domains can be divided into multi- and intracellular compartments (54,67). The multi-cellular compartment consists of the different cell types found within the heart, e.g., myocytes, smooth muscle cells, endothelial cells, fibroblasts, conductive cells and blood.

Although high-energy phosphates levels differ between the extracellular compartments, initially the bulk (>85%) is localized within the myocytes (67). This implies that changes in ATP and PCr concentrations of >15% should reflect differences within the myocytes alone. However, changes of metabolites normally present in low concentrations (uM) should be considered more carefully. For example, a small release of adenosine or one of its catabolites could be attributed to endothelial cells (60), and a fraction of the P_i measured by nuclear magnetic resonance (NMR) arises from the blood phosphate (14). Furthermore, different populations of myocytes are thought to exist within the heart (e.g., subepi- vs. subendocardial cells), which could give rise to non-uniform high-energy phosphate distribution. Finally, if redistribution of myocardial flow occurs (regional ischemia), large regional differences of high-energy phosphates can be expected within the heart (51).

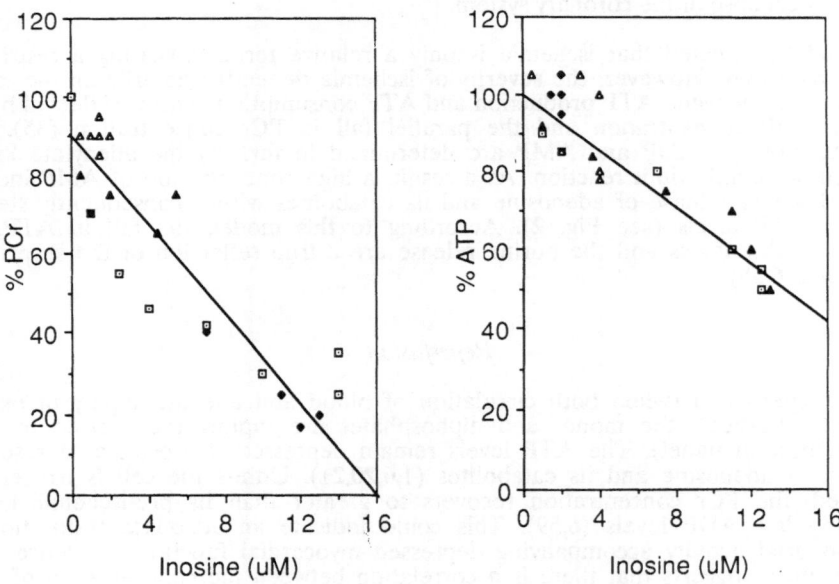

Fig. 2. Correlation between intracellular phosphocreatine (PCr) and inosine release (left panel) and intracellular ATP and inosine release (right panel) from isolated perfused rat hearts during a 90% flow reduction. PCr and ATP were measured by ^{31}P-NMR, and inosine concentrations in the coronary effluent were assayed by HPLC. □ = hearts made ischemic with pacing at 300 beats/min; ◆ = hearts made ischemic without pacing; △ = hearts made hypoxic with pacing at 300 beats/min; and ▲ = hearts made hypoxic without pacing (35). n = 5-9.

The intracellular compartments can be subdivided in three separate entities. First, there are physically distinct subcellular structures with different high-energy phosphate levels, e.g., mitochondria (16,28,30,66). It has been shown that there are marked differences in high-energy phosphate concentrations in the cytosol and mitochondrion. ATP and PCr are considered to be primarily cytosolic, whilst free ADP levels are higher in the mitochondria (4,16,66). Secondly, there is a bound compartment. Nucleotides readily bind to proteins, and in myocytes about 80% of the ADP is bound to actin (61). Thirdly, enzymes can form multi-enzyme complexes, in which the product of one enzyme is channeled to another enzyme, which uses it as a substrate. In the myocyte, adenylate kinase and ATPases are all in close proximity of each other (39,70). Furthermore, ATP produced by membrane-bound glycolytic enzymes and/or glycogen is used predominantly by ionic pumps (13,71). For these reasons one should be careful in the interpretation of the results of gross high-energy phosphate and catabolite determinations.

Techniques

To determine high-energy phosphates and P_i concentrations in heart, two fundamentally different methods can be used. The first method is to freeze the heart rapidly in order to arrest metabolism, and then, after appropriate extraction, to assay the compound of interest. One consequence of this method is that the sample is destroyed. The second method is to monitor the ATP, PCr and P_i levels with ^{31}P-NMR. This can be done in-vivo, and is a non-destructive and non-invasive method.

Determination of ATP in Extracted Tissue

Nucleotides are very unstable compounds in biological matrices. This means that during tissue sampling, ATP and PCr levels can fall rapidly with a concomitant rise in ADP and P_i levels. Wollenberger et al. (72) used aluminum tongs precooled in liquid nitrogen to "freeze-clamp" the tissue of interest within seconds. Hearse also repeatedly stressed the need for fast tissue enzyme deactivation (36). The frozen tissue can be extracted with perchloric acid (PCA) or trichloroacetic acid (TCA), with or without alcohol and EDTA. PCA is removed by precipitation of its potassium salt, or by extraction with an amine dissolved in freon. TCA is removed by ether extraction. Recoveries of 80 - 100% for both procedures have been obtained (34).

It should be stressed that slow freeze-clamping induces temporal tissue hypoxia with consequently lower PCr concentrations and hence increased ADP, AMP and P_i levels. The extent of this artefact depends on the rate of ATP turnover in the tissue (34,47). A high rate of ATP prior to freeze-clamping results in larger errors in ADP, AMP and P_i levels, than those from a tissue with a low ATP turnover rate. Freeze-clamping and subsequent acid-extraction result in the measurement of total nucleotides in cardiac tissue, independent of the form in which they exist and irrespective of their cellular location. To assess the nucleotide distribution within the cell, Sobel and Bünger isolated mitochondria from cells in a non-aqueous medium, to prevent ATP-hydrolysis, followed by an acid extraction (66). Mild extraction of freeze-clamped heart tissue with salt buffers can give an estimate of the free ADP available (61).

In neutralized samples, high-energy phosphates can be measured by enzymatic asays (10,50), bioluminescence (49,50), isotachophoresis (24) or high-performance liquid chromatography (HPLC, refs. 15,35). The differences between these techniques are not essential. The advantage of both HPLC and isotachophoresis is that, within a single run, several compounds can be determined. In principle, this leads to a more precise determination of metabolite ratios. Enzymatic determinations, however, can be more selective. Bioluminescence is specifically used for its sensitivity (50). In earlier days, the purity of the luciferine/luciferase enzymes caused problems.

Nowadays, with increased purification of these enzymes and inhibitors of enzymes such as adenylate kinase, the results of these assays are more reliable.

Adenosine and its catabolites in the coronary effluent are primarily measured by HPLC and enzymatic assays (10,33). Again, with these assays, the sample preparation is the most crucial step of the whole procedure. A small increase in ATP breakdown as a result of improper sample handling can result in a massive increase in adenosine and its catabolites (33).

Determination of High-Energy Phosphates with ^{31}P-NMR

Nuclear magnetic resonance is a spectroscopic method which utilizes the magnet spin properties of an atomic nucleus (e.g., ^{31}P, spin = 1/2). By applying a varying magnetic field and detecting the energy transitions in the radio frequence range, the quantity of a particular nucleus, its chemical micro-enviroment (e.g., the phosphate component in either ATP, PCr or P_i) and its macro-enviroment (whether a molecule is free or bound) can be monitored (26). This technique is completely non-invasive and can be applied to in-vivo systems (27,38). ^{31}P-NMR spectra have been obtained from isolated perfused hearts (6,29,35), from hearts in open-chested or intact animals by means of an internal or external surface coil (2,14,32,42,51) and from human myocardium using a spatial localization technique (12).

The sensitivity of this technique is, in part, dependent on the magnetogyric ratio and relaxation times of the nucleus under investigation and the strength of the magnetic field. The selectivity is dependent on the strength of the magnetic field. In a beating heart, the clearly observable peaks are P_i, PCr and the three phosphates of ATP with a very small contribution from ADP. The chemical shift of the P_i peak can determine the pH (26,29). In a blood-perfused heart, the P_i peak is contaminated with the 2,3-diphosphoglycerate from blood. It is possible to distinguish between the two by a spectral editing method (14).

Although the NMR technique does not have the sensivity and selectivity of the previously mentioned analytical techniques, its advantages are obvious. With this technique, it is possible to monitor biochemistry and physiology of the heart simultaneously. Furthermore, within one preparation, continuous measurements can be made. Also with NMR it is feasible to "magnetically label" specific phosphate groups and follow their transfer into another molecule as a result of an enzymic reaction (saturation transfer, refs. 26,64). This is a unique method for determining in vivo enzyme kinetics. Creatine kinase and ATPase activity have been studied extensively, and the results have helped clarify the role of these enzymes in the cell (7,9,25,45,48,52,64).

Conclusions

In the introduction, ideas concerning the importance of the regulation of ATP and its catabolites are described. Although there is still controversy about the exact mechanism of regulation of high-energy phosphate metabolism, there is a general consensus concerning the importance of measuring ATP and catabolite concentrations. One should, however, be aware of four problems:

1. In order to understand the regulatory properties of a compound or an enzyme (complex), knowledge of local subcellular concentrations is essential. At present, these concentrations cannot be determined.

2. Analytical assays in a tissue sample give an average measure of the compound in different compartments. In freeze-clamped and acid-extracted tissue, total (bound and soluble) nucleotides are determined, whilst with ^{31}P-NMR, only the soluble components are determined. Thus, determination of ADP and P_i by these two techniques

will generate very different values owing to their significant binding in vivo and breakdown of ATP and PCr on sample preparation.

3. Despite certain advantages in determining metabolite ratios (either in an extract or with an NMR spectrum), this method can be misleading. Energy charge calculations from extracts of freeze-clamped tissue, using total ADP levels, can lead to serious underestimation. Low PCr/ATP ratios can be used to indicate ischemia but, alternatively, a high PCr/ATP ratio can reflect either a normoxic tissue with high PCr levels, or a post-ischemic tissue with low ATP levels. Thus, absolute values of ATP are needed to distinguish between these two possibilities.

4. Relationships exist between biochemical parameters, the severity of ischemia and onset of reperfusion. Although the mechanisms linking the biochemical parameters and physiological functions are not always clear, some parameters (concentrations or ratios) can be used as indicators of ischemia and reperfusion damage. To truly evaluate the condition of a heart, some type of metabolic stress need to be used.

Acknowledgements

The critical comments of Drs Henk ter Keurs, Wim Leijendekker and Peter de Tombe from the Medical Department of the University of Calgary are greatly appreciated. Clinton Ho from the Biochemistry Department is thanked for the help with the drawings. This work was funded by the Medical Research Council, British Heart Foundation and the Wellcome Trust.

References

1. Achterberg PW, Harmsen E, De Tombe P, De Jong JW: Balance of purine nucleotide and catabolites in the isolated, ischemic rat heart. Adv Exp Biol Med 165B:483-486, 1984
2. Ackerman JJH, Grove TH, Wong GG, Gadian DG, Radda GK: Mapping of metabolites in whole animals by ^{31}P-NMR using surface coils. Nature 283:167-170, 1980
3. Allen DG, Morris PG, Orchard CH, Pirolo JS: A nuclear magnetic resonance study of metabolism in the ferret heart during hypoxia and inhibition of glycolysis. J Physiol (Lond) 361:185-204, 1985
4. Asimakis GK, Sordahl LA: Intramitochondrial adenine nucleotides and energy-linked factors of heart mitochondria. Am J Physiol 241:H672-H678, 1981
5. Atkinson D: Cellular energy metabolism and its regulation. New York: Acad Press, 1977
6. Bailey IA, Seymour A-ML, Radda GK: A ^{31}P-NMR study of the effects of reflow on the ischaemic rat heart. Biochim Biophys Acta 637:1-7, 1981
7. Bailey IA, Williams SR, Radda GK, Gadian DG: Activity of phosphorylase in total global ischaemia in the rat heart. Biochem J 196:171-178, 1981
8. Balaban RS, Kantor HL, Katz LA, Briggs RW: Relation between work and phosphate metabolite in the in vivo paced mammalian heart. Science 232:1121-1123, 1986
9. Bittl JA, Balschi JA, Ingwall JS: Contractile failure and high-energy phosphate turnover during hypoxia: ^{31}P-NMR surface coil studies in living rat. Circ Res 60:871-878, 1987
10. Bergmeyer H-U, ed: Methods of enzymatic analysis. Vol VII. Weinheim: Verlag Chemie, 1983
11. Bessman SP, Carpenter CL: The creatine-creatine phosphate energy shuttle. Annu Rev Biochem 54:831-862, 1985
12. Blackledge MJ, Rajagopalan B, Oberhaensli RD, Bolas NM, Styles P, Radda GK: Quantitative studies of human cardiac metabolism by ^{31}P rotating-frame NMR. Proc Natl Acad Sci (USA) 84:4283-4287, 1987
13. Bricknell OL, Daries PS, Opie LH: A relationship between adenosine triphosphate, glycolysis and ischaemic contracture in the isolated heart. J Mol Cell Cardiol 13:941-945, 1981
14. Brindle KM, Rajagopalan B, Bolas NM, Radda GK: Editing of ^{31}P NMR spectra of heart in vivo. J Magn Res 74:356-365, 1987
15. Brown PR, Krstulovic AM: Practical aspects of reversed-phase liquid chromatography applied to biochemical and biomedical research. Anal Biochem 99:1-21, 1979
16. Bünger R, Soboll S: Cytosolic adenylates and adenosine release in perfused working heart. Eur J Biochem 159:203-213, 1986

17. Chance B, Williams GR: Respiratory enzymes in oxidative phosphorylation. J Biol Chem 217:385-393, 1955
18. Clarke K, O'Connor AJ, Willis RJ: Temporal relationship between energy metabolism and myocardial function during ischemia and reperfusion. Am J Physiol 253:H412-H421, 1987
19. Cooke R, Pate E: The effects of ADP and phosphate on the contraction of muscle fibers. Biophys J 48:789-798, 1985
20. DeBoer LWV, Ingwall JS, Kloner RA, Braunwald E: Prolonged derangements of canine myocardial purine metabolism after a brief coronary artery occlusion not associated with anatomic evidence of necrosis. Proc Natl Acad Sci (USA) 77:5471-5475, 1980
21. De Jong JW. Biochemistry of the acutely ischemic myocardium. In: Schaper W, ed: Pathophysiology of myocardial perfusion. Amsterdam: Elsevier/North-Holland Biomed Press, 1979:719-750
22. Denton RM, McCormack JG: Ca^{2+} transport by mammalian mitochondria and its role in hormone action. Am J Physiol 249:E543-E554, 1985
23. Erecinska M, Wilson DF: Regulation of cellular energy metabolism. J Membr Biol 70:1-14, 1982
24. Everaerts FM, Beckers JL, Verheggen TPEM: Isotachophoresis. Theory, instrumentation and applications. Amsterdam: Elsevier Sci Publ, 1976
25. From AHL, Petein MA, Michurski SP, Zimmer SD, Ugurbil KM: ^{31}P-NMR studies of respiratory regulation in the intact myocardium. FEBS Lett 206:257-261, 1986
26. Gadian DG: Nuclear magnetic resonance and its applications to living systems. Oxford: Clarendon Press, 1982
27. Gadian DG, Radda GK: NMR studies of tissue metabolism. Annu Rev Biochem 50:69-83, 1981
28. Garlick PB, Brown TR, Sullivan RH, Ugurbil K: Observation of a second phosphate pool in the perfused heart by ^{31}P NMR; is this the mitochondrial phosphate? J Mol Cell Cardiol 15:855-858, 1983
29. Garlick PB, Radda GK: Seeley PJ: Studies of acidosis in the ischaemic heart by phosphorus nuclear magnetic resonance. Biochem J 184:547-554, 1979
30. Geisbuhler T, Altschuld RA, Trewyn RW, Ansel AZ, Lamka K, Brierley GP: Adenine nucleotide metabolism and compartmentalization in isolated adult rat heart cells. Circ Res 54:536-546, 1984
31. Gibbs C: The cytoplasmic phosphorylation potential. Its possible role in the control of myocardial respiration and cardiac contractility. J Mol Cell Cardiol 17:727-731, 1985
32. Grove TH, Ackerman JJH, Radda GK, Bore PJ: Analysis of rat heart in vivo by phosphorus nuclear magnetic resonance. Proc Natl Acad Sci (USA) 77:299-302, 1980
33. Harmsen E, De Jong JW, Serruys PW: Hypoxanthine production by ischemic heart demonstrated by high pressure liquid chromatography of blood purine nucleosides and oxypurines. Clin Chim Acta 115:73-84, 1981
34. Harmsen E, De Tombe PP, Hegge JAJ, De Jong JW: Determination of adenine-nucleotides and creatine phosphate in various mammalian tissues by high-performance liquid chromatography. In: Krstulovic AM, ed: Handbook of chromatography. Roca Baton: CRC Press, 1987 (in press)
35. Harmsen E, Hogan G, Radda GK, Seymour A-ML: Simultaneous determination of intracellular purine efflux and intracellular high-energy phosphates in an isometric paced rat heart. A ^{31}P-NMR study. Proc Soc Magn Res Med 1:475-476, 1985 (Abstr)
36. Hearse DJ: Microbiopsy metabolites and paired flow analysis. A new rapid procedure for homogenisation, extraction and analysis of high-energy phosphates and other intermediates without any errors from tissue loss. Cardiovasc Res 18:384-390, 1984
37. Hearse DJ: Oxygen deprivation and early myocardial contractile failure. Am J Cardiol 44:1115-1121, 1979
38. Hoult DI, Busby SJW, Gadian DG, Radda GK, Richards RE, Seeley PJ: Observation of tissue metabolites using ^{31}P nuclear magnetic resonance. Nature 252:285-287, 1974
39. Jacobus WE: Respiratory control and the integration of heart high-energy phosphate metabolism by mitochondrial creatine kinase. Annu Rev Physiol 47:707-725, 1985
40. Jacobus WE: Theoretical support for the heart phospho-creatine energy transport shuttle based on the intracellular diffusion limited mobility of ADP. Biochem Biophys Res Commun 133:1035-1041, 1985
41. Kammermeier H, Schmidt P, Jungling E: Free energy change of ATP-hydrolysis: a causal factor of early hypoxic failure of the myocardium? J Mol Cell Cardiol 14:267-277, 1982
42. Kantor HL, Briggs RW, RS Balaban: In vivo ^{31}P nuclear magnetic resonance measurements in canine heart using a catheter-coil. Circ Res 55:261-266, 1984

43. Kapel'ko VI, Corina MS, Novikova NA: Comparative evaluation of contraction and relaxation of isolated heart muscle with decreased calcium concentration in the perfusate, acidosis, and metabolic blockade. J Mol Cell Cardiol 14, Suppl 3:21-27, 1982

44. Kentish JC: The effect of inorganic phosphate and creatine phosphate on force production in skinned muscle from rat ventricle. J Physiol (Lond) 370:585-604, 1986

45. Kingsley-Hickman PB, Sako EY, Mohanakrishnan P, Robitaille PML, From AHL, Foker JE, Ugurbil K: ^{31}P NMR studies of ATP synthesis and hydrolysis kinetics in the intact myocardium. Biochemistry (in press)

46. Klingenberg M, Pfaff E: Structural and functional compartmentation in mitochondria. In: Tager JM, Papa S, Quagliariello E, Slater E, eds: Regulation of metabolic processes in mitochondria. New York: Elsevier, 1966:180-210

47. Kusmerick MJ, Meyer RA: Chemical changes in rat leg muscle by phosphorus nuclear magnetic resonance. Am J Physiol 248:C542-C549, 1985

48. LaNoue KF, Jeffries FMH, Radda GK: Kinetic control of mitochondrial ATP synthesis. Biochemistry 25:7667-7675, 1986

49. Lundberg R, Logren T: Bioluminescent determination of ATP in ischemic rat heart. Anal Lett 15:115-121, 1982

50. Lust WD, Feussner GK, Barbehenn EK, Passeneau JV: The enzymatic measurements of adenine nucleotides and P-creatine in picomole amounts. Anal Biochem 110:258-266, 1981

51. Malloy CR, Matthews PM, Smith MB, Radda GK: Influence of propranolol on acidosis and high-energy phosphates in ischaemic myocardium of the rabbit. Cardiovasc Res 20:710-720, 1986

52. Matthews PM, Bland JL, Gadian DG, Radda GK: The steady state of ATP synthesis in the perfused rat heart measured by ^{31}P NMR saturation transfer. Biochem Biophys Res Commun 103:1052-1059, 1981

53. Matthews PM, Taylor DJ, Radda GK: Biochemical mechanisms of acute contractile failure in the hypoxic rat heart. Cardiovasc Res 20:13-19, 1986

54. Moyer JD, Henderson JF: Compartmentation of intracellular nucleotides in mammalian cells. CRC Crit Rev Biochem 19:45-61, 1985

55. Nayler WG: Protection of the myocardium against post-ischemic reperfusion damage. J Thorac Cardiovasc Surg 84:897-905, 1982

56. Neely JR, Grotyohann LW: Role of glycolytic products in damage to ischemic myocardium. Circ Res 55:816-824, 1984

57. Reibel DK, Rovetto MJ: Myocardial adenosine salvage rates and restoration of ATP content following ischemia. Am J Physiol 237:H247-H252, 1979

58. Saks VA, Kupriyanov VV, Elizarova GV, Jacobus WE: Studies of energy transport in heart cells. J Biol Chem 255:755-763, 1980

59. Schaper J, Mulch J, Winkler B, Schaper W: Ultrastructural, functional and biochemical criteria for estimation of reversibility of ischemic injury: A study on the effects of global ischemia on the isolated dog heart. J Mol Cell Cardiol 11:521-541, 1979

60. Schrader J, Gerlach E: Compartmentation of cardiac adenine nucleotides and formation of adenosine. Pflügers Arch 367:129-135, 1976

61. Serayadan K, Mommaerts WFHM, Wallner A: The amount and compartmentalisation of adenosine diphosphate in muscle. Biochem Biophys Acta 65:443-460, 1962

62. Seymour A-ML, Bailey IA, Radda GK: A protective effect of insulin on reperfusing the ischemic rat heart shown using ^{31}P-NMR. Biochim Biophys Acta 762:525-530, 1983

63. Seymour A-ML, Harmsen E, Radda GK: Is ADP the primary regulator of respiration in the heart? Biochem Soc Trans 15:710, 1987 (Abstr)

64. Shoubridge EA, Jeffry FMH, Keogh JM, Radda GK, Seymour A-ML. Creatine kinase kinetics, ATP turnover, and cardiac performance in hearts depleted of creatine with the substrate analogue beta-guanidinepropionic acid. Biochim Biophys Acta 847:25-32, 1985

65. Slater EC, Rosing J, Mol A: The phosphorylation potential generated by respiring mitochondria. Biochim Biophys Acta 292:543-553, 1973

66. Soboll S, Bünger R: Compartmentation of adenine nucleotides in the isolated working guinea pig heart, stimulated by noradrenalin. Hoppe-Seyler's Z Physiol Chem 362:125-132, 1981

67. Van der Laarse A, Bloys van Treslong CHF, Vliegen HW, Ricciardi L: Relation between ventricular DNA content and number of myocytes and non-myocytes in hearts of normotensive and spontaneously hypertensive rats. Cardiovasc Res 21:223-229, 1987

68. Vary TC, Angelakos ET, Schaffer SW: Relationship between adenine nucleotide metabolism and irreversible ischemic tissue damage in isolated perfused rat heart. Circ Res 45:218-225, 1979
69. Veech RL, Lawson JWR, Cornell NW, Krebs HA: Cytosolic phosphorylation potential. J Biol Chem 254:6538-6547, 1979
70. Walker EJ, Dow EJ: Localisation and properties of two isoenzymes of cardiac adenylate kinase. Biochem J 203:361-369, 1982
71. Weiss JN, Lang ST: Glycolysis preferentially inhibits ATP-sensitive K channels in isolated guinea pig cardiac myocytes. Science 238:67-69, 1987
72. Wollenberger A, Ristau O, Schoffa G: Eine einfache Technik der extrem schnellen Abkühlung grösserer Gewebestücke. Pflügers Arch 270: 399-412, 1960
73. Zweier JL, Jacobus WE: Substrate-induced alterations of high energy phosphate metabolism and contractile function in the perfused heart. J Biol Chem 262:8015-8021, 1987

Chapter 12

Cardiac Energy Metabolism Probed with Nuclear Magnetic Resonance

C.J.A. van Echteld, Interuniversity Cardiology Institute
of the Netherlands and Department of Cardiology,
University Hospital, Utrecht, The Netherlands

Nuclear magnetic resonance (NMR) spectroscopy possesses great potential for studying myocardial energy metabolism. To ensure that the observed NMR signal predominantly originates from the heart, localization is required, which can be achieved by excision or exposure of the heart, or by means of sophisticated NMR localization techniques. A number of different atomic nuclei have been employed. ^{1}H NMR has been mainly used to follow lactate accumulation in ischemic or anoxic hearts. ^{13}C NMR has been applied to study the fate of different substrates in the citric acid cycle and amino acid pools, and the role of glycogen metabolism in ischemia or anoxia. ^{19}F, ^{23}Na and ^{39}K have been employed to investigate the consequences of altered energy metabolism for myocardial intracellular concentrations of Ca^{2+}, Na^+ and K^+. The most abundantly used nucleus for studying myocardial energy metabolism is ^{31}P. Numerous contributions have been made to the investigation of ischemia and reperfusion, protection of the heart against the consequences of ischemia and reperfusion, contractile failure, variation of high-energy phosphate levels over the cardiac cycle, regulation of oxidative phosphorylation and intracellular enzyme kinetics of both isolated perfused hearts and hearts in situ. Even human myocardial metabolism can be assessed by ^{31}P NMR, which is on the verge of becoming a clinical tool for investigating heart disease.

Introduction

Nuclear magnetic resonance (NMR) spectroscopy offers unique possibilities to study non-destructively in vivo and in vitro myocardial energy metabolism. The technique relies on the response of atomic nuclei, which are polarized by a strong magnetic field, to a radiofrequency pulse. Many atomic nuclei behave, due to a property called "spin", as magnetic dipoles. In a magnetic field the tiny nuclear magnets will orient themselves either parallel or anti-parallel to the magnetic field (when the nucleus has a spin quantum number ± 1/2). Due to an unequal occupation of both orientations, a net magnetization results. With low-energy radiofrequency pulses, transitions between the two orientations can be induced, allowing for manipulation of the net magnetization. The magnitude of this net magnetization can be detected as an emitted radiofrequency (RF) signal when after the pulse the nuclei relax to their equilibrium orientation.

The so-called resonance frequency of the emitted radiowaves is characteristic of the observed nucleus and directly proportional to the magnetic field strength. Table 1 shows the resonance frequencies and some properties of nuclei, commonly observed in in vivo and in vitro NMR. When a particular nucleus is observed, the exact resonance frequency is determined by its chemical environment. This property, known as chemical shift, allows us to discriminate between identical nuclei in different molecules or in molecules in different environments. The chemical shift δ is normally expressed as $\delta = (v_i - v_r) \cdot 10^6 / v_r$ ppm (parts per million) with v_i = resonance frequency and v_r = resonance frequency of a reference compound. The NMR signal from biological tissue is often a complex pattern of radiowaves, which is Fourier transformed into a frequency spectrum by computer.

Since the emitted signals are generally very weak, the measurement has to be repeated a number of times (signal averaging) to obtain a sufficient signal-to-noise ratio for reliable quantitative measurements. The signal intensity is in principal proportional to the number of observed nuclei, and higher at higher magnetic field strengths. However, upon signal averaging, signal intensity can be affected by other factors such as the spin-lattice relaxation time T_1 and the measurement repetition time. The relaxation time T_1 is a time constant which characterizes the return of the nuclei to their equilibrium distribution. When the next scan is performed before complete relaxation has occurred, the subsequent signal will be less than in the first measurement. T_1-relaxation times of nuclei in small molecules and ions in biological tissue normally range from a few hundreds of milliseconds to a few seconds.

Localization

For a meaningful determination of myocardial energy metabolism with NMR, some form of localization is required to ensure that the observed signal predominantly originates from the heart. The most straightforward and until now most commonly used method is the excision of the heart. The Langendorff perfused or working heart preparation is placed in a glass tube and brought in the magnet center. The RF-transmitter and receiver coil surrounds the tube and detects signal from the entire heart. The possibilities of this approach have first been demonstrated by Gadian et al. (36) in 1976.

Hearts in situ have been studied by surgically introducing NMR coils to fit around the heart of mechanically ventilated animals (41), thereby unavoidably picking up signal from blood. Other invasive approaches to study myocardial energy metabolism in vivo with NMR include chronically implanted coils (61) and small surface coils introduced as a catheter coil (53). The surface coil normally is a flat, circular coil detecting signal, as a rule of thumb, from two halve spheres on either side of the coil with a radius equal to the coil radius. This approach is easily clinically applicable (103), although the signal from the heart will be contaminated with signal from blood, bone and skeletal muscle.

Table 1. Properties of commonly observed nuclei in in vivo and in vitro NMR spectroscopy

Nucleus	Natural abundance (%)	Relative sensitivity (%)[*]	Resonance frequency (MHz) at 1 Tesla magnetic field strength
^{1}H	99.98	100.0	42.58
^{19}F	100.0	83.0	40.05
^{31}P	100.0	6.6	17.24
^{23}Na	100.0	9.3	11.26
^{13}C[**]	1.1	1.6	10.71
^{39}K	93.1	0.051	1.99

[*] Relative sensitivity is the NMR sensitivity of nuclei relative to that of an equal number of protons.

[**] The most abundant carbon isotope ^{12}C (98.9%) is not magnetically active, whereas the non-radioactive isotope ^{13}C is.

The surface coil method has been refined by combination with NMR-imaging derived slice selection (15,16). This method exploits the fundamental dependence of resonance frequency on magnetic field strength by applying a linear magnetic field gradient perpendicular to the plane of the surface coil, together with an RF pulse of limited bandwidth. In this way only a slice of tissue is excited instead of the entire half spherical volume, resulting in one dimensional localization. Other one dimensional localization methods make use of the inherent gradient of the RF field emitted by a surface coil (13).

Most elegant localization methods are the so-called volume selective excitation techniques (5) in which three orthogonal magnetic field gradients are used. Originally these methods were developed for coils that surround the subject and require high power RF-pulses. However, since safety considerations and technical limitations preclude their use for investigating human heart, alternative approaches have been developed. One of the most promising of these techniques is the so-called ISIS method (87), a three dimensional localization method which requires only low power RF-pulses and can be used with surface coils, and consequently is ideally suitable for studying human myocardial metabolism.

Myocardial Metabolism

A number of different nuclei has been employed to study myocardial metabolism with NMR. Each nucleus has its own characteristics and is associated with particular molecules commonly probed with it. Generally, narrow resonance peaks are only obtained from nuclei in molecules that are fairly mobile. Nuclei in large molecules give rise to very broad resonance lines underlying the narrow peaks from mobile metabolites or ions.

^1H

Despite its relatively high sensitivity, ^1H NMR has rarely been used to study myocardial energy metabolism. This is mainly due to the enormous signal arising from H_2O, which tends to conceal other resonances, and the ubiquity of protons in cellular components, which often results in very complicated spectra. Increasing the magnetic field strength not only increases sensitivity but also spectral dispersion which facilitates the interpretation of complicated spectra.

By saturating the H_2O resonance by selective irradiation with a very narrow frequency bandwidth prior to data acquisition, Ugurbil et al. (101) have been able to detect in isolated perfused rat hearts resonances corresponding with taurine, carnitine, creatine and phosphocreatine, glycerides, and especially in ischemic hearts, lactate. Lately, sophisticated methods for water suppression (14,47) and separation of overlapping peaks by editing techniques, which also frequently use selective irradiation (44,105), have been developed that allow a more reliable determination of tissue lactate in brain and skeletal muscle.

Only recently these methods have been applied to follow the time course of lactate accumulation in ischemic and anoxic isolated rat hearts (91). With a time resolution of 5 min, lactate was detected after 5 min in the ischemic hearts and after 10-20 min in anoxic hearts. Unfortunately, these measurements were performed at a relatively low field strength of 2 Tesla. At higher field strengths, a better time and spectral resolution can be obtained with a practical detection limit around 0.1 mM.

$$^{13}C$$

The chemical shift range of ^{13}C is much wider than that of ^1H (ca. 250 vs. ca. 15 ppm). Therefore, ^{13}C-spectra can be easier interpreted than ^1H-spectra. Unfortunately ^{13}C is far less sensitive than ^1H and has a natural abundance of only 1.1% (see Table 1). Therefore, ^{13}C in vivo and in vitro NMR spectroscopy is limited to molecules that have been specifically enriched with ^{13}C or are present at relatively high intracellular concentrations. In addition, nuclei which are close together by chemical bond or in space, can show homonuclear and heteronuclear interactions or couplings. As is the case with ^1H-^{13}C interactions, this often results in splitting of the peaks in doublets, triplets etc. By simultaneous irradiation with ^1H-resonance frequencies, this interaction can be "decoupled", which increases signal intensity and simplifies the spectra. However, decoupling requires the presence of a second coil for transmission of ^1H-frequencies.

Nevertheless, ^{13}C NMR has been successfully applied to study metabolism of both isolated, perfused hearts and hearts in situ. The first natural-abundance ^{13}C spectrum of isolated rat hearts was reported by Bailey et al. (6). It showed only resonances attributable to lipids, which are present in sufficiently high concentrations. When [2-^{13}C]acetate was offered as a substrate instead of glucose to investigate citric acid cycle activity, labeling of the C-2, C-3 and C-4 carbons of glutamate was observed. Since the C-3 carbon of alpha-ketoglutarate and hence glutamate (via transamination) can only be labeled in a second turn of the tricarboxylic acid cycle of previously labeled citric acid and since scrambling of the label occurs between the C-2 and C-3 of succinate, the steady state enrichment of glutamate at C-3 occurs at a slower rate than that at C-4. From this difference in incorporation rate, the relative rates of cycling in the tricarboxylic acid cycle and the transamination to glutamate can be estimated. When acetate in the perfusate was partly replaced by glucose, label also appeared in C-2 and C-3 of aspartate. This may reflect changes in transamination due to increased activity of the malate-aspartate shuttle. From coupling patterns between ^{13}C at C-4 of glutamate with neighboring ^{13}C at C-3 (which could only be observed in $HClO_4$ extracts of the heart, due to better resolution), it was concluded that exogenous acetate provided all of the substrate oxidized in acetate-perfused hearts. Finally, Bailey et al. (6) failed to detect any incorporation of label in alanine via pyruvate formed by malic enzyme from malate, although considerable malic enzyme activity has been reported in rat heart (93). They suggested that the malic enzyme is either kinetically controlled to feed substrate effectively unidirectionally into the citric acid cycle, or is inhibited in vivo, or cannot convert the labeled malate pool due to compartmentation.

The observation that only glutamate carbons become detectably labeled in [2-^{13}C]acetate-perfused rat hearts was essentially confirmed in a recent study (72), where also a minor additional resonance from acetylcarnitine was observed. About 90% of substrate oxidized was acetate. When propionate was added, which enters the tricarboxylic acid cycle via anaplerotic pathways, aspartate and malate appeared in the ^{13}C spectrum, whereas glutamate resonances decreased. It appeared that under these conditions only 50% of the substrate oxidized was acetate. In guinea-pig hearts perfused with [3-^{13}C]pyruvate as substrate, resonances were detected from glutamate, aspartate, alanine, citrate, malate, lactate and acetylcarnitine, whereas in [3-^{13}C]lactate-perfused hearts only glutamate resonances were observed. This most likely reflects a difference in the malate-aspartate shuttle activity, since lactate can only be utilized effectively when sufficient NAD^+ is provided for oxidation to pyruvate (95):

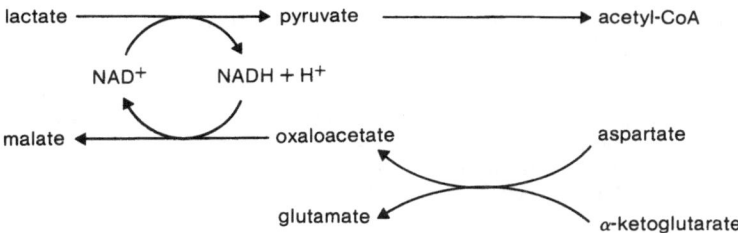

During anoxia in guinea-pig hearts, which had been preperfused with [3-^{13}C]pyruvate, substantial labeling of succinate occurred, whereas the intensities of labeled glutamate and aspartate simultaneously decreased. These results were interpreted as anaerobic metabolism of glutamate and aspartate to succinate yielding nucleoside triphosphates which may provide a supplemental mechanism to glycolysis for production of high-energy phosphates during anoxia or ischemia (18). Chance et al. (25) have presented a sophisticated mathematical analysis of ^{13}C labeling of tricarboxylic acid cycle intermediates, which allows calculation of total citric acid cycle flux.

Other authors have used D-[1-^{13}C]glucose as substrate to study myocardial glycogen synthesis and subsequent glycogenolysis, and appearance of label in lactate during anoxia (83,84) or ischemia (66). Glycogen is one of those rare exceptions in that all of the ^{13}C-carbons are NMR-visible, despite the very large mol. wt. as has been demonstrated for liver and heart glycogen (83,97). Apparently the molecule possesses high internal mobility.

It is clear that ^{13}C NMR offers unique possibilities and has distinct advantages over conventional ^{14}C-labeling techniques to study simultaneously the contribution of competing metabolic pathways and to elucidate specific pathways from preferential labeling. One may therefore expect that, despite the drawbacks mentioned before, ^{13}C NMR will become an important tool to study myocardial energy metabolism.

[19]F, [23]Na and [39]K

[19]F, [23]Na and [39]K are being increasingly used to study myocardial intracellular ion concentrations. ^{23}Na and [39]K NMR (31,90) are employed to directly measure intracellular $[Na^+]$ and $[K^+]$ in conjunction with so-called shift reagents. These paramagnetic reagents weakly interact with the extracellular ions, thereby changing the resonance frequency of these ions and separating the signals from intra- and extracellular ions. Intracellular $[Ca^{2+}]$ can be measured with ^{19}F NMR by loading the cell with a fluorinated Ca^{2+}-chelator (98). (Direct estimation of $[Ca^{2+}]$ -and $[Mg^{2+}]$ for that matter- using the magnetically active isotopes [43]Ca and [25]Mg is hardly feasible due to low sensitivity and low natural abundance.) Although ^{19}F, [23]Na and [39]K do not probe myocardial energy metabolism directly, they may become a powerful tool to study the consequences of altered energy metabolism for intracellular ion concentrations. Unless gated to the cardiac cycle, the time resolution of these measurements in isolated, perfused hearts, which varies from about 1 min for [23]Na to several min for [39]K and ^{19}F, is not sufficient to detect variations in intracellular concentrations during the cardiac cycle. However, it has been possible to observe substantial differences in intracellular $[Na^+]$ and $[Ca^{2+}]$ at different levels of cardiac performance (21,73,90,99) and after 10 min of global ischemia (73,99).

^{31}P

^{31}P is the most abundantly used nucleus for studying myocardial energy metabolism. Figure 1 shows a ^{31}P NMR spectrum of an isolated Langendorff-perfused rabbit heart, illustrating the ^{31}P resonances one may find in heart. Since protonation of phosphate groups changes the chemical environment of the phosphorus nucleus, a number of the peaks in Fig. 1 show a pH-dependent chemical shift. Alternatively, the resonance frequency can be used to determine intracellular pH. Inorganic phosphate (P_i) is a good candidate for intracellular pH-measurement (80), which is illustrated by the P_i-peaks corresponding with an extracellular pH of 7.4 and an intracellular pH of 7.0 in Fig. 1.

Intracellular pH measurement of hearts in situ is often complicated by the (partially) overlapping resonances of 2,3-diphosphoglycerate (2,3-DPG) from blood and P_i. Brindle et al. (19) have used an editing technique, which exploits the [^{1}H-^{31}P]-couplings in 2,3-DPG and the absence of these couplings in P_i, to effectively suppress the resonances of 2,3-DPG. The P_i resonance thus measured indicated a pH of approximately 7.3. Therefore they concluded that the P_i signal obtained from normoxic hearts in vivo predominantly originates from blood, although spectra obtained from isolated heart at high workloads can show significant resonances from intracellular P_i (35).

Another external factor which may influence chemical shifts is binding of metal ions. Especially the gamma- and beta-ATP resonances are sensitive to the binding of

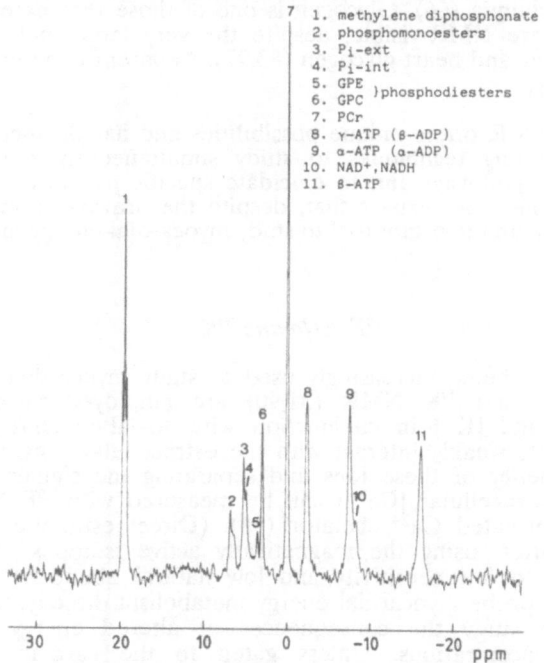

1. methylene diphosphonate
2. phosphomonoesters
3. P_i-ext
4. P_i-int
5. GPE)
6. GPC) phosphodiesters
7. PCr
8. γ-ATP (β-ADP)
9. α-ATP (α-ADP)
10. NAD+,NADH
11. β-ATP

Fig. 1. 81.0 MHz (4.7 Tesla) ^{31}P NMR spectrum of an isolated rabbit heart obtained from 128 scans with an interpulse time of 10 s. Methylene diphosphonate is an external reference, phosphomonoesters include sugar phosphates and AMP, P_i = inorganic phosphate, ext = extracellular, int = intracellular, GPE = glycerophosphoethanolamine, GPC = glycerophosphocholine, PCr = phosphocreatine, ATP = adenosine 5'-triphosphate, ADP = adenosine 5'-diphosphate, NAD(H) = nicotinamide adenine dinucleotide.

Mg^{2+} (42,48). From the spectrum in Fig. 1, it can be derived that >92% of ATP is complexed to Mg^{2+}-ions and, assuming the [ATP] to be 8 mM (37), the intracellular $[Mg^{2+}]$ can be estimated (42) to be 0.5 mM. Only high resolution ^{31}P NMR spectra of heart extracts show ADP resonances resolved from ATP resonances. In in vivo spectra from intact hearts, the peaks of alpha- and beta-ADP unfortunately coincide with the peaks of alpha- and gamma-ATP, resp. The beta-ATP peak, however, is unique for ATP, so strictly speaking, the difference between beta- and gamma-ATP signal intensities should yield the [ADP]. This approach is not realistic since ATP levels are normally two orders of magnitude higher than ADP levels. Nevertheless, free cytosolic [ADP] can be derived from the creatine kinase (CK) reaction. Combined measurement of O_2-consumption and fluxes through the CK reaction (using magnetization transfer methods, which will be discussed below) have shown that CK fluxes in normoxic, perfused heart exceed the mitochondrial ATP synthesis rate by at least a factor of 3 (12,76,77) over a wide range of cardiac performance. This demonstrates that the CK reaction is sufficiently rapid to maintain the cytosolic reactants near their equilibrium concentration. It is therefore justified to derive $[ADP]_{free}$ from the CK equilibrium as $[ADP] = ([ATP] [Cr])/([PCr] [H^+] \cdot K_{eq})$ in which $K_{eq} = 1.66 \cdot 10^9$ M^{-1} (at pH 7.2, $[Mg^{2+}]_{free} = 1$ mM and an ionic strength of 0.25 M) (68). [ATP], [PCr] and $[H^+]$ can be measured from ^{31}P NMR spectra and the creatine (Cr) concentration can be derived assuming a constant total pool of Cr + PCr of about 26 mM (37,69,106). Depending on the workload and the carbon substrate being oxidized, $[ADP]_{free}$ has been reported in the range from 10 um up to 132 um (35,77,106).

Ischemia and reperfusion. One of the first topics addressed with ^{31}P NMR was intracellular acidosis during ischemia and subsequent recovery upon reperfusion (39,45,51), which soon became combined with simultaneous measurement of left ventricular pressure in paced hearts (46). The metabolic consequences of global ischemia, which were investigated in these and many subsequent studies, comprise a fast decrease of the PCr level at the onset of ischemia, followed by a slower decrease of the ATP level and a concomitant decrease of intracellular pH and increase of the P_i level. Garlick et al. (38) have suggested that glycogenolysis during ischemia is the major H^+-producing reaction. They observed an attenuation of intracellular acidosis in glycogen-depleted hearts. However, as pointed out by Gevers (40), glycogenolysis and lactate-producing glycolysis is not associated with a net generation of H^+-ions, and acidification is rather the result of ATP hydrolysis. Bailey et al. (7) have pretreated glucose-perfused rat hearts with insulin resulting in increased glycogen content and increased accessibility of glycogen to phosphorylase during ischemia. They observed a decrease of intracellular pH to 5.7 as compared to 6.2 for untreated hearts, and a slower decline of ATP levels. The extremely low intracellular pH contradicts the commonly believed control of glycolysis by pH at the level of phosphofructokinase. Probably cessation of glycolysis in untreated hearts at pH 6.2 is rather a consequence of reduced accessibility (7). Seymour et al. (94) observed a better mechanical recovery upon reperfusion of insulin-pretreated ischemic heart, indicating that ATP maintenance is more important to recovery than maintenance of intracellular pH is. In protecting isolated, perfused rabbit hearts by hypothermia and cardioplegia against the consequences of global ischemia, Flaherty et al. (29) also found better functional recovery with higher ATP content at the end of the ischemic period, although intracellular acidosis and increased P_i levels correlated inversely with recovery of postischemic ventricular structure and function.

Numerous other interventions have been applied to protect the isolated heart against the consequences of ischemia or hypoxia and reperfusion, including inosine infusion (49), adenosine deaminase inhibitors (27), perfluorocarbon cardioplegia (28), free radical scavenging (3,4) and treatment with prostaglandin (88), beta-blockers and calcium antagonists (58,65,67,71,81,85,89,92). The beneficial effect of treatment with calcium antagonists in most cases relied on an energy-sparing effect.

However, we recently showed that pretreatment of rats with anipamil, a new calcium antagonist, did neither alter left ventricular developed pressure (LVDP) of the isolated hearts under normoxic conditions, nor the rate and extent of ATP and PCr depletion during ischemia, as can be seen in Fig. 2, whereas intracellular acidosis was attenuated. Upon reperfusion, hearts from anipamil pretreated animals recovered significantly better than untreated hearts, with respect to high-energy phosphate levels, return to low P_i levels (see Fig. 2), and left ventricular function and coronary flow. Intracellular pH also recovered rapidly to pre-ischemic levels, whereas in untreated hearts a complex intracellular P_i-peak indicated areas of different pH within the myocardium. Protection in this case cannot be attributed to an energy-sparing effect during ischemia, since a negative inotropic effect was absent during normoxia.

Contractile failure. The decline of contractile function of isolated rabbit hearts during the early phases of ischemia has been studied by Jacobus et al. (50). From comparison with respiratory acidosis, they concluded that intracellular acidification early in the ischemic period accounts for at the most a 50% reduction in LVDP and that other factors must also play a role. Since ATP levels did not change appreciably during their experiments, these cannot fulfil this role. Only when the [ATP] falls below about 35% of the normal level, contractility is affected (62,82).

Early contractile failure during hypoxia has recently been studied by several investigators using ^{31}P NMR. Allen et al. (1) have found in isolated, perfused ferret hearts that also during hypoxia with normal glycolysis rates, the decreased pH only partially accounted for the decreased LVDP. In this situation the amplitude of the calcium transients was unaffected (2). Therefore, they suggested intracellular P_i to be the major other factor that contributes to early contractile failure. When glycolysis was inhibited, no pH decrease occurred, but in this situation a decline in calcium transients has been previously reported (2). At the same time a decrease of 20% of the [ATP] during the first 5 min of hypoxia was observed. Using the CK

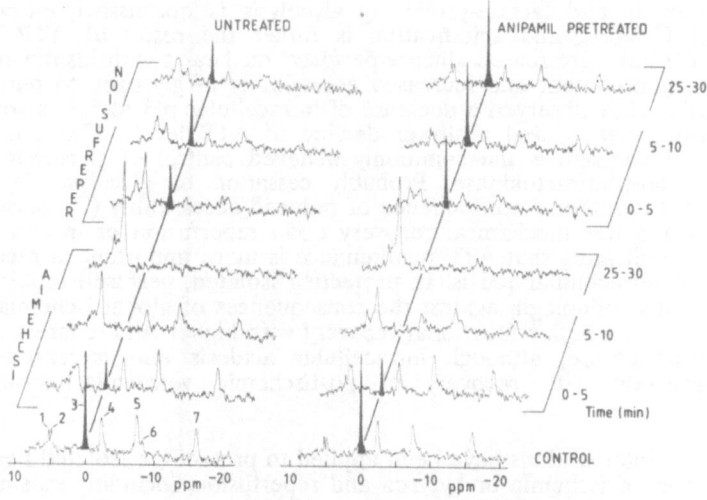

Fig. 2. Serial ^{31}P NMR spectra obtained from untreated (left) and anipamil-pretreated (right) rat hearts during control perfusion, 30 min of total ischemia, and 30 min of reperfusion. Numbered peaks include: 1) extracellular P_i; 2) intracellular P_i; 3) PCr; 4, 5 and 7) gamma-, alpha- and beta-ATP, resp.; 6) NAD(H).

equilibrium, the [ADP] could be derived and was used to calculate the free energy change of ATP hydrolysis: ΔG_{ATP}. As originally suggested by Kammermeier et al. (52), Allen et al. (1) discussed the decline in ΔG_{ATP} and consequent lack of energy to maintain sarcoplasmic reticulum calcium pump activity as a causal factor in the rapid and total contractile failure during hypoxia with inhibited glycolysis. However, a similar decrease in ΔG_{ATP} was observed during hypoxia with normal glycolytic activity.

Using a method for eliciting and measuring maximal Ca^{2+}-activated tension in ferret hearts (74), Kusuoka et al. (64) concluded that during the early phase of hypoxia [P_i] is responsible for the depression of contractility and that pH plays only a minor role. Matthews et al. (78) suggested that the myocardial inotropic state may be directly responsive to the ambient pO_2, although their results are qualitatively similar to those of Allen et al. (1). Interestingly, they also reported that no significant decrease in the ATP-utilization rate occurred, despite an 80% reduction in mechanical activity during hypoxia.

Using a surface coil and an open-chest model (41), Bittl et al. (11) have investigated the effects of hypoxia in the heart of living rats. Unfortunately, due to the previously mentioned overlap of 2,3-DPG and P_i resonances, measurements of myocardial intracellular pH and [P_i] were not possible. However, based on the decrease of the steady state [high-energy phosphate] at different levels of hypoxia, they calculated with the data of Kusuoka et al. (64) at the most a 20% reduction in maximal Ca-activated pressure caused by P_i accumulation. On the other hand, they observed a 45% decrease in systolic pressure at 8% O_2 ventilation, suggesting that other mechanisms than the [P_i] are involved too. They measured flux through the CK reaction and found a close match between this flux and cardiac performance at different degrees of hypoxia, whereas ATP levels fell only slightly during hypoxia. They suggested that the decreased turnover of ATP and PCr reflects the combination of impaired ATP synthesis and utilization and may be the energetic basis for hypoxic cardiac failure. However, their decreased ATP turnover is not consistent with the unchanged ATP utilization in isolated hypoxic hearts (78).

Regional ischemia. The surface coil approach with both open and closed thorax has been used by other investigators to study the effects of ligation of the left anterior descending coronary artery (LAD). ^{31}P NMR spectra of ischemic regions in cat hearts showed similar changes as were observed in global ischemic rat or rabbit hearts (100). In rabbits ATP levels decreased together with PCr levels, although at a slower rate. After about 30 min of ligation, substantial amounts of high-energy phosphate were still present (71). In dog hearts the decrease of high-energy phosphates was relatively limited (17,43). The different results obtained for different species may well reflect differences in collateral flow. In dog hearts a correlation was found between regional contractile function and both pH and P_i, whereas ATP levels correlated with recovery of regional contractile function (43). A closed-chest dog model allowed for long-term evaluation of the effects of permanent LAD occlusion (17). The results were quite diverse. In some dogs P_i and pH returned to normal within 30 min, whereas in others PCr/P_i ratios remained depressed for up to 5 days. In all dogs myocardial pH returned to normal within 15 h.

Cardiac Cycle. The variation of high-energy phosphate and P_i levels during the cardiac cycle has also been investigated with ^{31}P NMR. Using an NMR method, in which the data acquisition was gated to the heart beat, Fossel et al. (33) observed a considerable variation in high-energy phosphate and P_i levels during the heart cycle of an isolated working heart. They found both PCr and ATP levels to be maximal at minimal aortic pressure and minimal at maximal aortic pressure, whereas

P_i showed a direct phase relationship with the aortic pressure curve. However, using a saturation transfer method, Matthews et al. (76) demonstrated that in an isolated Langendorff-perfused rat heart only 2.5% of the total ATP is turned over during one heart cycle compared to about 35% in the other study (33). An explanation for the observed discrepancy may be a poor nutrient and/or oxygen supply in the isolated, working rat heart.

Gated ^{31}P NMR spectra of rat (61) and dog (54) hearts in situ did not reveal any variation over the heart cycle. Assuming that the total Cr pool remains constant, also during the cardiac cycle, variation of the [ADP] over the cardiac cycle cannot be concluded, although the [ADP]$_{free}$ is believed to be a key regulatory factor in oxidative phosphorylation (24).

Respiratory regulation. The regulation of oxidative phosphorylation at different steady state levels of cardiac performance has been studied using ^{31}P NMR in isolated rat hearts perfused with different substrates (35) and in hearts in situ of different species. For isolated rat hearts perfused with pyruvate (+ glucose), ADP levels were generally very low and varied with myocardial O_2 consumption, which suggests a regulation of oxidative phosphorylation through ADP availability. However, with glucose as carbon substrate in the presence or absence of insulin, ADP levels were much higher and the phosphorylation potential, ADP/ATP ratio and [ADP] were relatively constant over the range of workloads examined. It has been previously reported that, even at low workloads, levels of citric acid cycle intermediates, pyruvate and mitochondrial NADH are at least an order of magnitude lower with glucose as carbon substrate than with pyruvate (59). These observations suggest that when glucose is used as substrate, carbon substrate delivery to mitochondria is the rate-limiting step in mitochondrial NADH generation and therefore oxidative phosphorylation. This suggestion has been substantiated by a combined ^{31}P NMR and NADH fluorescence study in which increased O_2 consumption was associated with increased mitochondrial NAD(P)H levels and not with P_i, PCr, ATP or ADP (55).

When varying the rate pressure product over a wide range from $5 \cdot 10^3$ to $25 \cdot 10^3$ mmHg/min, Balaban et al. (8) also did not observe a change in PCr/ATP ratio in in vivo paced dog hearts using ^{31}P NMR surface coil spectroscopy with a catheter probe. From the constant PCr/ATP ratio a constant [ADP] is inferred. Chance et al. (23) studied transfer functions, which relate work and substrate concentrations of (possible) control chemicals, in various tissues in situ. They observed in dog hearts a very steep transfer function when relating work and [ADP], in line with other studies in this species (8,70). These results again suggest a respiratory regulation by mitochondrial NADH. However, this steep transfer function is not found in hearts in general since in neoate lamb (23) and cat (70) hearts a far less steep relationship between work and [ADP] is observed. The authors suggest that these species differences may be related to adaptation to different life-styles, i.e., an endurance-performance adaptation for dogs vs. a stalking and sprinting adaptation for cats.

When comparing the results in refs. 8 and 35, it is found that the [ADP] in dog hearts in vivo is comparable to the [ADP] in isolated rat hearts perfused with glucose/insulin. From et al. (35) presented preliminary data showing that fatty-acid perfused hearts have ADP levels that are lower than those in glucose-perfused hearts, but not sufficiently low to become rate limiting as is the case in pyruvate-perfused hearts. Since fatty acids are considered to be the major carbon source for myocardial energy metabolism in vivo, a similar situation may be found for the heart in situ. The fact that different substrates may lead to different ratios of high-energy phosphates and P_i has been recognized before (79), but in addition, Zweier and Jacobus (106) observed an increase in contractility when switching from glucose to pyruvate as substrate.

Magnetization transfer. Where [high-energy phosphates] do not change upon increased cardiac performance, there is still a well-established coupling between cardiac performance and myocardial O_2 consumption and therefore net ATP synthesis (59). As mentioned before, magnetization transfer methods can be be used to measure fluxes through the enzymatic reactions which are involved in energy production, like the CK reaction:

$$PCr + ADP + H^+ \underset{k\text{-}1}{\overset{k1}{\rightleftharpoons}} ATP + Cr$$

and the ATP synthetase reaction: $P_i + ADP \underset{k\text{-}2}{\overset{k2}{\rightleftharpoons}} ATP$.

The magnetization transfer method (30) will be briefly outlined for the simplest case of a two-site exchange process, illustrated by the exchange of phosphate between PCr and gamma-ATP. In this case the CK reaction is treated as a pseudo first-order reaction by assuming [ADP], [H$^+$] and [Cr] to be constant, resulting in:

$$PCr \underset{k_{rev}}{\overset{k_{for}}{\rightleftharpoons}} gamma\text{-}ATP$$

in which $k_{for} = k_1 \cdot [ADP] \cdot [H^+]$ and $k_{rev} = k_2 \cdot [Cr]$. The flux in the forward direction is defined as $F_f = k_{for} \cdot [PCr]$ and in the reverse direction as $F_r = k_{rev} \cdot [gamma\text{-}ATP]$. The equilibrium magnetizations of PCr and gamma-ATP are symbolized as M°_{PCr} and $M^\circ_{gamma\text{-}ATP}$. When perturbed, the magnetization of, e.g., PCr (M_{PCr}) returns to its equilibrium situation, when we assume the chemical exchange to be absent, as described by the following equation:

$$dM_{PCr}/dt = (M^\circ_{PCr} - M_{PCr})/(T_{1\ PCr})$$

in which t = time and $T_{1\ PCr}$ is the intrinsic relaxation time T_1 of PCr that would be observed in the absence of exchange. In the presence of exchange, this equation must be modified. PCr loses magnetization through transfer of phosphate groups to ADP to form ATP and gains magnetization from the reverse process which results in:

$$dM_{PCr}/dt = (M^\circ_{PCr} - M_{PCr})/(T_{1\ PCr}) - k_{for} \cdot M_{PCr} + k_{rev} \cdot M_{gamma\text{-}ATP}.$$

By applying selective irradiation of a very narrow bandwidth to the position of the gamma-ATP resonance for a certain period of time, as illustrated in Fig. 3, it is possible to selectively saturate or nullify the gamma-ATP magnetization, resulting in:

$$dM_{PCr}/dt = (M^\circ_{PCr} - M_{PCr})/(T_{1\ PCr}) - k_{for} \cdot M_{PCr}.$$

In a steady-state situation, where $dM_{PCr}/dt = 0$, it therefore appears that

$$M^s_{PCr} = (M^\circ_{PCr})/(1 + k_{for} \cdot T_{1\ PCr}),$$

which means that in the presence of saturation of the gamma-ATP resonance the magnetization or signal intensity of PCr (M^s_{PCr}) is no longer equal to M°_{PCr}, but is reduced by a factor $1/(1 + k_{for} \cdot T_{1\ PCr})$ as can be seen in Fig. 3. After perturbation M_{PCr} returns to its equilibrium value according to an effective T_1, given by:

$$T_{1\ eff\ PCr} = (T_{1\ PCr}) \cdot M^s_{PCr}/M^\circ_{PCr}.$$

The rate constant k_{for} can then be derived from:

$$k_{for} = (1 - M^s_{PCr}/M^o_{PCr})/(T_{1\ eff\ PCr}).$$

$T_{1\ eff\ PCr}$ and M^s_{PCr}, the magnetization of PCr in the presence of long saturation of gamma-ATP, can by obtained from a so-called progressive saturation T_1-measurement (34) in which T_1 is derived from a series of measurements with repetitive pulses with varying interpulse delays, in the presence of irradiation of gamma-ATP. M^o_{PCr} can be obtained from a control spectrum in the absence of saturation.

k_{rev} can be obtained in an analogous way by selective irradiation of PCr and substitution of PCr by gamma-ATP and vice versa in the above equations and considerations. Since the complete measurement of both k_{for} and k_{rev} with T_1's is a lengthy procedure, which requires at least 30 min, depending on magnetic field strength and heart size, intracellular enzyme kinetics can only be studied in hearts in a metabolic steady state.

With magnetization transfer methods, fluxes in the forward direction were found to be larger than in the reverse direction (77,86). Initially, these results were interpreted as an indication for compartmentation (86) and support for the "shuttle"

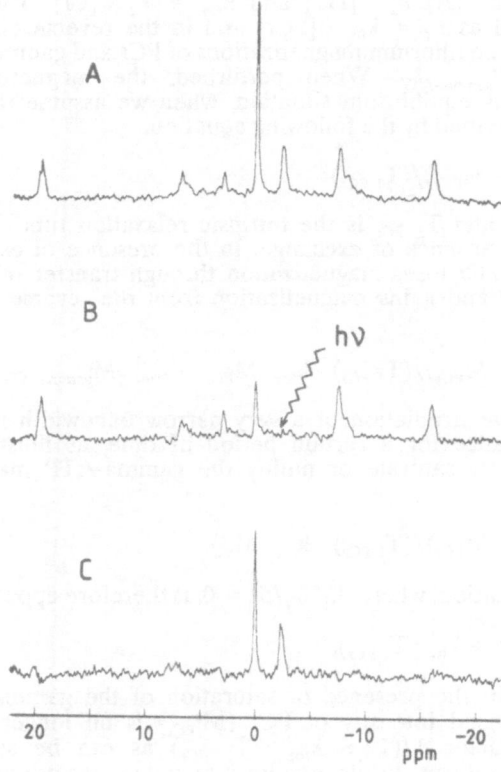

Fig. 3. ^{31}P NMR spectra of an isolated rabbit heart. Spectrum A represents 128 scans with a recycle time of 10 s. Spectrum B represents 128 scans with selective irradiation on gamma-ATP during 9.6 s in between scans. Spectrum C represents the difference between A and B.

model of the CK reaction (9). However, as pointed out by several investigators (26,77), PCr is a substrate unique for CK, but ATP participates in many more reactions. This leads to an underestimation of the reverse flux. When pyruvate or glucose/insulin was used as substrate at high workloads, substantial differences in forward and reverse fluxes were present, whereas with glucose alone at lower workloads fluxes were equal (102). By using a multiple saturation transfer method in which both PCr and P_i were saturated, including all reactions in which ATP hydrolysis is involved, Ugurbil et al. (102) could fully account for the observed discrepancies. However, Koretsky et al. (60) mentioned that this approach is not succesful in rat hearts in situ. In addition, they measured the influence of the adenylate kinase reaction by simultaneous irradiation of the beta-ATP resonance, but concluded that the activity of this reaction is too low to account for the observed discrepancy in forward and reverse fluxes in the heart in situ. Therefore, they suggested that for the heart in situ, ATP compartmentation is still involved.

Regulation of CK fluxes in relation to cardiac performance is still a matter of controversy. In isolated rat hearts , Matthews et al. (77) found little change in CK forward flux upon increasing cardiac performance, but observed a considerable change in PCr/ATP ratio, which was compensated by changes in k_{for}. In contrast, Bittl and Ingwall (12) reported a 3.5-fold increase in forward CK flux when changing from K^+-arrest to a rate-pressure product (RPP) of $44.7 \cdot 10^3$ mmHg/min for isolated rat heart. Yet they observed very little change in PCr/ATP ratio, but increased flux was primarily caused by increased k_{for}. Also in the heart in situ, they observed a comparable relation between forward flux and RPP (10). Kupriyanov et al. (63) studied the same range of RPP (12), but found only a 1.5-fold increase in forward flux in isolated rat hearts, in agreement with earlier results (77). Using isolated rabbit hearts, we recently did not observe significant dependence of forward CK flux on RPP although necessarily a smaller RPP range was studied than in the rat heart studies. Since experimental conditions vary between the different studies cited, no firm conclusions can yet be reached concerning the relation of CK flux and cardiac performance.

Bittl and Ingwall (12) have interpreted their results as evidence for a direct coupling between energy production and energy transfer. However, Shoubridge et al. (96) reached a different conclusion. They depleted rat hearts of PCr by feeding rats a creatine analogue, which is not a substrate for CK. These hearts could perform normally despite a 90% depletion of PCr. Therefore, PCr does not seem an obligate intermediate of energy transduction in the heart. Fossel and Hoefeler (32) found with a similar approach (CK inhibition) that only at high workload isolated rat hearts were unable to maintain their peak-systolic and end-diastolic pressures.

ATP-synthesis rates have also been studied using magnetization transfer methods (76), despite the greater technical difficulty due to the relatively low P_i content of the heart. Kingsley-Hickman et al. (56) compared ATP synthesis rate and O_2 consumption and found a surprisingly high P:O ratio of 6.3. They explained their results by invoking a unidirectional ATP synthesis which exceeded the net ATP synthesis by a factor of 2. However, Brindle and Radda (20) suggested that the glycolytic enzymes glyceraldehyde-3-phosphate dehydrogenase and phosphoglycerate kinase could also catalyze a significant P_i-ATP exchange in addition to the mitochondrial ATP-synthase. In a subsequent study in which these enzymes were inhibited, Kingsley-Hickman et al. (57) found a P:O ratio of 2.36 and concluded that the mitochondrial ATP-synthase works essentially unidirectionally.

Studies on human heart. Studies of human cardiac metabolism have been few and merely demonstrate the possibility to obtain localized ^{31}P NMR spectra of human heart (13,16). Only one study has been reported where ^{31}P NMR is used for diagnosis and therapeutic evaluation of a pediatric case of cardiomyopathy (103).

Since the technical difficulties in obtaining ^{31}P NMR spectra of human skeletal muscle are far less than for cardiac muscle, a number of investigators has studied skeletal muscle metabolism in patients with congestive heart failure (22,75,104). They found metabolic alterations which did not appear to be merely related to reduced blood flow, but also to impaired substrate availability and altered biochemistry. However, in the near future, ^{31}P NMR will also be used to study *myocardial* metabolism of patients with congestive heart failure and other heart diseases and will become an increasingly important tool for clinical investigation.

Acknowledgements

I would like to thank Prof. T.J.C. Ruigrok, PhD, and J.H. Kirkels, MD, for helpful discussions and correcting the manuscript, and Mrs. A.I. Diepeveen for preparing the manuscript.

References

1. Allen DG, Morris PG, Orchard CH, Pirolo JS: A nuclear magnetic resonance study of metabolism in the ferret heart during hypoxia and inhibition of glycolysis. J Physiol (Lond) 361:185-204, 1985
2. Allen DG, Orchard CH: The effect of hypoxia and metabolic inhibition on intracellular calcium in mammalian heart muscle. J Physiol (Lond) 339:102-122, 1983
3. Ambrosio G, Weisfeldt ML, Jacobus WE, Flaherty JT: Evidence for a reversible, oxygen radical mediated component of reperfusion damage: reduction by recombinant human superoxide dismutase administered at the time of reflow. Circulation 75:282-291, 1987
4. Ambrosio G, Zweier JL, Jacobus WE, Weisfeldt ML, Flaherty JT: Improvement of postischemic myocardial function and metabolism induced by administration of deferoxamine at the time of reflow: the role of iron in the pathogenesis of reperfusion injury. Circulation 76:906-915, 1987
5. Aue WP, Müller S, Cross TA, Seelig J: Volume-selective excitation. A novel approach to topical NMR. J Magn Reson 56:350-354, 1984
6. Bailey IA, Gadian DG, Matthews PM, Radda GK, Seeley PJ: Studies of metabolism in the isolated, perfused rat heart using ^{13}C NMR. FEBS Lett 123:315-318, 1981
7. Bailey IA, Radda GK, Seymour A-ML, Williams SR: The effects of insulin on myocardial metabolism and acidosis in normoxia and ischemia. A ^{31}P NMR study. Biochim Biophys Acta 720:17-27, 1982
8. Balaban RS, Kantor HL, Katz LA, Briggs RW: Relation between work and phosphate metabolites in the in vivo paced mammalian heart. Science 232:1121-1123, 1986
9. Bessman SP, Geiger PJ: Transport of energy in muscle: the phosphorylcreatine shuttle. Science 211:448-452, 1981
10. Bittl JA, Balschi JA, Ingwall JS: Effects of norepinephrine infusion on myocardial high energy phosphate content and turnover in the living rat. J Clin Invest 79:1852-1859, 1987
11. Bittl JA, Balschi JA, Ingwall JS: Contractile failure and high-energy phosphate turnover during hypoxia: ^{31}P NMR surface coil studies in living rat. Circ Res 60:871-878, 1987
12. Bittl JA, Ingwall JS: Reaction rates of creatine kinase and ATP synthesis in the isolated rat heart. J Biol Chem 260:3512-3517, 1985
13. Blackledge MJ, Rajagopalan B, Oberhaensli RD, Bolas NM, Styles P, Radda GK: Quantitative studies of human cardiac metabolism by ^{31}P rotating-frame NMR. Proc Natl Acad Sci (USA) 84:4283-4287, 1987
14. Blondet P, Decorps M, Albrand JP, Benabid AL, Remy C: Water-suppressing pulse sequence for in vivo ^{1}H NMR spectroscopy with surface coils. J Magn Reson 69:403-409, 1986
15. Bottomley PA: Noninvasive study of high-energy phosphate metabolism in human heart by depth-resolved ^{31}P NMR spectroscopy. Science 229:769-772, 1985
16. Bottomley PA, Foster TB, Darrow RD: Depth resolved surface-coil spectroscopy (DRESS) for in vivo ^{1}H, ^{31}P and ^{13}C NMR. J Magn Reson 59:338-347, 1984
17. Bottomley PA, Smith LS, Brazzamano S, Hedlund LW, Redington RW, Herfkens RJ: The fate of inorganic phosphate and pH in regional myocardial ischemia and infarction: a noninvasive ^{31}P NMR study. Magn Reson Med 5:129-142, 1987

18. Brainard JR, Hoekenga DE, Hutson JY: Metabolic consequences of anoxia in the isolated, perfused guinea pig heart: anaerobic metabolism of endogenous amino acids. Magn Reson Med 3:637-684, 1986

19. Brindle KM, Rajagopalan B, Bolas NM, Radda GK: Editing of ^{31}P NMR spectra of heart in vivo. J Magn Reson 74:356-365, 1987

20. Brindle KM, Radda GK: ^{31}P NMR saturation transfer measurements of exchange between P$_i$ and ATP in the reactions catalysed by glyceraldehyde-3-phosphate dehydrogenase and phosphoglycerate kinase in vitro. Biochim Biophys Acta 928:45-55, 1987

21. Burstein D, Fossel ET: Nuclear magnetic resonance studies of intracellular ions in perfused frog heart. Am J Physiol 252:H1138-H1146, 1987

22. Chance B, Clark BJ, Nioka S, Subramanian H, Maris JM, Argov Z, Bode H: Phosphorus nuclear magnetic resonance spectroscopy in vivo. Circulation 72, Suppl IV:103-110, 1985

23. Chance B, Leigh JS, Kent J, McCully K, Nioka A, Clark BJ, Maris JM, Graham T: Multiple controls of oxidative metabolism in living tissues as studied by phosphorus magnetic resonance. Proc Natl Acad Sci (USA) 83:9458-9462, 1986

24. Chance B, Williams GR: Respiratory enzymes in oxidative phosphorylation. Kinetics of oxygen utilization. J Biol Chem 217:383-393, 1955

25. Chance EM, Seeholzer SH, Kobayashi K, Williamson JR: Mathematical analysis of isotope labeling in the citric acid cycle with applications to ^{13}C NMR studies in perfused hearts. J Biol Chem 258:13785-13794, 1983

26. Degani H, Laughlin M, Campbell S, Shulman RG: Kinetics of creatine kinase in heart: a ^{31}P NMR saturation and inversion-transfer study. Biochemistry 24:5510-5516, 1985

27. Dhasmana JP, Digerness SB, Geckle JM, Ng TC, Glickson JD, Blackston EH: Effect of adenosine deanimase inhibitors on the heart's functional and biochemical recovery from ischemia: a study utilizing the isolated rat heart adapted to ^{31}P nuclear magnetic resonance. J Cardiovasc Pharmacol 5:1040-1047, 1983

28. Flaherty JT, Jaffin JH, Magovern GJ, Kanter KR, Gardner TJ, Miceli MV, Jacobus WE: Maintenance of aerobic metabolism during global ischemia with perfluorocarbon cardioplegia improves myocardial preservation. Circulation 69:585-592, 1984

29. Flaherty JT, Weisfeldt ML, Bulkley BH, Gardner TJ, Gott VL, Jacobus WE: Mechanisms of ischemic myocardial cell damage assessed by phosphorus-31 nuclear magnetic resonance. Circulation 65:561-571, 1982

30. Forsèn S, Hoffman RA: Study of moderately rapid chemical exchange reactions by means of nuclear magnetic double resonance. J Chem Phys 39:2892-2901, 1963

31. Fossel ET, Hoefeler H: Observation of intracellular potassium and sodium in the heart by NMR: a major fraction of potassium is invisible. Magn Reson Med 3:534-540, 1986

32. Fossel ET, Hoefeler H: Complete inhibition of creatine kinase in isolated perfused rat hearts. Am J Physiol 252:E124-E130, 1987

33. Fossel ET, Morgan HE, Ingwall JS: Measurement of changes in high energy phosphates in the cardiac cycle by using gated ^{31}P nuclear magnetic resonance. Proc Natl Acad Sci (USA) 77:3654-3658, 1980

34. Freeman R, Hill HDW. Fourier transform study of NMR spin-lattice relaxation by "progressive saturation". J Chem Phys 54:3367-3377, 1971

35. From AHL, Petein M, Michurski SP, Zimmer SD, Ugurbil K: ^{31}P NMR studies of respiratory regulation in the intact myocardium. FEBS Lett 206:257-261, 1986

36. Gadian DG, Hoult DI, Radda GK, Seeley PJ, Chance B, Barlow C: Phosphorus nuclear magnetic resonance studies on normoxic and ischemic cardiac tissue. Proc Natl Acad Sci (USA) 73:291-332, 1976

37. Gard JK, Kichura GM, Ackerman JJH, Eisenberg JD, Billadello JJ, Sobel BE, Gross RW: Quantitative ^{31}P nuclear magnetic resonance analysis of metabolite concentrations in Langendorff-perfused rabbit hearts. Biophys J 48:803-813, 1985

38. Garlick PB, Radda GK, Seeley PJ: Studies of acidosis in the ischaemic heart by phosphorus nuclear magnetic resonance. Biochem J 184:547-554, 1979

39. Garlick PB, Radda GK, Seeley PJ, Chance B: Phosphorus NMR studies on perfused heart. Biochem Biophys Res Commun 74:1256-1262, 1977

40. Gevers W: Generation of protons by metabolic processes in heart cells. J Mol Cell Cardiol 9:867-874, 1977

41. Grove TH, Ackerman JJH, Radda GK, Bore PJ: Analysis of rat heart in vivo by phosphorus nuclear magnetic resonance. Proc Natl Acad Sci (USA) 77:299-302, 1980

42. Gupta RJ, Moore RD: ^{31}P NMR studies of intracellular free Mg^{2+} in intact frog skeletal muscle. J Biol Chem 255:3987-3993, 1980

43. Guth BD, Martin JF, Heusch G, Ross J: Regional myocardial blood flow, function and metabolism using phosphorus-31 nuclear magnetic resonance spectroscopy during ischemia and reperfusion in dogs. J Am Coll Cardiol 10:673-681, 1987

44. Hetherington HP, Avison MJ, Shulman RG: ^{1}H homonuclear editing of rat brain using semiselective pulses. Proc Natl Acad Sci (USA) 82:3115-3118, 1985

45. Hollis DP, Nunnally RL, Jacobus WE, Taylor GJ: Detection of regional ischemia in perfused beating hearts by phosphorus nuclear magnetic resonance. Biochem Biophys Res Commun 75:1086-1091, 1977

46. Hollis DP, Nunnally RL, Taylor GJ, Weisfeldt ML, Jacobus WE: Phosphorus nuclear magnetic resonance studies of heart physiology. J Magn Reson 29:319-330, 1978

47. Hore P: Solvent suppression in Fourier transform nuclear magnetic resonance. J Magn Reson 55:283-300, 1983

48. Hoult DI, Busby SJW, Gadian DG, Radda GK, Richards RE, Seeley PJ: Observation of tissue metabolites using ^{31}P nuclear magnetic resonance. Nature 252:285:287, 1974

49. Ingwall JS: Phosphorus nuclear magnetic resonance spectroscopy of cardiac and skeletal muscles. Am J Physiol 242:H729-744, 1982

50. Jacobus WE, Pores IH, Lucas SK, Weisfeldt ML, Flaherty JT: Intracellular acidosis and contractility in the normal and ischemic heart as examined by ^{31}P NMR. J Mol Cell Cardiol 14, Suppl 3:13-20, 1982

51. Jacobus WE, Taylor G, Hollis DP, Nunnally RL: Phosphorus nuclear magnetic resonance of perfused working rat hearts. Nature 265:756-758, 1977

52. Kammermeier H, Schmidt P, Jüngling E: Free energy change of ATP hydrolysis: a causal factor of early hypoxic failure of the myocardium? J Mol Cell Cardiol 14:267-277, 1982

53. Kantor HL, Briggs RW, Balaban RS: In vivo ^{31}P nuclear magnetic resonance measurements in canine heart using a catheter coil. Circ Res 55:261-266, 1984

54. Kantor HL, Briggs RW, Metz KR, Balaban RS: Gated in vivo examination of cardiac metabolites with ^{31}P nuclear magnetic resonance. Am J Physiol 251:H171-H175, 1986

55. Katz LA, Koretsky AP, Balaban RS: Respiratory control in the glucose perfused heart. A ^{31}P NMR and NADH fluorescence study. FEBS Lett 221:270-276, 1987

56. Kingsley-Hickman P, Sako EY, Andreone PA, St. Cyr JA, Michurski S, Foker JE, From AHL, Petein M, Ugurbil K: ^{31}P NMR measurement of ATP synthesis rate in perfused intact rat hearts. FEBS Lett 198:159-163, 1986

57. Kingsley-Hickman PB, Sako EY, Mohanakrishnan P, Robitaille PML, From AHL, Foker JE, Ugurbil K: ^{31}P NMR studies of ATP synthesis and hydrolysis kinetics in the intact myocardium. Biochemistry 1988, in press

58. Kirkels JH, Ruigrok TJC, Van Echteld CJA, Meijler FL: Protective effect of pretreatment with anipamil on the ischemic-reperfused rat myocardium; a phosphorus-31 nuclear magnetic resonance study. J Am Coll Cardiol 1988, in press

59. Kobayashi K, Neely JR: Control of maximum rates of glycolysis in rat cardiac muscle. Circ Res 44:166-175, 1979

60. Koretsky AP, Wang S, Klein MP, James TL, Weiner MW: ^{31}P NMR saturation transfer measurements of phosphorus exchange reactions in rat heart and kidney in situ. Biochemistry 25:77-84, 1986

61. Koretsky AP, Wang S, Murphy-Boesch J, Klein MP, James TL, Weiner MW: ^{31}P NMR spectroscopy of rat organs, in situ, using chronically implanted radiofrequency coils. Proc Natl Acad Sci (USA) 80:7491-7495, 1983

62. Kupriyanov VV, Lakomkin VL, Kapelko VI, Steinschneider AY, Ruuge EK, Saks VA: Dissociation of adenosine triphosphate levels and contractile function in isovolumic hearts perfused with 2-deoxyglucose. J Mol Cell Cardiol 19:729-740, 1987

63. Kupriyanov VV, Steinschneider AY, Ruuge EK, Kapel'ko VI, Zuera MY, Lakomkin VL, Smirnov VN, Saks VA: Regulation of energy flux through the creatine kinase reaction in vitro and in perfused rat heart. Biochim Biophys Acta 805:319-331, 1984

64. Kusuoka H, Weisfeldt ML, Zweier JL, Jacobus WE, Marban E: Mechanism of early contractile

failure during hypoxia in intact ferret heart: evidence for modulation of maximal Ca^{2+}-activated force by inorganic phosphate. Circ Res 59:270-282, 1986

65. Lavanchy N, Martin J, Giacomelli M, Rossi A: Evaluation by ^{31}P NMR of the effects of acebutolol on the ischaemic isolated rat heart. Eur J Pharmacol 125:341-351, 1986

66. Lavanchy N, Martin J, Rossi A: Glycogen metabolism: a ^{13}C NMR study on the isolated perfused rat heart. FEBS Lett 178:34-38, 1984

67. Lavanchy N, Martin J, Rossi A: Effects of diltiazem on the energy metabolism of the isolated rat heart submitted to ischemia: a ^{31}P NMR study. J Mol Cell Cardiol 18:931-941, 1987

68. Lawson IWR, Veech RL: Effects of pH and free Mg^{2+} on the K_{eq} of the creatine kinase reaction and other phosphate hydrolyses and phosphate transfer reactions. J Biol Chem 254:6528-6537, 1979

69. Lee YCP, Visscher MB: Perfusate cations and contracture and Ca, Cr, PCr and ATP in rabbit myocardium. Am J Physiol 219:1637-1641, 1970

70. Ligeti L, Osbakken MD, Clark BJ, Schnall M, Bolinger L, Subramanian H, Leigh JS, Chance B: Cardiac transfer function relating energy metabolism to workload in different species as studied with ^{31}P NMR. Magn Reson Med 4:112-119, 1987

71. Malloy CR, Matthews PM, Smith MB, Radda GK: Influence of propranolol on acidosis and high energy phosphates in ischaemic myocardium of the rabbit. Cardiovasc Res 20:710-720, 1986

72. Malloy CR, Sherry AD, Jeffrey FMH: Carbon flux through citric acid cycle pathways in perfused heart by ^{13}C NMR spectroscopy. FEBS Lett 212:58-62, 1987

73. Marban E, Kitakaze M, Kusuoka H, Porterfield JK, Yue DT, Chacko VP: Intracellular free calcium concentrations measured with ^{19}F NMR spectroscopy in intact ferret hearts. Proc Natl Acid Sci (USA) 84:6005-6009, 1987

74. Marban E, Kusuoka H, Yue DT, Weisfeldt ML, Wier WG: Maximal Ca^{2+}-activated force elicited by tetanization of ferret papillary muscle and whole heart: mechanism and characteristics of steady contractile activation in intact myocardium. Circ Res 59:262-269, 1986

75. Massie BM, Conway M, Yonge R, Frostick S, Sleight P, Ledingham J, Radda G, Rajagopalan B: ^{31}P nuclear magnetic resonance evidence of abnormal skeletal muscle metabolism in patients with congestive heart failure. Am J Cardiol 60:309-315, 1987

76. Matthews PM, Bland JL, Gadian DG, Radda GK: The steady state rate of ATP synthesis in the perfused rat heart measured by ^{31}P NMR saturation transfer. Biochem Biophys Res Commun 103:1052-1059, 1981

77. Matthews PM, Bland JL, Gadian DG, Radda GK: A ^{31}P NMR saturation transfer study of the regulation of creatine kinase in the rat heart. Biochim Biophys Acta 721:312-320, 1982

78. Matthews PM, Taylor DJ, Radda GK: Biochemical mechanisms of acute contractile failure in the hypoxic rat heart. Cardiovasc Res 20:13-19, 1986

79. Matthews PM, Williams SR, Seymour A-M, Schwartz A, Dube G, Gadian DG, Radda GK: A ^{31}P NMR study of some metabolic and functional effects of the inotropic agents epinephrine and ouabain, and the ionophore R02-2985 (X573A) in the isolated, perfused rat heart. Biochim Biophys Acta 720:163-171, 1982

80. Moon RB, Richards JH: Determination of intracellular pH by 31P magnetic resonance. J Biol Chem 248:7276-7278, 1973

81. Nakazawa M, Katano Y, Imai S, Matsushita K, Ohuchi M: Effects of l- and d-propranolol on the ischemic myocardial metabolism of the isolated guinea pig heart, as studied by ^{31}P NMR. J Cardiovasc Pharmacol 4:700-704, 1982

82. Neely JR, Grotyohann LW: Role of glycolytic products in damage to ischemic myocardium. Dissociation of adenosine triphosphate levels and recovery of function of reperfused ischemic hearts. Circ Res 55:816-824, 1984

83. Neurohr KJ, Barrett EJ, Shulman RG: In vivo carbon-13 nuclear magnetic resonance studies of heart metabolism. Proc Natl Acad Sci (USA) 80:1603-1607, 1983

84. Neurohr KJ, Gollin G, Neurohr JM, Rothman DL, Shulman RG: Carbon-13 nuclear magnetic resonance studies of myocardial glycogen metabolism in live guinea pigs. Biochemistry 23:5029-5035, 1984

85. Nunnally RL, Bottomley PA: Assessment of pharmacological treatment of myocardial infarction by phosphorus-31 NMR with surface coils. Science 211:177-180, 1981

86. Nunnally RL, Hollis DP: Adenosine triphosphate compartmentation in living hearts: a phosphorus nuclear magnetic resonance saturation transfer study. Biochemistry 18:3642-3646, 1979

87. Ordidge RJ, Connelly A, Lohman JAB: Image-selected in vivo spectroscopy (ISIS). A new technique for spatially selective NMR spectroscopy. J Magn Reson 66:283-294, 1986
88. Pieper GM, Todd GL, Wu ST, Salhany JM, Clayton FC, Eliot RS: Attenuation of myocardial acidosis by propranolol during ischaemic arrest and reperfusion: evidence with ^{31}P nuclear magnetic resonance. Cardiovasc Res 14:646-653, 1980
89. Pieper GM, Wu ST, Salhany JM: A polymeric prostaglandin (PGB) attenuates adenine nucleotide loss during global ischemia and improves myocardial function during reperfusion. J Mol Cell Cardiol 17:775-783, 1985
90. Pike MM, Frazer JC, Dedrick DF, Ingwall JS, Allen PH, Springer CS, Smith TW: ^{23}Na and ^{39}K NMR studies of perfused rat hearts: discrimination of intra- and extracellular ions using a shift reagent. Biophys J 48:159-173, 1985
91. Richards TL, Terrier F, Sievers RE, Lipton MJ, Moseley ME, Higgins CB: Lactate accumulation in ischemic- and anoxic-isolated rat hearts assessed by H-1 spectroscopy. Invest Radiol 22:638-641, 1987
92. Ruigrok TJC, Van Echteld CJA, De Kruijff B, Borst C, Meijler FL: Protective effect of nifedipine in myocardial ischemia assessed by phosphorus-31 nuclear magnetic resonance. Eur Heart J 4, Suppl C:109-113, 1983
93. Saito T, Tomita K: Biochem J (Tokyo) 72:807-815, 1972
94. Seymour A-ML, Bailey IA, Radda GK: A protective effect of insulin on reperfusing the ischaemic rat heart shown using ^{31}P NMR. Biochim Biophys Acta 762:525-530, 1983
95. Sherry AD, Nunnally RL, Peshock RM: Metabolic studies of pyruvate- and lactate-perfused guinea-pig hearts by ^{13}C NMR. J Biol Chem 260:9272-9279, 1985
96. Shoubridge EA, Jeffry FMH, Keogh JM, Radda GK, Seymour A-ML: Creatine kinase kinetics, ATP turnover and cardiac performance in hearts depleted of creatine with the substrate analogue beta-guanidinopropionic acid. Biochim Biophys Acta 847:25-32, 1985
97. Sillerud LO, Shulman RG: Structure and metabolism of mammalian liver glycogen monitored by carbon-13 nuclear magnetic resonance. Biochemistry 22:1087-1094, 1983
98. Smith GA, Hesketh RT, Metcalfe JC, Feeney J, Morris PG: Intracellular calcium measurements by ^{19}F NMR of fluorine-labeled chelators. Proc Natl Acad Sci (USA) 80:7178-7182, 1983
99. Steenbergen C, Murphy E, Levy L, London RE: Elevation in cytosolic free calcium concentration early in myocardial ischemia in perfused rat heart. Circ Res 60:700-707, 1987
100. Stein PD, Goldstein S, Sabbah HN, Liu ZQ, Helpern JA, Ewing JR, Lakier JB, Chopp M, LaPenna WF, Welch KMA: In vivo evaluation of intracellular pH and high-energy phosphate metabolites during regional myocardial ischemia in cats using ^{31}P nuclear magnetic resonance. Magn Reson Med 3:262-269, 1986
101. Ugurbil K, Petein M, Maidan R, Michurski S, Cohn JN, From AH: High resolution proton NMR studies of perfused hearts. FEBS Lett 167:73-78, 1984
102. Ugurbil K, Petein M, Maidan R, Michurski S, From AHL: Measurement of an individual rate constant in the presence of multiple exchanges: application to myocardial creatine kinase reaction. Biochemistry 25:100-107, 1986
103. Whitman GJR, Chance B, Bode H, Maris J, Haselgrove J, Kelly R, Clark BJ, Harken AH: Diagnosis and therapeutic evaluation of a pediatric case of cardiomyopathy using phosphorus-31 nuclear magnetic resonance spectroscopy. J Am Coll Cardiol 5:745-749, 1985
104. Wiener DH, Fink LI, Maris J, Jones RA, Chance B, Wilson JR: Abnormal skeletal muscle bioenergetics during exercise in patients with heart failure: role of reduced muscle blood flow. Circulation 73:1127-1136, 1986
105. Williams SR, Gadian DG, Proctor E, Sprague DB, Talbot DF, Young IR, Brown F: Proton NMR studies of muscle metabolites in vivo. J Magn Reson 63:406-412, 1985
106. Zweier JL, Jacobus WE: Substrate-induced alterations of high energy phosphate metabolism and contractile function in the perfused heart. J Biol Chem 262:8015-8021, 1987

Chapter 13

Positron Emission Tomography and Myocardial Biochemistry

W. Wyns and J.A. Melin, Positron Emission Tomography Laboratory,
University of Louvain Medical School, Brussels, Belgium

Positron emission tomography (PET) is a new method for the external quantification of regional myocardial blood flow, substrate fluxes and biochemical reaction rates. It takes advantage of physiologic tracers, tracer kinetic principles and the quantitative imaging capabilities of the tomograph. To date, the technique has been used primarily for the study of ischemic heart disease and cardiomyopathies. This report summarizes experimental and clinical investigations on substrate metabolism during the process of ischemic injury. Uptake of C-11 palmitate and F-18 2-fluoro-2-deoxyglucose appear to be well-suited markers for the study of myocardial fatty acid and glucose metabolism, respectively. Observations with PET have shown noninvasively in vivo the known impairment of fatty acid oxidation and the enhanced glycolytic flux during ischemia. Some distinct findings unexpected from existing biochemical knowledge will be highlighted.

Introduction

Positron emission tomography (PET) offers the opportunity for studying noninvasively regional myocardial tissue function. The technique owes its unique capabilities to several features which will be briefly outlined. For more detail on physics and instrumentation, the reader is referred to excellent recent reviews (16,29). This imaging device relies on the emission (in diametrically opposed directions) of two 511 keV photons during positron annihilation. Registration and localization of the annihilation event by a ring of external detectors occur through coincidence detection which allows electronic collimation and correction for photon attenuation. The cross-sectional images produced by the tomograph after appropriate reconstruction reflect a quantitative measurement of regional radioactivity.

The physiologic tracers used with PET are compounds labeled with positron-emitting isotopes of elements that are abundantly present in vivo, such as carbon-11, fluorine-18, nitrogen-13 or oxygen-15. Labeled substrates, substrate analogs and drugs can be synthesized in high specific activity without disturbing their biological properties. The local tissue concentrations of the tracers are in the pico- to nanomolar range and, therefore, they do not interfere with the process to be studied.

Combining these physiologic tracers with the quantitative capabilities of PET allows one to image the distribution, uptake and turnover of tracer in the myocardium. If a kinetic model correctly describing the tissue kinetics of the tracer is available, quantitative measurements of regional blood flow in ml/min/g tissue or myocardial substrates fluxes in mmol/min/g tissue can be obtained noninvasively. A successful example is the application to the myocardium of the operational equation initially developed to measure exogenous glucose utilization rates in the brain with the glucose analog F-18 2-fluoro-2-deoxyglucose or FDG (30,48). This agent traces the transmembraneous exchange of glucose and the hexokinase-mediated phosphorylation to glucose-6-phosphate. After phosphorylation to FDG-6-phosphate, the tracer becomes trapped in the cell. The kinetics of FDG in tissue are described by a three-compartment model, validated in isolated perfused rabbit myocardium (20), in canine hearts (31), and more recently in humans (42). As shown in Fig. 1, the estimates of exogenous glucose utilization determined externally with PET and the FDG model agreed well with those calculated by the Fick method (31).

Thus, PET appears potentially useful as an in vivo biochemical assay technique that will permit examination and validation of functional processes that have been described primarily with in vitro experiments. Although there is continuing interest in receptor binding studies (49,50) and quantification of regional myocardial blood flow (2,12,35,46), we will limit this review to the most intriguing application of PET which in our view relates to the study of myocardial substrate metabolism. The reader interested in the entire scope of PET studies of the heart is referred to recent detailed monographs (1,34,37).

Normal Alterations in Substrate Metabolism Demonstrated by PET

Unlike other organs, such as the brain, the heart meets its energy demands oxidizing several substrates (17,18). Selection of fatty acids (FA), glucose, lactic acid, ketone bodies and amino acids for oxidation depends largely upon their arterial concentrations which in turn are regulated by the dietary state, by physical activity and by hormones. For example, plasma FA levels are high in the fasting state (0.9 mM). Oxidation of FA therefore accounts for about half of the myocardial oxygen consumption. Arterial lactate levels often rise from 0.9 mM at rest up to 9.5 mM during exercise; lactate then becomes the major substrate for oxidation. Uptake and utilization rates of various substrates rapidly respond to changing demand in order to meet instantaneously changes in energy requirements of the cardiac muscle.

The heart's ability to increase substrate oxidation in response to greater demand and to resort to alternate substrates have been demonstrated noninvasively in vivo with PET. Fasted subjects were studied first at control and then again during states

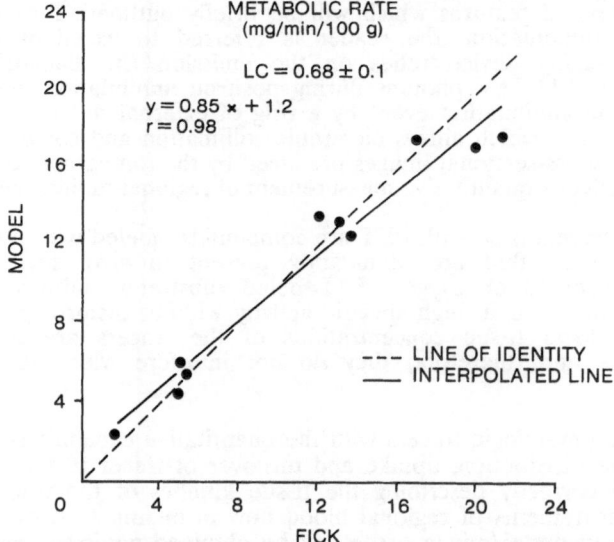

Fig. 1. Utilization of exogenous glucose by canine myocardium measured by the Fick method and externally with the fluorodeoxyglucose (FDG) model and PET. LC = lumped constant. From Ratib et al. (31), with permission.

of increased cardiac work induced by rapid atrial pacing while plasma substrate concentrations remained unchanged (14). Compared to control, the expected increase in FA oxidation was observed using serial PET imaging and C-11 palmitic acid (CPA) as a tracer for FA metabolism. Although a tracer kinetic model for CPA is not available, its tissue kinetics have been extensively documented (21,22,38,39). Normal myocardium avidly extracts this long chain FA (16 carbon); the first transit retention fraction averages 65% (38,56). Maximal myocardial C-11 activity tissue concentrations are attained within 3 to 4 min. Thereafter, the activity typically clears from normal myocardium in a biexponential fashion, indicating fractional distribution of the tracer between two major pools of different sizes and with different turnover rates (Fig. 2). This is consistent with the known dual fate of FA (19,53): the late slow turnover pool is related to the endogenous lipid pool, consisting mainly of triglycerides. The early rapid phase of the time - activity curve relates to the release of C-11 carbon dioxide which is the end product of CPA oxidation (56). In the study by Henze et al. (14), during pacing at rates above 100 beats/min, the relative size of the early rapid phase increased while the relative size of the late slow phase declined. Similarly, the steepness of the slope of the early rapid phase increased. Thus, when demand was high, more CPA entered oxidative pathways and was oxidized more rapidly.

The second entity demonstrated in vivo by PET studies is the heart's normal ability to alternate between substrates in response to changes in substrate concentrations in plasma. With combined use of tracers of glucose (FDG) and FA (CPA) metabolism, the UCLA group has documented the effect of substrate availability on substrate selection by normal myocardium. This was shown initially in open-chest dogs studied first after an overnight fast and then again after glucose/insulin infusion (32). Despite comparable blood flows and myocardial oxygen consumption, the fractional distribution of CPA in cardiac muscle changed in response to altered plasma substrate concentration. For example, the relative size of the early rapid clearance phase and its clearance rate declined or this curve component disappeared entirely when plasma FA levels were lowered and glucose and lactic acid concentrations increased by a factor two (14). This was accompanied by lesser amounts of C-11 carbon dioxide released from myocardium, indicating that less CPA oxidized. Observations in human myocardium, though still small in number, are similar (14) and consistent with the notion that as FA availability declines, the myocardium switches to alternate substrates (e.g., glucose and lactic acid).

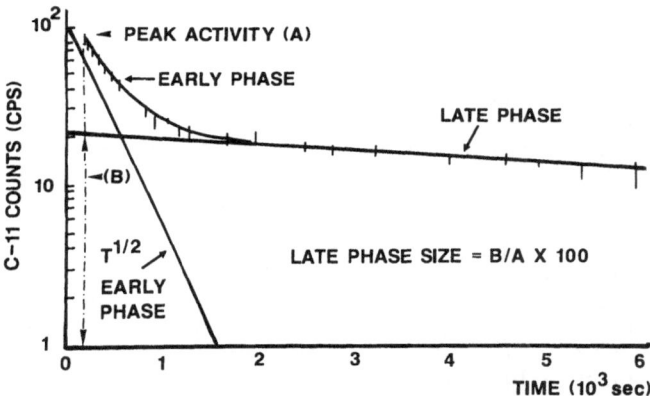

Fig. 2. C-11 time activity curve derived from serial PET images after intravenous C-11 palmitic acid (CPA) injection. Note the peak activity (A) and the subsequent biexponential tissue clearance. The late phase is extrapolated back to the time (A) and from the intercept B, the fraction of CPA entering the early and late phase can be estimated. Biexponential fitting also provides the halftime (T 1/2) of each curve component.

Metabolic Studies in Myocardial Ischemia and Postischemic Dysfunction

The sequence of metabolic events initiated by ischemia has been extensively studied in invasive animal models or in vitro experimental systems (23,27,28). It appears that the amount of residual blood flow determines whether residual metabolic activity is maintained or whether ischemia will proceed to irreversible tissue injury. As oxygen availability decreases, the flux of FA through the beta-oxidative spiral falls. This provides substrate for esterification of FA with alpha-glycerophosphate resulting in increased triglyceride synthesis. Thus, the onset of ischemia is usually associated with enhanced glycolysis. The glucose is derived from endogenous (i.e., glycogen) and/or from exogenous sources. As ischemia progresses and tricarboxylic acid cycle activity decreases, pyruvate is converted to lactate and released from myocardium. Anaerobic glycolysis remains then the only source of ATP. The rate of lactic production and removal from the cell determines how long and at what rate glycolysis can be maintained.

Because beta-oxidation is most sensitive to oxygen deprivation, initial studies with PET used CPA as a tracer for identifying ischemia.

In animal experiments, both low-flow and high-demand ischemia affected the fractional distribution of CPA in tissue (22,33,39). The fraction immediately entering oxidative pathways decreased, its clearance rate from myocardium declined and a greater fraction was deposited in the endogenous lipid pool. These changes were seen as a consequence of an impaired transfer of palmitoyl-CoA into mitochondria and of greater "trapping" of CPA in the triglyceride pool. Also, in patients with high-demand ischemia induced by moderate atrial pacing, abnormal kinetics of CPA were observed in segments supplied by stenosed coronary arteries (13). The expected increase in tracer entering the rapid curve component and its clearance rate observed in normal myocardium, was attenuated in segments subtended by stenosed coronary arteries. Interestingly, these metabolic abnormalities occurred without electrocardiographic, or in some instances, functional (by two-dimensional echocardiography) correlates of acute myocardial ischemia.

These studies show the ability of the CPA approach to externally detect expected changes in FA metabolism as well as the sensitivity of the technique. Yet, some limitations should be recognized.

First of all, as discussed earlier, in the absence of an appropriate tracer kinetic model, studies with CPA remain largely qualitative.

Second, the initial uptake of CPA depends on the delivery, i.e., on blood flow. Thereafter, the activity clears rapidly from myocardium. This limits the usefulness of this tracer in severe ischemia because regional blood flow is then markedly reduced. Limited uptake of the tracer will preclude derivation of statistically adequate data, needed for construction of reliable tissue time - activity curves.

Third, Fox et al. (10) have shown that during severe ischemia and hypoxia, about 40 to 50% of the C-11 activity released from myocardium into the coronary sinus is represented by unaltered non-metabolized CPA. Back-diffusion competes with metabolic sequestration of the tracer, suggesting a block of the energy-requiring activation step of CPA to palmitoyl-CoA. This factor will confound the use of the tissue clearance-rate of C-11 activity during the early rapid phase of CPA kinetics.

These limitations prompted many investigators to resort to FDG for the detection of exogenous glucose utilization which is enhanced in ischemia and which presents itself on the cross-sectional image as a positive signal.

Increases in FDG uptake were demonstrated in the isolated arterially perfused rabbit septum (24) as well as noninvasively in dogs with coronary stenoses and rapid atrial pacing (36). Findings in patients are similar. Camici et al. (6) studied patients with known coronary disease during and after exercise induced ischemia. FDG uptake was preserved in myocardial segments that were inadequately perfused during exercise. When injected after exercise, absolute FDG uptake was increased (compared to normal segments). The authors suggested that glucose utilization remained high in postischemic myocardium as a consequence of restoration of glycogen stores (7).

Following on this observation, several groups examined the recovery of metabolism from transient ischemic insult (11,45,55). Short periods of transient ischemia such as, in dogs, a complete coronary occlusion for 20 min or a critical stenosis for one hour resulted in prolonged recovery of FA metabolism despite return to baseline of myocardial blood flow (Figs. 3 and 4). Interestingly, the recovery of metabolism paralleled the slow recovery of regional function, a concept frequently referred to as "myocardial stunning" (3,15). Recent observations in postischemic myocardium using the tissue clearance of C-11 acetate as a marker for tricarboxylic cycle activity showed impaired FA oxidation in segments with nearly normal C-11 acetate kinetics (4,5). These findings in stunned myocardium suggest not only that repair of the biochemical machinery after ischemic damage takes time but also that different metabolic pathways may recover at different rates. Why glucose utilization remains predominant at the expense of FA oxidation during reflow is unclear but may reflect a prolonged "stunning" of beta-oxidation or the residual inhibitory influence of FA intermediates.

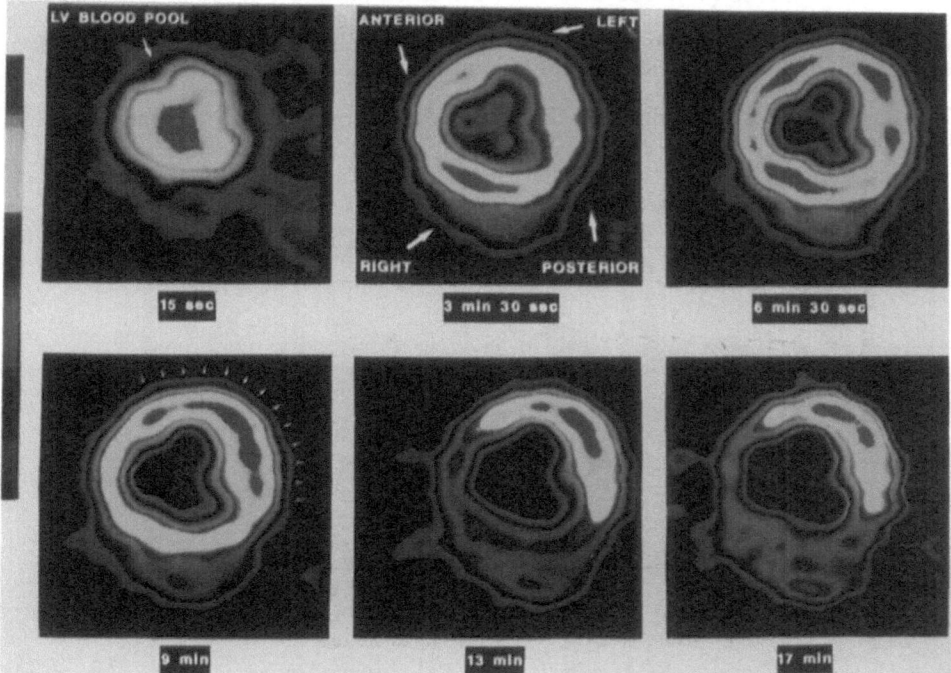

Fig. 3. PET imaging with C-11 palmitic acid (CPA) in a dog 24 h after a transient ischemic insult (1 h critical circumflex stenosis reducing transmural blood flow to 40% of control). Note the delayed clearance of activity from the lateral and posterior walls (arrows). The time of acquisition after intravenous CPA injection is indicated under each image.

Myocardial Infarction and Reperfusion: Metabolism as an Indicator of Tissue Viability

Estimates of size and mass of infarcted myocardium were obtained with PET and CPA in dog experiments (54) and in patients with chronic infarction (47,51). Tomographic estimates of infarct size by planimetry of the segmental defect in CPA uptake correlated well (r = 0.92) with enzymatic estimates by serial creatine kinase serum sampling.

More recently, the widespread use of thrombolytic agents for restoring blood flow in evolving myocardial infarction raised the interest in the metabolic fate of reperfused tissue. Early identification of reversible versus irreversible injury is critical for patient management.

In a canine model of reperfused infarction, ischemic but viable myocardium was characterized by a persistent glucose uptake which was absent in necrotic tissue, as shown in Fig. 5 (26). Schwaiger et al. (44) examined the recovery of metabolic abnormalities over a 4-wk period in dogs after a 3-h occlusion of the left anterior descending coronary artery followed by reperfusion. Enhanced glucose utilization and impaired fatty acid oxidation persisted for hours to days. Again, as these abnormalities disappeared, segmental function improved. In order to elucidate the mechanisms of increased glucose utilization in reperfused myocardium, the relative contribution of oxidative and non-oxidative glucose metabolism was studied by arterio-venous sampling for C-14 carbon dioxide and C-14 lactate after infusion of C-14 labeled glucose (43). The data show increased glucose utilization of which 75% entered glycolysis, most of it being released as lactate. This suggests that the energy produced from non-oxidative glucose metabolism is sufficient to maintain cellular viability but insufficient to allow a normal mechanical function.

Encouraged by these experimental findings, several groups started to submit patients with recent myocardial infarction to PET studies (8,9,25,40,41).

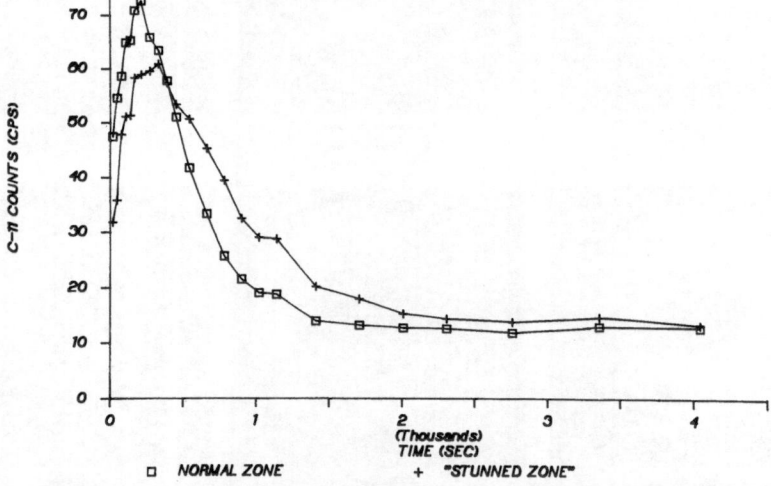

Fig. 4. Uptake and clearance of C-11 palmitic acid (CPA) in stunned myocardium, 24 h after critical stenosis of dog circumflex. The two curves were obtained from the study shown in Fig. 3. As compared with the normal septal and anterior walls, clearance of C-11 activity is delayed in the posterior and lateral walls. The halftime of the early phase is increased from 4.2 min in the normal zone to 7.2 min in the stunned zone. The tracer fraction entering the late phase is increased to 21% in the stunned as compared to 16% in the normal zone.

Marshall et al. (25) compared regional glucose utilization and blood flow in 15 patients studied an average of two weeks after myocardial infarction. Different metabolic patterns were defined. In normal myocardium, blood flow and FDG uptake match closely. A marker of blood flow (i.e., N-13 ammonia) is used to identify the abnormal segment, and FDG to distinguish between presence and absence of metabolism in segments with reduced blood flow. In infarcted segments, flow and metabolism are concordingly decreased. By contrast, ischemic myocardium is characterized by a "mismatch", i.e., an increase in exogenous glucose utilization relative to blood flow. Segments with "mismatches" were present mainly in patients with post-infarction angina and more extensive disease. They corresponded in location to the sites of electrocardiographic changes and wall motion abnormalities occurring during anginal episodes. Consequently, 12 patients were studied <72 h after onset of symptoms (40,41). The same "mismatch" pattern was observed in about half of the segments indicating the presence of injured but viable myocardium in the infarction area for prolonged periods after the acute event.

These are important findings as PET could help in selecting patients for more aggressive therapeutic approaches. Reperfusion of an area of scar is not likely to be of much benefit; revascularization of ischemic tissue should improve segmental left ventricular function, as demonstrated recently in a study on the effects of coronary artery bypass grafting (52).

Conclusions

This review attempts to describe some of the features of PET as a tool for probing myocardial regional biochemistry. Observations by PET made in human heart are consistent with the knowledge on biochemistry derived from experimental in vitro systems. For the first time, PET offers the opportunity to test the validity of this knowledge directly in the living human heart. On the other hand, observations in patients showed some distinct differences to experimental findings. For example, augmented glycolytic flux persists in ischemically injured myocardium for longer periods than expected. Also, transient reductions in oxygen delivery to myocardium do promote prolonged impairments in FA utilization despite reoxygenation. The detection of tissue viability appears nowadays as a potentially clinically useful application of PET. However, these clinical investigations have remained largely qualitative and major aspects of myocardial metabolism such as protein synthesis and tricarboxylic acid cycle activity await further investigation.

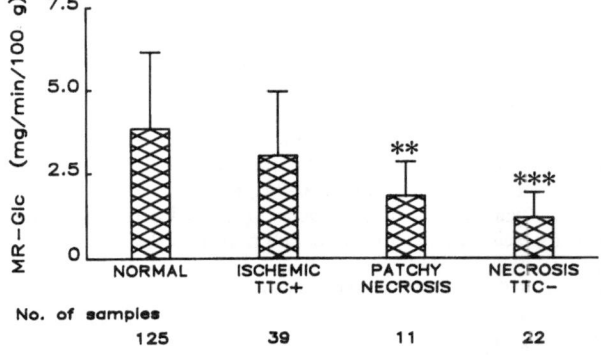

Fig. 5. Myocardial glucose utilization rates (MR-Glc) are calculated by the fluorodeoxyglucose (FDG) model in tissue samples characterized as normal, ischemic but viable, or necrotic by histochemistry, electron microscopy and blood flow. Measurements are performed 4 h after reperfusion in dogs subjected to a 2-h period of coronary occlusion. Depressed glucose uptake indicates irreversible injury while normal FDG uptake is associated with viable tissue (**$p < 0.01$; ***$p < 0.001$). From Melin et al. (26), with permission. TTC = triphenyl tetrazolium chloride.

Acknowledgments

Supported in part by grants from the "Fonds National de la Récherche Scientifique et Médicale" 3.4550-85 and 3.4551-87. The authors appreciate the expert secretarial assistance of Mrs. Ch. Proumen and I. Moschion as well as the continuous support of the Damman Foundation.

The support of Siemens Gammasonics B.V., Uithoorn, The Netherlands is gratefully acknowledged.

References

1. Bergmann SR, Fox KAA, Geltman EM, Sobel BE: Positron emission tomography of the heart. Progr Cardiovasc Dis 28:165-194, 1985
2. Bergmann SR, Fox KAA, Rand AL, McElvany KD, Welch MJ, Markham J, Sobel BE: Quantification of regional myocardial blood flow in vivo with $H_2^{15}O$. Circulation 70:724-733, 1984
3. Braunwald E, Kloner RA: The stunned myocardium: Prolonged, post-ischemic ventricular dysfunction. Circulation 66:1146-1149, 1982
4. Brown M, Marshall DR, Sobel BE, Bergmann SR: Delineation of myocardial oxygen utilization with carbon-11 labeled acetate. Circulation 76:687-696, 1987
5. Brown M, Myears DW, Herrero P, Bergmann SR: Disparity between oxidative and fatty acid metabolism in reperfused myocardium assessed with positron emission tomography. Circulation 76, Suppl IV:IV-4, 1987 (Abstr)
6. Camici P, Araujo LI, Spinks T, Lammertsma AA, Kaski JC, Shea MJ, Selwyn AP, Jones T, Maseri A: Increased uptake of F-18 fluorodeoxyglucose in postischemic myocardium of patients with exercise-induced angina. Circulation 74:81-88, 1986
7. Camici P, Bailey IA: Time course of myocardial glycogen repletion following acute transient ischemia. Circulation 70, Suppl II:II-85, 1984 (Abstr)
8. De Landsheere CM, Raets D, Pierard LA, Berthe C, Legrand V, Crochelet L, Chapelle JP, Lemaire C, Del Fiore G, Guillaume M, Lamotte D, Kulbertus HE, Rigo P: Thrombolysis in anterior myocardial infarction: Effect on regional viability studied with positron emission tomography. Circulation 76, Suppl IV:IV-5, 1987 (Abstr)
9. De Landsheere CM, Raets D, Pierard LA, Materne P, Del Fiore G, Lemaire C, Quaglia L, Guillaume M, Peters JM, Lamotte D, Kulbertus HE, Rigo P: Fibrinolysis and viable myocardium after an acute infarction: A study of regional perfusion and glucose utilization with positron emission tomography. Circulation 72, Suppl III:III-393, 1985 (Abstr)
10. Fox KAA, Abendschein DR, Ambos HD, Sobel BE, Bergmann SR: Efflux of metabolized and nonmetabolized fatty acid from canine myocardium. Implications for quantifying myocardial metabolism tomographically. Circ Res 57:232-243, 1985
11. Fox KAA, Abendschein DR, Sobel BE, Bergmann SR: Detection of persistent metabolic impairment despite reperfusion after transient (20 min) myocardial ischemia. Fed Proc 43:902, 1984 (Abstr)
12. Gould KL, Goldstein RA, Mullani NA, Kirkeeide RL, Wong WH, Tewson TJ, Berridge MS, Bolomey LA, Hartz RK, Smalling RW, Fuentes F, Nishikawa A: Noninvasive assessment of coronary stenoses by myocardial perfusion imaging during pharmacologic coronary vasodilation. VIII. Clinical feasibility of positron cardiac imaging without a cyclotron using generator-produced rubidium-82. J Am Coll Cardiol 7:775-789, 1986
13. Grover-McKay M, Schelbert HR, Schwaiger M, Sochor H, Guzy PM, Krivokapich J, Child JS, Phelps ME: Identification of impaired metabolic reserve by atrial pacing in patients with significant coronary artery stenosis. Circulation 74:281-292, 1986
14. Henze E, Grossman EJ, Huang SC, Barrio JR, Phelps ME, Schelbert HR: Myocardial uptake and clearance of C-11 palmitic acid in man: Effects of substrate availability and cardiac work. J Nucl Med 23:P212, 1982 (Abstr)
15. Heyndrickx GR, Millard RW, McRitchie RJ, Maroko PR, Vatner SF: Regional myocardial functional and electrophysiological alterations after brief coronary artery occlusion in conscious dogs. J Clin Invest 56:978-985, 1975
16. Hoffman EJ, Phelps ME: Positron emission tomography: Principles and quantitation. In: Phelps M, Mazziotta J, Schelbert H, eds: Positron emission tomography and autoradiography: Principles and applications for the brain and heart. New York: Raven Press, 1986:237-286
17. Keul J, Doll E, Steim H, Fleer U, Reindell H: Ueber den Stoffwechsel des menschlichen Herzens. III. Der oxidative Stoffwechsel des menschlichen Herzens unter verschiedenen Arbeitsbedingungen. Pflügers Arch 282:43-53, 1965

18. Keul J, Doll E, Steim H, Homburger H, Kern H, Reindell H: Ueber den Stoffwechsel des menschlichen Herzens. I. Substratversorgung des gesunden Herzens in Ruhe, während und nach körperlicher Arbeit. Pflügers Arch 282:1-27, 1965
19. Klein MS, Goldstein RA, Welch MJ, Sobel BE: External assessment of myocardial metabolism with C-11-palmitate in rabbit hearts. Am J Physiol 237:H51-H58, 1979
20. Krivokapich J, Huang SC, Phelps ME, Barrio JR, Watanabe CR, Selin CE, Shine KJ: Estimation of rabbit myocardial metabolic rate for glucose using fluorodeoxyglucose. Am J Physiol 243:H884-H895, 1982
21. Lerch RA, Ambos HD, Bergmann SR, Welch MJ, Ter-Pogossian MM, Sobel BE: Localization of viable, ischemic myocardium by positron-emission tomography with C-11 palmitate. Circulation 64:689-699, 1981
22. Lerch RA, Bergmann SA, Ambos HD, Welch MJ, Ter-Pogossian MM, Sobel BE: Effect of flow-independent reduction of metabolism on regional myocardial clearance of C-11 palmitate. Circulation 65:731-738, 1982
23. Liedtke AJ: Alterations of carbohydrate and lipid metabolism in the acutely ischemic heart. Progr Cardiovasc Dis 23:321-336, 1981
24. Marshall RC, Nash WW, Shine KI, Phelps ME, Ricchiuti N: Glucose metabolism during ischemia duw to excessive oxygen demand or altered coronary flow in the isolated arterially perfused rabbit septum. Circ Res 49:640-648, 1981
25. Marshall RC, Tillisch JH, Phelps ME, Huang SC, Carson RC, Henze E, Schelbert HR: Identification and differentiation of resting myocardial ischemia and infarction in man with positron computed tomography F-18 labeled fluorodeoxyglucose and N-13 ammonia. Circulation 64:766-778, 1981
26. Melin JA, Wyns W, Keyeux A, Gurné O, Cogneau M, Michel C, Bol A, Robert A, Charlier AA, Pouleur H: Assessment of thallium-201 redistribution versus glucose uptake as predictors of viability after coronary occlusion and reperfusion. Circulation 77:000-000, 1988
27. Myears DW, Sobel BE, Bergmann SR: Substrate use in ischemic and reperfused canine myocardium: quantitative considerations. Am J Physiol 253:H107-H114, 1987
28. Opie LH: Effects of regional ischemia on metabolism of glucose and fatty acids. Circ Res 38, Suppl I:52-68, 1976
29. Phelps ME: Emission computed tomography. Semin Nucl Med 7:337-365, 1977
30. Phelps ME, Huang SC, Hoffmann EJ, Selin C, Sokoloff L, Kuhl DE: Tomographic measurement of local cerebral glucose metabolic rate in humans with [F-18] 2-fluoro-2-deoxy-D-glucose: Validation of method. Ann Neurol 6:371-388, 1979
31. Ratib O, Phelps ME, Huang SC, Henze E, Selin CE, Schelbert HR: Positron tomography with deoxyglucose for estimating local myocardial glucose metabolism. J Nucl Med 23:577-586, 1982
32. Schelbert HR, Henze E, Schon HR, Keen R, Hansen HW, Selin C, Huang SC, Barrio JR, Phelps ME: C-11 palmitate for the noninvasive evaluation of regional myocardial fatty acid metabolism with positron computed tomography. III. In vivo demonstration of the effects of substrate availability on myocardial metabolism. Am Heart J 105:492-504, 1983
33. Schelbert HR, Henze E, Schon HR, Najafi A, Hansen H, Huang SC, Barrio JR, Phelps ME: C-11 palmitic acid for the noninvasive evaluation of regional fatty acid metabolism with positron computed tomography. IV. In vivo demonstration of impaired fatty acid oxidation in acute myocardial ischemia. Am Heart J 106:736-750, 1983
34. Schelbert HR, Neely JR, Phelps ME, Heiss HW: Regional myocardial metabolism by positron tomography. In: Foundation for advances in clinical medicine, ed: Advances in clinical cardiology (vol III). New Jersey: Mahwah, 1987
35. Schelbert HR, Phelps ME, Huang SC, MacDonald NS, Hansen H, Selin C, Kuhl DE: N-13 ammonia as an indicator of myocardial blood flow. Circulation 63:1259-1272, 1981
36. Schelbert HR, Phelps ME, Selin C, Marshall RC, Hoffman EJ, Kuhl DE: Regional myocardial ischemia assessed by F-18 2-deoxyglucose and positron emission computed tomography. In: Heiss HW, ed: Advances in clinical cardiology (vol I): Quantification of myocardial ischemia. New York: Witzstrock Publ, 1980:437-447
37. Schelbert HR, Schwaiger M: PET studies of the heart. In: Phelps M, Mazziotta J, Schelbert H, eds: Positron emission tomography and autoradiography: Principles and applications for the brain and heart. New York: Raven Press, 1986:581-661
38. Schon HR, Schelbert HR, Najafi A, Robinson G, Huang SC, Barrio J, Phelps ME: C-11 labeled palmitic acid for the noninvasive evaluation of regional myocardial fatty acid metabolism with

positron computed tomography: I. Kinetics of C-11 palmitic acid in normal myocardium. Am Heart J 103:532-547, 1982

39. Schon HR, Schelbert HR, Najafi A, Hansen H, Robinson GR, Huang SC, Barrio J, Phelps ME: C-11 labeled palmitic acid for the noninvasive evaluation of regional myocardial fatty acid metabolism with positron computed tomography. II. Kinetics of C-11 palmitic acid in acutely ischemic myocardium. Am Heart J 103:548-561, 1982

40. Schwaiger M, Brunken R, Grover-McKay M, Krivokapich J, Child J, Tillisch JH, Phelps ME, Schelbert HE: Regional myocardial metabolism in patients with acute myocardial infarction assessed by positron emission tomography. J Am Coll Cardiol 8:800-808, 1986

41. Schwaiger M, Brunken R, Krivokapich J, Child J, Tillisch JH, Phelps ME, Schelbert HR: Beneficial effect of residual antegrade flow on tissue viability as assessed by positron emission tomography in patients with myocardial infarction. Eur Heart J 8:981-988, 1987

42. Schwaiger M, Huang SC, Krivokapich J, Phelps ME, Schelbert HR: Myocardial glucose utilization measured noninvasively in man by positron tomography. J Am Coll Cardiol 1:688, 1983 (Abstr)

43. Schwaiger M, Neese R, Wyns W, Wisneski J, Grover-McKay M, Phelps ME, Schelbert HR, Gertz E: Glucose metabolism in post-ischemic canine myocardium. J Nucl Med 27:890-891, 1986 (Abstr)

44. Schwaiger M, Schelbert HR, Ellison D, Hansen H, Yeatman L, Vinten-Johansen J, Selin C, Barrio J, Phelps ME: Sustained regional abnormalities in cardiac metabolism after transient ischemia in the chronic dog model. J Am Coll Cardiol 6:336-347, 1985

45. Schwaiger M, Schelbert HR, Keen R, Vinten-Johansen J, Hansen H, Selin C, Barrio J, Huang SC, Phelps ME: Retention and clearance of C-11 palmitic acid in ischemic and reperfused canine myocardium. J Am Coll Cardiol 6:311-320, 1985

46. Shah A, Schelbert HR, Schwaiger M, Henze E, Hansen H, Selin C, Huang SC: Measurement of regional myocardial blood flow with N-13 ammonia and positron emission tomography in intact dogs. J Am Coll Cardiol 5:92-100, 1985

47. Sobel BE, Weiss ES, Welch MJ, Siegel BA, Ter-Pogossian MM: Detection of remote myocardial infarction in patients with positron emission transaxial tomography and intravenous C-11 palmitate. Circulation 55:853-857, 1977

48. Sokoloff L, Reivich M, Kennedy C, Des Rosiers MH, Patlak CS, Pettigrew KD, Sakurada O, Shinohara M: The C-14 deoxyglucose method for the measurement of local cerebral glucose utilisation: Theory, procedure and normal values in the conscious and anesthetized albino rat. J Neurochem 28:897-916, 1977

49. Syrota A, Comar D, Paillotin G, Davy JM, Aumont MC, Stulzaft O, Maziere B: Muscarinic cholinergic receptor in the human heart evidenced under physiological conditions by positron emission tomography. Proc Natl Acad Sci (USA) 82:584-588, 1985

50. Syrota A, Paillotin G, Davy JM, Aumont MC: Kinetics of in vivo bindings of antagonist to muscarinic cholinergic receptor in the human heart studied by positron emission tomography. Life Sci 35:937-945, 1984

51. Ter-Pogossian MM, Klein MS, Markham J, Roberts R, Sobel BE: Regional assessment of myocardial metabolic integrity in vivo by positron emission tomography with C-11 labeled palmitate. Circulation 61:242-255, 1980

52. Tillisch JH, Brunken R, Schwaiger M, Mandelkern M, Phelps ME, Schelbert HR: Reversal of cardiac wall motion abnormalities predicted by using positron tomography. N Engl J Med 314:884-888, 1986

53. Ward BJ, Gloster JA, Harris P: The incorporation and distribution of H-3 oleic acid in the isolated perfused guinea-pig heart: A biochemical and EM autoradiographic study. Tissue Cell 11:793-801, 1979

54. Weiss ES, Ahmed SA, Welch MJ, Williamson JR, Ter-Pogossian MM, Sobel BE: Quantification of infarction in cross sections of canine myocardium in vivo with positron emission transaxial tomography and C-11 palmitate. Circulation 55:66-73, 1977

55. Wyns W, Melin JA, Keyeux A, Bol A, Michel C, Cogneau M, Heyndrickx GR: Metabolic imaging in canine myocardium stunned by one hour critical coronary artery stenosis. Circulation 76, Suppl IV:IV-115, 1987 (Abstr)

56. Wyns W, Schwaiger M, Huang SC, Keen R, Phelps ME, Schelbert HR: Effects of inhibition of fatty acid oxidation on myocardial kinetics of C-11 palmitate. Circulation 72, Suppl III:III-337, 1985 (Abstr)

Chapter 14

Energy Metabolism and Transport in Neonatal Heart Cells in Culture

A. Pinson and *T. Huizer, Laboratory for Myocardial Research,
Hebrew University-Hadassah Medical School, Jerusalem, Israel;
*Cardiochemical Laboratory, Thoraxcenter, Rotterdam, The Netherlands

Heart cells in culture are increasingly being used in studies at the cellular level in many fields of cardiac biology and pathology. This chapter reviews the metabolism of the energy-producing substrates - glucose and fatty acids (FA's) - pinpointing the findings in cultured heart cells and their contribution to understand uptake and subsequent distribution of these metabolites. The roles of glucose and FA's in energy production and balance have been determined mainly from studies of $^{14}CO_2$ production from labelled substrates. Oxidation of glucose and FA's is incomplete if either of these substrates is present in excess, indicating that substrate utilization is controlled by a cellular mechanism. In addition, FA's in cellular reserves seem to be preferentially utilized over newly incorporated ones, implying that cellular lipids undergo continuous renewal. The recently developed technique for measuring oxygen consumption without disrupting the cultures will undoubtedly allow further progress to be made in elucidating the energy demands of the cardiomyocyte. The regulation of high-energy phosphate production and its transport from the mitochondria to the myofibrils via the creatine kinase system are also discussed in this chapter. Cultured heart cells have been used for studying oxygen deprivation. Under suitable conditions, such systems provide information on the sequence of events at the cellular level during anoxia. Kinetic studies of high-energy phosphate depletion and concomitant nucleoside and oxypurine formation during anoxic injury are presented. The repair mechanism via reoxygenation and reperfusion deserves detailed investigation.

1. Introduction

The relationship between cellular chemical energy and mechanical work in muscles is one of the most fascinating links between the classical disciplines of biochemistry and physiology. This chapter will be limited to studies on energy metabolism in cultures of neonatal heart cells, pinpointing findings that have furthered understanding of cardiac energy metabolism. The first part will deal with substrate utilization; the second, with certain aspects of high-energy phosphate metabolism and transport.

The heart-cell culture system is a stable preparation, in which both hormonal influences and diffusion barriers are absent. In addition, the composition of the extracellular environment may be controlled in this system, which allows investigations at the cellular level to be carried out. The cells in culture must, however, adapt to the new substratum and a continually changing environment owing to frequent changes of medium, with catabolic products that are never completely removed from the vicinity of the cells. In addition, cultured heart cells, which clearly resemble those in situ in their ability to contract spontaneously and rhythmically, are not required to act against the frictional resistance of blood vessels. Therefore, their energy requirements correspond to those for isotonic contractions. Consequently, there are less mitochondria in cultured cells than in cells in vivo. In spite of the reservations mentioned above, the basic metabolism, structure and electrophysiological properties of cardiomyocytes resemble those of the intact heart. Thus, studies with cultured heart cells, in which the fate of a substrate can easily be followed, have certainly extended our knowledge of cardiac physiology and biochemistry.

2. Lipid Metabolism

There is a well-established dual role for lipids in the heart: a) they are the major functional-structural component of the sarcolemma, and b) they also serve as the major energy source of the myocardium, primarily an oxidative organ. For this reason, storage of lipids is much more important than that of glycogen, since they provide a much more efficient energy source (15,45).

Quantitative studies of heart metabolism in situ were greatly advanced when cardiac catheterization techniques were devised. Cardiac O_2 consumption was shown to be too high to be accounted for by the oxidation of glucose, lactate and pyruvate (4,6). Non-esterified fatty acids (NEFA's) were shown to be the major energy-providing substrate in the heart (5,33).

Fatty Acid Uptake and Distribution

Fatty acids (FA's) are transported in the plasma in the form of complexes with albumin, which has several high-affinity FA binding sites. Less than 1% of the FA's circulates in the free state or in loosely-bound forms (110,132,135).

In cultured heart cells (28,76,77,101), two phases occur in the uptake of labelled FA's: a) Passive diffusion, predominating at FA levels of >200 uM; b) A saturable specific binding process (K_m about 10 uM). FA uptake in cardiomyocytes - as in liver cells - is dependent on the total [FA] (both free and albumin bound refs. 101,132). The various long-chain FA's compete for uptake. Since a specific binding protein could not be demonstrated, the transport system is probably located on the cell surface (101). Maximal uptake of palmitate occurs at palmitate:albumin ratios ranging from 7 to 10, differing markedly from the in vivo value of about 2. Albumin enhances both uptake and subsequent esterification of palmitate, but reduces that of oleate (76). This is due to binding of most palmitate molecules to low- and medium-affinity binding sites in albumin. With oleate (an unsaturated FA), albumin acts as an alternative uptake site to cardiac cells, thus lowering oleate uptake into cardiomyocytes (76). The uptake process is not energy dependent, but is probably facilitated (77).

Following uptake, FA's are immediately metabolized in the sarcoplasmic reticulum (20,113). A cytosolic protein of mol. wt. 12,000 that binds FA acyl-CoA has been demonstrated (63,64). It suppresses the detergent-like properties of acyl-CoA.

Detailed kinetic investigations of FA pathways have recently been carried out by our group (8). We found that most of the FA's were immediately esterified, NEFA only accounting for 5 to 10% of the the total FA incorporated. Diglycerides did not accumulate in this system, indicating that the FA-conversion rate exceeds uptake. On depletion of the cells of energy-providing substrates and uncoupling oxidative phosphorylation, FA esterification was inhibited by 80% but not accompanied by FA accumulation in the cells. Impairment of esterification led to export of both diglycerides and FA's. Under normal conditions, FA incorporation into triglycerides and phospholipids was about equal. Most of the triglycerides were located in the

100,000 g supernatant fraction, whereas phospholipids were generally associated with the particulate and membranous fractions.

Fatty Acid Oxidation

The mitochondrial route is usually considered the major pathway for the metabolism of exogenous FA, indicating that exogenous, and not endogenous FA's, are preferentially used for energy supply. However, some reports suggest that this is not the case - most FA's were found to be in the esterified form shortly after uptake (8,113). Also in skeletal muscle, FA's are not oxidized directly, but are first converted to the esterified form and generally stored in the vicinity of the mitochondria as lipid droplets (18,111,141).

Based on studies with perfused hearts, Shipp et al. (108) suggested that endogenous lipids provide energy, although it was not clear whether phospholipids (108) or triglycerides (72,73) are preferred. We demonstrated in cultured heart cells that endogenous lipids are preferentially used over FA's that have just entered the cell (82,83).

To undergo oxidation acyl-CoA thioesters are transported into the mitochondria after conversion to acylcarnitine at the outer mitochondrial membrane. On reaching the inner membrane, the carnitine is exchanged with the inner mitochondrial membrane carnitine by a transferase, the acyl group then being tranferred to the mitochondrial matrix CoA and finally, beta-oxidation occurs (89,90,92).

In cardiomyocytes there are conflicting data regarding the importance of FA oxidation as compared to that of other substrates. According to Ross and McCarl (97), glucose, pyruvate and lactate contribute more to energy production than palmitate. Others (112) reported that more energy is derived from glutamine than from either glucose or FA's. The shift from lipid to carbohydrate metabolism in aging cultures (32) is thought to be due to low intracellular levels of the enzymes involved in FA metabolism (37,38) or to a postulated hypoxic glycolytic pathway in such cells (95). However, according to our data (30,82), FA's are the major energy source for cardiac myocytes; the decrease in FA oxidation upon aging could be eliminated by preincubating the cells with palmitate. The variation in cell populations in culture as a function of age may also be of significance (136). Clearly, this topic warrants reexamination in the light of these findings.

FA degradation occurs primarily via beta-oxidation involving four sequential reactions (47). FA acyl-CoA's of chain-lengths from C_4 to C_{18} undergo oxidation via the "oxidase". Short FA intermediates of palmitate and stearate oxidation could only be found under special conditions, e.g., O_2 restriction (60,114). There were no differences in the oxidation rates (i.e., the amount of $^{14}CO_2$ released) from ^{14}C-palmitate labelled at either at C 1, 6 or 11, fed to intact rats (131) or given to perfused hearts (27,88).

Long-chain FA's are completely oxidized to CO_2 in cultured heart cells (101). However, in agreement with earlier findings (88), we have reported a striking difference in the production of $^{14}CO_2$ from either 1-^{14}C- or 16-^{14}C-palmitate in cultured heart cells (82,83), the release being several fold higher from the former. This implies that the two termini of the palmitate molecule have different catabolic fates (see ref. 39).

In hepatocytes total oxidation of 1-^{14}C- or 16-^{14}C-palmitate was almost identical, but considerably less $^{14}CO_2$ was produced from 16-^{14}C-palmitate (16). These data imply the preferential incorporation of omega-C_2 units into ketone bodies (see ref.

9). If this also holds for heart cells, reutilization of the ketone bodies would be expected, since these are the energy source of choice in the myocardium (133).

Our studies revealed considerable release into the medium of water-soluble products from the ^{14}C-methyl terminus of palmitate (82,83,84). This extracellular radioactivity persisted even after long incubation periods, indicating that the released product did not undergo further metabolism. The metabolite was identified as 3-hydroxybutyrate (83,84), and by isomer separation (61), 80% of the radioactivity was shown to be associated with the L(+)-isomer.

The CoA derivative of L(+)-3-hydroxybutyrate is an intermediate in long-chain FA oxidation (128). It is thought that the L(+)-isomer of 3-hydroxybutyrate participates in beta-oxidation attached to the enzyme complex as the CoA derivative, while free 3-hydroxybutyrate released into the body fluids is the D(-)form. Although some studies indicated that mammalian tissues could metabolize L(+)-3-hydroxybutyrate, this route was thought to be unimportant (54,55). Furthermore, mitochondrial 3-hydroxybutyrate dehydrogenase, which oxidizes 3-hydroxybutyrate to acetoacetate, has an absolute stereospecificity for the D(-)-isomer (56). Based on the uneven distribution of radioactivity in both the ^{14}CO$_2$ and in the water-soluble 3-hydroxybutyrate, we suggested that L(+)-hydroxybutyryl-CoA is detached and deacylated to form the free acid, which is transported across the membrane, contributing to the water-soluble, radioactive 3-hydroxybutyrate in the extracellular medium. It may, however, be metabolized at a very slow rate, as indicated by the ^{14}C-DL-3-hydroxybutyrate disappearance from the medium (83). Furthermore, other radioactive metabolites were also released into the medium, particularly from 1-^{14}C-palmitate. However, under such conditions, the total excreted radioactivity was considerably lower and was distributed among several compounds, one of which was identified as D-3-hydroxybutyrate. This is consistent with earlier reports (27,13) of high [3-hydroxybutyrate] in the extracellular compartment in perfused hearts. It is also possible that some of the early reports in which 3-hydroxybutyrate was chemically determined, in fact unknowingly dealt with the L(+)-isomer.

Abnormal Fatty Acid Metabolism

Erucic acid (EA, delta-13 docosenoic acid; C$_{22:1}$) is the major fatty acid contained in rapeseed oils. An EA-rich diet causes an accumulation of EA-rich triglycerides in heart-muscle cells, which leads to cardiac lipidosis if fed to rats and other animals. We have shown that cardiac cells have a decreased ability to oxidize EA, as studied by using EA labelled at both the C-14 (on the double bond) and the carboxyl carbons. ^{14}CO$_2$ originating from the carboxyl group appears rapidly, whereas that derived from C-14 lagged behind for about 12 h (82,85,86). Thus, apparently EA cannot be directly oxidized in the mitochondrion, but undergoes first chain shortening. This occurs in two phases: first to C$_{20:1}$ and then to C$_{18:1}$ (86). The C$_{18:1}$ produced may then be undergo beta-oxidation. Chain shortening to a lesser degree also occurs with other FA's, as lower labelled homologues, such as C$_{14:0}$, C$_{12:0}$ and C$_{10:0}$, derived from 16-^{14}C-palmitate could be detected in heart cell cultures (Pinson et al., unpublished).

3. Carbohydrate Metabolism

Glucose Uptake

Glucose-6-phoshate is at the crossroads of glucose metabolism, several different pathways being available from that point: glycolysis yielding pyruvate (and lactate) usually followed by complete oxidation via the citric acid cycle; glycogen synthesis via uridine diphosphoglucose; and the pentose-phosphate shunt.

In the aerobic heart, glucose does not constitute an important energy source. Even under extremely high work load with glucose as the only substrate in the extracellular fluid, glucose oxidation only accounts for 40% of the energy required for contraction, the rest being derived from intracellular triglycerides (70,73).

In the only extensive study on glucose transport in cultured heart cells, Paris et al. (75) compared the metabolism of 3-O-methylglucose, which enters the cell but is not further metabolized, and 2-deoxyglucose, which undergoes phosphorylation by hexokinase but may not be subsequently used. They showed that the rate of transport is several-fold higher than the phosphorylation rate. The latter is limited by the rate of transport into the cell, reaching a steady state within a few min. It is dependent on the extracellular [carbohydrate]. The intracellular [glucose] is below the K_m for hexokinase and, in turn, determines the phosphorylation rate, which is increased by countertransport and decreased by cytochalasin B. Thus, any changes in carbohydrate metabolism are reflected in the rate of 2-deoxyglucose phosphorylation. Glucose transport is not affected by FA's in the medium or by preloading the cells with FA's (75). These findings conflict with those reported by Neely et al. (68), who found that in the working perfused heart palmitate inhibited glucose uptake. This discrepancy is possibly due to the contribution of a regulation system in cultured heart cells, which responds to extracellular hexoses (75). The decreased transport in cultured cells following incubation with cycloheximide suggests that the synthesis of a carrier protein is involved and that its activity may be modulated.

Other studies (29,30) have pointed to the importance of the composition of the extracellular medium for glucose uptake. Supplementation of the medium with 20% serum resulted in glucose uptake at a constant rate of 3 umol/h/mg protein. The medium thus became depleted of glucose after 10 to 12 h, well before the next medium change. In the absence of serum, however, glucose was taken up at about half this rate (30). Serum in the medium stimulated both glucose transport and hexokinase activity (29). These rates of glucose uptake are an order of magnitude higher than those found in both perfused (30 umol/h/g of tissue, ref. 66) and in fetal heart (66 umol/h/g, ref. 17). Other investigators obtained data with cultured cells (about 1.5 umol/h/mg protein, refs. 74,95,140) which were very similar to the findings in perfused hearts. The differences may be attributed to variations in the distribution of various cell types since cardiomyocytes take up glucose at a much higher rate than non-muscle cells (97).

Lactate Production

Glycolysis in the heart is regulated via the activities of phosphofructokinase and glyceraldehyde-3-phosphate dehydrogenase. The supply of glucose-6-phosphate, either following glucose uptake and phosphorylation or via glycogen degradation, is a crucial step in the pathway from glucose to pyruvate and lactate (69,90,92).

Under conditions of normal O_2 supply, the heart does not produce lactate, glucose either being used for glycogen synthesis or undergoing complete oxidation. The lactate dehydrogenase (LDH) H_4 isoenzyme pattern is present, and consequently lactate is oxidized to pyruvate. Lactate is only the end product of glucose degradation during O_2 deprivation, when the energy is supplied to the ischemic area via glycolysis or in the partially anaerobic fetal heart.

Studies with cardiomyocytes have shown that most of the glucose incorporated is rapidly converted to lactate which is then released into the extracellular space (29-31,97). However, lactate production is not a result of O_2 restriction in the culture medium (30). The released lactate is not derived from endogenous glycogen, but from newly incorporated glucose, and is not dependent on the [lactate] in the medium (30,31). However, serum in the medium promotes optimal glucose uptake and lactate production (30). Non-muscle cells in culture only oxidize glucose to lactate, while cardiomyocytes also metabolize some glucose via the citric acid cycle (97). Thus, the amount of lactate produced also varies with the initial seeding density (129). Indeed, it is difficult to compare the amount of glucose converted to lactate in various experiments, since different methods employed for preparing cell cultures by various research groups affect the proportion of myocytes and non-muscle cells in the system. Our group (31) and Ross and McCarl (97) have reported similar values for lactate production, 40 umol/mg protein (more than 65% of the incorporated glucose) and 0.88 umol/h/mg protein, representing 75% of the glucose taken up from the medium, respectively - values which are 4-fold higher than Orloff and McCarl's data (74).

Lactate production may serve as a mechanism for removing excess glucose from cultured heart cells, since only a small amount is used in energy production and a mere 5% is stored as glycogen (30,31). The shift in the LDH isoenzyme pattern from the heart (H_4) to the muscle type (M_4) may reflect this (11). However, changes in the relative proportion of the different cell types in the culture cannot be ruled out (120). Randle et al. reported that in the non-working perfused heart significant amounts of glucose and pyruvate are converted to lactate (91). These findings are possibly related to the fact that cells in culture, in direct contrast to those in vivo, are not engaged in work against resistance. Only after the medium has been completely depleted of glucose and >65% has been transformed to lactate, does the [lactate] in the medium decrease in a linear fashion, by oxidation or by transamination (30,31,97).

Glucose Oxidation Pathways

The growth of heart cells in culture occurs in two distinct phases: an initial phase of cell division, the duration of which depends upon the plating density, until confluency is reached, and a non-dividing phase with continued cell growth (136). These two phases are reflected in the glucose oxidation via two pathways in cultured cells: a) the pentose-phosphate pathway, and b) glucose oxidation via the citric acid cycle and oxidative phosphorylation. The importance of both pathways has been studied in cultured cells - the oxidation of glucose has been estimated by measuring the release of $^{14}CO_2$ from an appropriately labelled substrate, either directly in the culture flask (129), or by means of a simple multi-channel trapping device (82,87), or by the rather sophisticated "radiorespirometer" working along similar lines (98).

The *pentose-phosphate pathway* is very active in early embryonic heart and becomes less so as the heart approaches the adult form, when the division index decreases. The end product of this pathway, ribose-phosphate, is a precursor for nucleoside and subsequent DNA synthesis. Since ribose is formed from carbons 2 to 6 of the glucose molecule, the relative importance of each pathway has been studied by comparing the amount of $^{14}CO_2$ released from 1-^{14}C-glucose obtained via the pentose-phosphate shunt and the citric acid cycle to that produced from 6-^{14}C-glucose via the citric acid cycle alone. Warshaw and Rosenthal (129) reported enhanced glucose oxidation in conditions of low-density plating. This presumably reflects increased flow through the pentose-phosphate pathway, which is inhibited when the cells reach confluency (96,129). We have reported (30,31) similar findings, as well as a link between this pathway and ^{3}H-thymidine incorporation into DNA. Thus, DNA synthesis, in the wake of the pentose-phosphate pathway, increases in parallel with the flow through the pentose shunt, after a delay of several hours, which is consistent with the increased requirement for ribose-phosphate in proliferating cultures.

The level of *glucose oxidation* via the tricarboxylic acid cycle has been shown to remain constant during all the phases of heart-cell culture (31,129). However, conflicting results have been reported (103): rates of glucose utilization varied in cultures of different ages. These were interpreted as a postnatally-induced decline of glucose as an energy source.

Pyruvate and Lactate Oxidation

We have reported that following glucose uptake from the medium, lactate is released back into the medium and reutilized within the cell either via oxidation or in further transformations. Rosenthal and Warshaw (96) found a lag phase in the time course of $^{14}CO_2$ formation from 2-^{14}C-pyruvate as compared to that produced from 1-^{14}C-pyruvate, after which $^{14}CO_2$ production from both substrates increased linearly, although $^{14}CO_2$ formation from 1-^{14}C-pyruvate occurred at a higher rate. This may be explained if CO_2 derived from carbon 1 is produced directly by the action of pyruvate dehydrogenase, while CO_2 from carbons 2 and 3 is formed via incorporation into acetyl-CoA followed by the citric acid cycle reactions.

Ross and McCarl (97) found that pyruvate was taken up at a rate of 4.8 umol/min/mg protein. About 50% of the incorporated pyruvate is reexported as lactate, the remainder being oxidized. They also confirmed an earlier finding (96) that $^{14}CO_2$ release from 2-^{14}C-pyruvate lagged behind that formed from 1-^{14}C-pyruvate by about 10 min. The steady state for oxidation was reached within 20 min with the former substrate, and within 80 to 100 min with the latter. Lactate oxidation led to linear release of $^{14}CO_2$ from 1-^{14}C-lactate over a 2-h period. Subsequently the release rate fell (97). More $^{14}CO_2$ was also produced from 1-^{14}C-lactate than from 1-^{14}C-pyruvate under similar conditions. As yet no satisfactory explanation has been given for this finding. Furthermore, 15% of the total lactate was found to be incorporated into cellular components - 2/3 of this into water-soluble molecules and 1/3 into lipids and proteins.

In conclusion, the above results show that in cultured heart cells, pyruvate is either oxidized or converted to lactate, but is not transformed into other cellular components, while lactate undergoes either oxidation or transformation to other compounds. Similar findings have been reported in the perfused heart (71,91), indicating that the LDH isoenzyme profile does not play a significant role in the regulation of pyruvate-lactate formation under low work load conditions.

4. Oxygen Consumption

Oxygen consumption measurements are essential in studies of cellular metabolism. Although the current oxygraph may provide useful data, most of these experiments are carried out on cells that have been suspended by protease treatment and therefore consist of dying non-beating cells. Recently, Takehori et al. (115) have constructed a lucite attachment with a built-in O_2 electrode, which allows O_2 consumption to be measured in beating cultured heart cells attached to the substratum. With this device, they showed that O_2 consumption was 47% higher in beating than in quies- cent cells, and that these beating rates were comparable to those observed in intact heart. In addition, the O_2 consumption of quiescent cardiomyocytes was two-fold higher than in either cardiac non-muscle cells or in the L_6 muscle cell line - a clear indication of the high level of energy metabolism in cardiomyocytes.

In adult cardiomyocytes, O_2 consumption was two-fold higher than in neonatal cultured cells (10). It correlated with the higher mitochondrial creatine kinase (CK) in adult cells (100,118), implying that this CK plays a role in the coupling of energy production and utilization, presumably by preventing high-energy phosphate depletion during high cardiac work loads. However, the available data are scarce and much more research is required in this area.

5. High-Energy Phosphate Transport

Over the last few decades, attention has been directed toward elucidation of myocyte energy metabolism, focusing on the roles played by CK and the creatine/phosphocreatine ratio (Cr/PCr) in the maintenance of cellular energy levels and energy transport.

The finding that ADP undergoes phosphorylation by PCr (58) and the discovery of myosine ATPase (23) served to demonstrate the direct relationship between ATP utilization and mechanical work, and, consequently, the restricted role of PCr as a reservoir for high-energy phosphate for the coupling between energy utilization and production (12,44,58,65). However, in spite of the progress in this area, the mechanism of transformation of chemical energy into mechanical work is still not clear, although many of the biochemical events involved have been characterized (for review, see ref. 46).

Mitochondrial energy production is determined by the rate of ATP utilization, which is regulated via a cytosolic signal (53). Presumably, ADP formation and the Cr-PCr-CK system are involved in intracellular energy formation and in energy transport from the mitochondria to the myofibrils (100,104,106). CK bound to either myofibrils or membranes may be involved in phosphorylation of ATP, thus providing ATP at the sites where it is used (14,81,137,138,). According to this view, CK is one the key enzymes involved in cellular energetics.

Creatine kinase is a dimeric enzyme (19) consisting of two types of monomers - M and B. Of the three isozymes - MM, BB and MB - the MM form predominates in muscle (24,25). The existence of a fourth form of CK has been demonstrated in mitochondria (mCK, ref. 50). In rat heart, about 20% of the CK activity is located in the mitochondria, and about 75% in the cytosol (21). In human heart, these values are 10% and 85% (7). Electrophoresis (50), in situ histochemical staining (3,107), and immunological techniques showed that mCK is distinct from the cytoplasmic CK's (94). mCK is bound to the outer surface of the inner mitochondrial membrane and possibly via cardiolipin (36).

Mitochondrial CK plays a critical role in the regulation of respiration, enhancing both O_2 consumption and PCr synthesis in isolated mitochondria, whereas PCr reduces O_2 uptake owing to competition for ADP between mCK and oxidative phosphorylation (51,52,109,127). There are apparently several pools of ATP. The ATP pool which is preferentially used by mCK and the possible role played by the mitochondrial membrane in the regulation of ATP fluxes between the mitochondria and the cytoplasm remain to be elucidated (2,26,57,139).

Inhibition of oxidative phosphorylation with oligomycin in cultured heart cells leads to cessation of contraction within 30 to 60 min (106). This effect is correlated with the depletion in PCr levels, the ATP levels remaining virtually unchanged. In addition, when cells are cultured in Cr-containing medium, or are pulsed with Cr, PCr levels increase, while there is no change in ATP. Yet, there is no evidence for direct participation of PCr in contractile activity.

The inhibition of PCr synthesis by oligomycin suggests that it is dependent on oxidative phosphorylation, and that mCK is involved in the control of PCr levels in cultured myocardial cells. Indeed, this isozyme has been identified in rat myocardial cells. It is absent from cardiac non-muscle cells, and from the L_6 muscle cell line (118).

The relative proportion of the mCK isozyme increases upon aging, both in culture and in vivo. The BB form decreases under these conditions, and the MM isozyme predominates during development of myocardial cells. The relative isozyme distribution is similar in culture and in fresh tissue, although total CK activity is 4- to 5-fold higher in the latter: 1.5 U/mg protein in culture, 6.2 U/mg in fetal heart, and 7.3 U/mg in 1- to 6-d old heart (118).

Creatine appears to trigger mCK synthesis - if present in the culture medium, it leads to a 44% increase in mCK. It is reasonable to assume that myocardial energy demands increase postnatally. Energy production is regulated at the mitochondrial level by Cr concomitantly with the development of the Cr-PCr-mCK energy shuttle system (106,118). However, the cause-and-effect relationship of Cr in the coupling of energy production to energy utilization via resynthesis of PCr in the mitochondria remains to be demonstrated (62).

Analysis of the relative proportion of CK isoenzymes at various times before and after birth revealed striking changes - 14-d old fetal heart contains only BB isozymes, while in the 17-d old fetus, 80% of the enzyme activity is due to M forms (MB and MM isozymes). In mice, mCK is also undetectable in either fetuses or immediately after birth, and accounts for 1% and 9% of the total CK activity in 6-d old and adult animals, respectively. In the lamb, however, cardiac CK and Cr are detectable even prenatally (49), whereas the BB form is totally absent, and the level of the MB form is decreased from 24% to 5% during this period (25,48). In cultured striated muscle cells, transition from B to M forms is incomplete, with significant levels of MB and BB isozymes expressed throughout the period in culture (22,59,79,80). The M form is apparently characteristic of the post-fusion state (13,78). In addition, fusion and synthesis of contractile proteins can be stimulated by the supplementation of Cr to the culture medium (105,106).

6. Energy Metabolism during Oxygen Deprivation

Under conditions of O_2 restriction or deprivation, the heart immediately switches to anaerobic glycolysis. By this means, it briefly maintains the high-energy phosphate levels, prior to a rapid decline in ATP and CP production with a concomitant increase in degradation products (35,93,102,116). Cultured heart cells have also been used as a model system for studying O_2 deprivation at the cellular level (1,40,121). In some experiments, metabolic inhibitors were employed together with O_2 deprivation (41).

Recently, an alternative approach has been taken to assess the effects of O_2 deprivation in cultured cells. By combining anoxia with glucose deprivation, coupled with extreme reduction in the volume of extracellular medium, a true microenvironment with reduced flow is simulated, in which the products of cellular catabolism, such as lactate, remain in the vicinity of the cells and are not washed away (122,124,126). Indeed, anoxic injury, as reflected in higher rates of enzyme release and sarcolemmal transformations, is increased under these conditions (122-126).

Several studies relating to nucleotide and purine metabolism in anoxic heart cells in culture have shown that in chick fetal heart cells, anoxia induces a two-fold increase in production of adenosine and its metabolites. Surprisingly, the levels of ATP remained unchanged (67). In another investigation, it was demonstrated that hypoxanthine release precedes cellular enzyme depletion (119). Correlation between cellular ATP depletion and degradation of the cell membrane by phospholipase C has also been reported (42,43).

Recently, we studied the kinetics of high-energy phosphate depletion and the release of nucleotide degradation products in cultured heart cells under conditions of anoxia combined with volume restriction (122,123). Decrease in CP levels usually precedes that of ATP. There is marked acceleration of the decrease in the levels of both of these, when anoxia is combined with volume restriction of the extracellular medium. In the presence of glucose and without volume restriction, the glycolytic flux from glucose and the glycogen stores may provide sufficient ATP to delay its depletion for considerable periods. In perfused heart (69), under conditions of high flow, the glycolytic flux from cellular glycogen and supplementary glucose may even provide sufficient energy to maintain some contractile activity.

The irreversible stage in anoxic cells is presumably related to marked depletion in ATP content. Indeed, major cellular damage, as reflected in lysosomal enzyme release, only starts when 50% of the cellular ATP has been depleted (122). In addition, release of free arachidonate and palmitate was associated with ATP depletion - arachidonate accumulation started when the ATP levels reached <25% of the normal values (34). If the rate of ATP utilization exceeds the rate of ATP production in the absence of O_2 and glucose, ATP breakdown gives rise to intracellular adenosine, which may be recovered in the form of purine bases in the medium. The breakdown of ATP results in the formation of ADP and AMP, which is further deaminated to IMP or dephosphorylated to adenosine under anoxic conditions. Adenosine may be further degraded to inosine and then, in turn, to hypoxanthine, xanthine, and urate. The degradation of inosine presumably involves nucleoside phosphorylase which is located in endothelial cells (99). In our studies, adenosine was not detected in cultured cells during O_2 deprivation, inosine levels increased slightly, while the [hypoxanthine] in the external medium was even decreased. However, anoxia combined with volume restriction, led to high levels of purine release, although inosine levels remained low, and hypoxanthine, xanthine, and urate release reflected the decrease in cellular ATP levels (122,123).

7. Conclusions and Perspectives

Glucose metabolism in cultured heart cells has been termed "impaired" or "hypoxic", since these cells produce lactate. Careful examination of the literature on both cultured cardiomyocytes and perfused hearts reveals that lactate production should not be considered as an abnormal response, but rather as an expression of the cellular energy requirements. In this respect, cultured heart cells are more akin to the heart at rest or under low work loads, in which lactate production also occurs. Much remains to be learned about the mechanisms of carbohydrate metabolism. Clearly, the shift from the H_4 to the M_4 forms of lactate dehydrogenase is an inadequate explanation for the lactate production. Other crossroads of enzymatic control - the active and inactive forms of pyruvate dehydrogenase, governing the flow of pyruvate into the citric acid cycle, and the enzymes activating glycogen synthesis and degradation - deserve further study.

In cultured cardiomyocytes, fatty acids constitute the main source of metabolic energy, as in the heart in vivo. The use of these cultures has furthered our understanding of FA transport across membranes. In addition, FA's derived by the cleavage of intracellular reserves were shown to be preferentially oxidized over newly incorporated FA's. Another highly significant achievement made with cultured heart cells was the discovery of the pathway for incomplete oxidation of FA's leading to the release of L(+)-3-hydroxybutyrate, an intermediate in beta-oxidation. It will be interesting to examine the physiological role of this pathway, bearing in mind that L(+)-3-hydroxybutyrate, when injected into suckling rats, may be used in the synthesis of cholesterol and other lipids, chiefly in the brain (130).

Little mention has been made in this review of interactions between carbohydrates, ketone bodies and lipid metabolism in cultured heart cells - surprisingly this field has suffered from almost complete neglect. Cells in culture provide an ideal system for carrying out such investigations owing to the ease with which they may be manipulated and controlled.

At present, only a tentative scheme can be drawn up for the metabolic pathways of lipids within the cell. The flow of substrates between organelles, the sites where they are stored and metabolized, and the (hormonal) control of these events remain to be elucidated.

The energy shuttle requires the presence of creatine kinase bound to mitochondria and to myofibrils. Phosphocreatine, synthesized in the mitochondria by mCK, diffuses to the myofibrils where the ATP level is maintained by the myofibrillar CK. One may speculate that PCr is not susceptible to non-specific enzymic reactions, and that this constitutes a distinct advantage in energy transport. Regulation of energy metabolism and the energy shuttle by the Cr-PCr-mCK system occurs concomitantly with the continuous energy demand of the myocardium imposed upon a high resting metabolism. This system is not ubiquitous. It seems that its development in myocardial cells usually occurs postnatally. Further studies in this field together with O_2 consumption measurements in situ under various metabolic conditions should provide more information on the energy requirements of the cultured heart cell.

Finally, since the cultured cardiomyocyte is a rhythmically contracting cell, one may expect that ATP utilization and some enzymic activities fluctuate during cycles of cardiac contraction and relaxation. The cultured cell system is ideal for studies on this level, since the technical difficulties of freeze-clamping (117,134) should be easier to overcome than in the perfused heart.

Acknowledgements

Dr. Pinson's work reported in this chapter was supported by the Institut National de la Santé et la Recherche Médicale (INSERM), France; the Dutch Heart Foundation; the Netherlands Society for the Advancement of Pure Research (ZWO); the Ministry of Education and Sciences of the State of Niedersachsen; Mr. and Mrs. D. Vidal-Madjar (France); Mrs. F. Berk (Belgium); and Mrs. R. Missistrano (France).

References

1. Acosta D, Puckett M, McMillin R: Ischemic myocardial injury in cultured heart cells: leakage of cytoplasmic enzymes from injured cells. In Vitro 14:728-732, 1978
2. Altschuld RA, Brierley GP: Interaction between the creatine kinase of heart mitochondria and oxidative phosphorylation. J Mol Cell Cardiol 9:875-896, 1986
3. Baba N, Kim S, Farrell EC: Histochemistry of creatine kinase. J Mol Cell Cardiol 8:599-617, 1976
4. Bing RJ, Hammond MM, Handlesman JC, Powers SR, Spencer FC, Eckenhoff JE, Goodale WT, Hafenshiel JH, Kety SS: Measurement of coronary blood flow, oxygen consumption, and efficiency of the left ventricle in man. Am Heart J 38:1-12, 1949
5. Bing RJ, Siegel A, Ungar I, Gilbert M: Metabolism of the human heart. II. Studies on fat, ketone and amino acid metabolism. Am J Med 16:504-515, 1954
6. Bing RJ, Siegel A, Vitale A, Balboni P, Sparks E, Taeschler E, Klapper M, Edwards S: Metabolic studies on the human heart in vitro: studies on carbohydrate metabolism of the human heart. Am J Med 15:284-296, 1953
7. Blum HE, Weber B, Deus B, Gerok W: The mitochondrial isozyme from human heart muscle. In: Lang H, ed: Creatine kinase isoenzymes: pathophysiology and clinical application. Berlin: Springer-Verlag, 1981:19-31
8. Brandes R, Pinson A, Heller M: Transport and metabolism of free fatty acids in cultured heart cells. In: Freysz L, Gatt S, Dreyfus H, Massarelli R, eds: Enzymes of lipid metabolism. New York: Plenum Press, 1986:459-465

9. Brown GW Jr, Chapman DD, Matheson HR, Chaikoff IL, Dauben WG: Acetoacetate formation in the liver. III. On the mechanism of acetoacetate formation from palmitic acid. J Biol Chem 209:537-548, 1954

10. Burns AH, Ready WJ: Amino acid stimulation of oxygen and substrate utilization by cardiac myocytes. Am J Physiol 235:E461-E466, 1978

11. Cahn RD: Developmental changes in embryonic enzyme patterns: The effect of oxidative substrates on lactic dehydrogenase in beating chick embryonic heart cell cultures. Dev Biol 9:327-346, 1964

12. Cain DF, Davies RE: Breakdown of adenosine triphosphate during a single contraction of working muscle. Biochem Biophys Res Commun 8:361-366, 1962

13. Caravatti M, Perriard JC, Eppenberger HM: Developmental regulation of creatine kinase in myogenic cell cultures from chicken. Biosynthesis of creatine kinase subunits M and B. J Biol Chem 254:388-394, 1979

14. Carvahlo AP, Motta AM: The role of ATP and of bound phosphoryl group acceptor on Ca binding and exchangeability in SR. Arch Biochem Biophys 142:201-212, 1971

15. Christiansen K: Membrane-bound lipid particles from beef heart acylglycerol synthesis. Biochim Biophys Acta 380:390-402, 1975

16. Christiansen RZ: The effects of clofibrate feeding on hepatic fatty acid metabolism. Biochim Biophys Acta 530:314-324, 1978

17. Clark CM: Carbohydrate metabolism in isolated foetal rat heart. Am J Physiol 220:583-588, 1978

18. Dagenais GR, Tancredi RG, Zierler KL: Free fatty acid oxidation by forearm muscle at rest, and evidence for an intramuscular lipid pool in human forearm. J Clin Invest 58:421-431, 1976

19. Dawson DM, Eppenberger HM, Kaplan NO: Creatine kinase: evidence for dimeric structure. Biochem Biophys Res Commun 21: 346-353, 1965

20. Denton RM, Randle PJ: Concentration of glycerides and phospholipids in rat heart and gastrocnemius muscles. Biochem J 104:416-422, 1967

21. Desjardins PR, Pesclovitch R: Subcellular localization of human atypical creatine kinase. Clin Chim Acta 135:35-40, 1983

22. Dym H, Turner DC, Eppenberger HM, Yaffe D: Creatine kinase isoenzyme transition in actinomycin D-treated differentiating muscle cultures. Exp Cell Res 113:15-21, 1978

23. Engelhart VA, Ljubimova MN: Myosin and adenosine-triphosphatase. Nature 144:668-669, 1939

24. Eppenberger HM, Dawson DM, Kaplan NO: The comparative enzymology of creatine kinases: isolation and characterization from chicken and rabbit tissue. J Biol Chem 242:204-209, 1967

25. Eppenberger HM, Eppenberger M, Richterich R, Aebi H: The ontogeny of creatine kinase isozymes. Develop Biol 10:1-16, 1964

26. Erickson-Viitanen S, Viitanen P, Geiger PJ, Yang WCT, Bessman SP: Compartmentation of mitochondrial creatine phosphokinase. I. Direct demonstration of compartmentation with the use of labelled precursors. J Biol Chem 257:14395-14404, 1982.

27. Evans JR, Opie LH, Shipp JC: Metabolism of palmitic acid in perfused rat heart. Am J Physiol 205:766-770, 1963

28. Franchi A, Ailhaud G: Incorporation of a fatty acid containing a photosensitive group into lipids of cultured cardiac cells from chick embryo. Biochimie 59:813-817, 1977

29. Frelin C: The growth of heart cells in culture. Evidence for a multiple activation of the pleiotypic program. Biochimie 60:627-638, 1978

30. Frelin C, Pinson A, Athias P, Surville JM, Padieu P: Glucose and palmitate metabolism by beating rat heart cells in culture. Pathol Biol 27:45-50, 1979

31. Frelin C, Pinson A, Moalic JM, Padieu P: Energy metabolism of beating rat heart cell cultures. II. Glucose metabolism. Biochimie 56:1596-1602, 1974

32. Fujimoto A, Harary I: Studies in vitro on single beating rat-heart cells. IV. The shift from fat to carbohydrate metabolism in culture. Biochim Biophys Acta 86:74-80, 1964

33. Gordon RS Jr: Unesterified fatty acid in human blood plasma. II. The transport function of unesterified fatty acid. J Clin Invest 36:810-815, 1957

34. Gunn MD, Sen A, Chang A, Willerson JI, Buja LM, Chien KR: Mechanisms of accumulation of arachidonic acid in cultured myocardial cells during ATP depletion. Am J Physiol 249:H1188-H1194, 1985

35. Haider W, Eckersberger F, Wolner E: Preventative insulin administration for myocardial protection in cardiac surgery. Anesthesiology 60:422-429, 1984

36. Hall N, Addis P, De Luca M: Purification of mitochondrial creatine kinase: two interconvertible forms of active enzyme. Biochem Biophys Res Commun 76:950-956, 1977
37. Harary I, McCarl R, Farley B: Studies in vitro on single beating rat heart cells. IX. Restoration of beating by serum lipids and fatty acids. Biochim Biophys Acta 115:15-22, 1966
38. Harary I, Slater EC: Studies in vitro on single beating rat heart cells. VIII. The effect of oligomycin, dinitrophenol and ouabain on the beating rate. Biochim Biophys Acta 99:227-233, 1965
39. Harper RD, Saggerson ED: Factors affecting fatty acid oxidation in fat cells isolated from rat white adipose tissue. J Lipid Res 17:516-526, 1976
40. Higgins TJC, Allsopp D, Bailey PJ: The effect of extracellular calcium concentration and Ca-antagonist drugs on enzyme release and lactate production by anoxic heart cell cultures. J Mol Cell Cardiol 12:909-927, 1980
41. Higgins TJC, Allsopp D, Bailey PJ, D'Souza EDA: The relationship between fatty acid metabolism and membrane integrity in neonatal myocytes. J Mol Cell Cardiol 13:599-615, 1981
42. Higgins TJC, Bailey PJ, Allsopp D: The influence of ATP depletion on the action of phospholipase C on cardiac myocyte membrane phospholipids. J Mol Cell Cardiol 13:1027-1030, 1981
43. Higgins TJC, Bailey PJ, Allsopp D: Interrelationship between cellular metabolic status and susceptibility of heart cells to attack by phospholipase. J Mol Cell Cardiol 14:645-654, 1982
44. Hill AV: A challenge to biochemists. Biochim Biophys Acta 4:4-11, 1950
45. Hochachka PW, Neely JR, Driedzic WR: Integration of lipid utilization with Krebs cycle activity in muscle. Fed Proc 36:2009-2014, 1977
46. Homsher E, Kean CJ: Skeletal muscle energetics and metabolism. Annu Rev Physiol 40:93-131, 1978
47. Huxtable RJ, Wakil SJ: Comparative mitochondrial oxidation of fatty acid. Biochim Biophys Acta 239:168-177, 1971
48. Ingwall JS, Kramer MF, Friedman WF: Developmental changes in heart creatine kinase. In: Jacobus WE, Ingwall JS, eds: Heart creatine kinase. Baltimore: Williams & Wilkins, 1980:9-17
49. Ingwall JS, Kramer MF, Woodman D, Friedman WF: Maturation of energy metabolism in the lamb: changes in myosin ATPase and creatine kinase activities. Pediat Res 15:1128-1133, 1981
50. Jacobs HK, Heldt HW, Klingenberg M: High activity of creatine kinase in mitochondria from muscle and brain and evidence for a separate mitochondrial isoenzyme of creatine kinase. Biochem Biophys Res Commun 16:516-521, 1964
51. Jacobus WE: Theoretical support for the heart phosphocreatine energy transport shuttle based on the intracellular diffusion limited mobility of ADP. Biochem Biophys Res Commun 133:1035-1041, 1985
52. Jacobus WE, Lehninger AL: Creatine kinase of rat heart mitochondria. J Biol Chem 248:4803-4810, 1973
53. Jacobus WE, Vandegaer KM, Moreadith RW: Aspects of heart respiratory control by the mitochondrial isozyme of creatine kinase. Adv Exp Med Biol 194:169-191, 1985
54. Klee CB, Sokoloff L: Changes in the D(-)-beta-hydroxybutyric dehydrogenase activity during brain maturation in the rat. J Biol Chem. 242:3880-3883, 1967
55. Lehninger AL, Greville GD: The enzymatic oxidation of d- and l-beta-hydroxybutyrate. Biochim Biophys Acta 12:188-202, 1953
56. Lehninger AL, Sudduth HC, Wise JB: D-beta-Hydroxybutyric dehydrogenase of mitochondria. J Biol Chem 235:2450-2455, 1960
57. Lipskaya TY, Temple VI, Belousova LV, Molokova EV: Interaction between heart mitochondrial creatine kinase and oxidative phosphorylation. Biokhimiya (Engl Transl) 45:1015-1022, 1980
58. Lohman K: Ueber die enzymatische Aufspaltung der Kreatinphosphosaure; zugleich ein Beitrage zum Chemismus der Muskelkontraktion. Biochem Z 271:264-279, 1934
59. Lough J, Bischoff R: Differentiation of creatine phosphokinase during myogenesis: quantitative fractionation of isozymes. Develop Biol 29:410-418, 1972
60. Lynen F: Biosynthesis of saturated fatty acids. Fed Proc 20:941-951, 1961
61. McCann WP, Greville GD: D(-) and L(+)-beta-Hydroxybutyric acid. Biochem Prep 9:63-68, 1962
62. Meyer RA, Sweeney HL, Kushmerick NJ: A simple analysis of the "phosphocreatine shuttle". Am J Physiol 246:C365-C377, 1984
63. Mishkin S, Stein L, Gatmaitan Z, Arias IM: The binding of fatty acids to cytoplasmic proteins:

     Binding to Z protein in the liver and other tissues of the rat. Biochem Biophys Res Commun 47:997-1003, 1972

64. Mishkin S, Trucotte R: The binding of long chain fatty acid CoA to Z, a cytoplasmic factor present in the liver and other tissues of the rat. Biochem Biophys Res Commun 57:918-926, 1974

65. Mommaerts WFHM: Energetics of contraction. Physiol Rev 40:427-508, 1969

66. Morgan HE, Henderson MJ, Regen DM, Park CR: Regulation of glucose uptake in muscle. I. The effects of insulin and anoxia on glucose transport and phosphorylation in the isolated, perfused heart of normal rats. J Biol Chem 236:253-261, 1961

67. Mustafa SJ, Berne RM, Rubio R: Adenosine metabolism in cultured chick-embryo heart cells. Am J Physiol 228:1474-1478, 1975

68. Neely JR, Bowman RHM, Morgan HE: Interaction of fatty acid and glucose oxidation by cultured heart cells. Am J Physiol 216:804-811, 1969

69. Neely JR, Morgan HE: Substrate and energy metabolism in the heart. Annu Rev Physiol 36:414-459, 1974

70. Neely JR, Rovetto MJ, Oram JF: Myocardial utilization of carbohydrates and lipids. Progr Cardiovasc Dis 15:289-329, 1972

71. Neely JR, Whitmer KM, Mochizuki S: Effects of mechanical activity and hormones on myocardial glucose and fatty acid utilization. Circ Res 37:733-741, 1975

72. Olson RE, Hoeschen RJ: Utilization of endogenous lipids by the isolated perfused heart. Biochem J 103:796-801, 1967

73. Opie LH: Myocardial energy metabolism. Adv Cardiol 12:70-83, 1974

74. Orloff KJ, McCarl RL: The effect of metabolic inhibitors on cultured rat heart cells. J Cell Biol 57:225-228, 1973

75. Paris S, Pousségur J, Ailhaud G: Sugar transport in chick embryo cardiac cells in culture. Analysis by countertransport, relationship to phosphorylation and effect of glucose starvation: Biochim Biophys Acta 602:644-652, 1980

76. Paris S, Samuel D, Jacques Y, Gache C, Franchi A, Ailhaud G: The role of serum albumin in the uptake of fatty acids by cultured cardiac cells from chick embryo. Eur J Biochem 83:235-243, 1978

77. Paris S, Samuel D, Romey G, Ailhaud G: Uptake of fatty acids by cultured heart cells from chick embryo. Evidence for a facilitation process without energy dependence. Biochimie 61:361-367, 1979

78. Perriard JC: Developmental regulation of creatine kinase isoenzymes in myogenic cell culture from chicken. Levels of mRNA for creatine subunits M and B. J Biol Chem 254:7036-7041, 1979

79. Perriard JC, Caravatti M, Perriard E, Eppenberger HM: Quantitation of creatine kinase isoenzyme transitions in differentiating chick embryonic breast muscle and myogenic cell cultures by immunoadsorption. Arch Biochem Biophys 191:90-100, 1978

80. Perriard JC, Perriard E, Eppenberger HM: Detection and relative quantitation of mRNA for creatine kinase isoenzymes in RNA from myogenic cell cultures and in fibroblasts. J Biol Chem 252:6529-6535, 1978

81. Perry SV: Creatine phosphokinase and the enzymic and contractile properties of the isolated myofibrils. Biochem J 57:427-434, 1954

82. Pinson A: Métabolisme de l'acide palmitique et de l'acide érucique par les cellules cardiaques bathmotropes de rat en culture. Acad thesis, Dijon, 1975

83. Pinson A, Degrès J, Heller M: Partial and incomplete oxidation of palmitate by cultured beating cardiac cells from neonatal rats. J Biol Chem 254:8331-8335, 1979

84. Pinson A, Frelin C, Padieu P: Palmitate oxidation by beating heart cells. Rec Adv Stud Cardiac Struct Metab 12:667-676, 1978

85. Pinson A, Padieu P: Erucic acid metabolism by cultured beating heart cells of the postnatal rat. In: Fleckenstein A, ed: Myocardial cell damage. 6th Annu Meet Int Study Group Res Cardiac Metab. Freiburg i Br, 1973: no 91 (Abstr)

86. Pinson A, Padieu P: Erucic acid oxidation by beating heart cells in culture. FEBS Lett 39:88-90, 1974

87. Pinson A, Padieu P: Quelques aspects du métabolisme des acides gras à longue chaine par les cellules bathmotropes de coeur de rat en culture. Biochimie 56:1587-1596, 1974

88. Posner B, Matsuzaki F, Raben MS: Differential oxidation of C-1, C-11 and C-16 of palmitate by rat tissues. Fed Proc 23:269, 1964 (Abstr)

89. Ramsay RR, Tubbs PK: The mechanism of fatty acid uptake by heart mitochondria: an acylcarnitine-carnitine exchange. FEBS Lett 54:21-25, 1975
90. Randle PJ: Fuel selection in animals. Biochem Soc Trans 14:799-806, 1986
91. Randle PJ, Garland PB, Hales CN, Newsholme EA, Denton RM, Pogson CI: Interactions of metabolism and the physiological role of insulin. Rec Progr Horm Res 22:1-48, 1966
92. Randle PJ, Tubbs PK: Carbohydrate and fatty acid metabolism. In: Berne RM, Sperelakis N, Geiger SR, eds: Handbook of Physiology. Vol I, Sect 2. The cardiovascular system. Bethesda: American Physiological Society, 1979:805-844
93. Reibel DK, Rovetto MJ: Myocardial adenosine salvage rates and restoration of ATP content following ischemia. Am J Physiol 237:H247-H252, 1979
94. Roberts R, Grace AM: Purification of mitochondrial creatine kinase. J Biol Chem 255:2870-2877, 1980
95. Rosenthal MD, Warshaw JB: Interactions of fatty acid and glucose oxidation by cultured heart cells. J Cell Biol 58:332-339, 1973
96. Rosenthal MD, Warshaw JB: Fatty acid and glucose oxidation by cultured rat heart cells. J Cell Physiol 93:31-40, 1977
97. Ross PD, McCarl RL: Oxidation of carbohydrate and palmitate by intact cultured neonatal rat heart cells. Am J Physiol 246:H389-H397, 1984
98. Ross PD, McCarl RL, Hartzell CR: Radiorespirometry and metabolism of cultured heart cells attached to petri dishes. Anal Biochem 112:378-386, 1981
99. Rubio R, Berne RM: Localization of purine and pyrimidine nucleoside phosphorylase in heart, kidney and liver. Am J Physiol 239:H721-H730, 1980
100. Saks VA, Chernousova GB, Vornkov II, Smirnov VN, Chasov EI: Study of the energy transport in myocardial cells. Circ Res 34, Suppl III:138-149, 1974
101. Samuel D, Paris S, Ailhaud G: Uptake and metabolism of fatty acids by cardiac cells from chick embryo. Eur J Biochem 64:583-595, 1976
102. Schoutsen B, De Jong JW, Harmsen E, De Tombe PP, Achterberg PW: Myocardial xanthine oxidase/dehydrogenase. Biochim Biophys Acta 762:519-524, 1983
103. Schroedl NA, Hartzell CR, Ross PD, McCarl RL: Glucose metabolism, insulin effects, and developmental age of cultured neonatal rat heart cells. J Cell Physiol 113: 231-239, 1982
104. Seraydarian MW, Abbott BC: The role of the creatine-phosphocreatine system in muscle. J Mol Cell Cardiol 8:741-746, 1976
105. Seraydarian MW, Artaza L: Regulation of energy metabolism by creatine in cardiac and skeletal muscles in culture. J Mol Cell Cardiol 8:669-678, 1976
106. Seraydarian MW, Artaza L, Abbott BC: Creatine and control of energy metabolism in cardiac and skeletal muscle cells in culture. J Mol Cell Cardiol 6:405-413, 1974
107. Sharov VA, Saks VA, Smirnov VN, Chasov EI: An electron microscopic histochemical investigation of the localization of creatine kinase in heart cells. Biochim Biophys Acta 468:495-501, 1977
108. Shipp JC, Thomas J, Crevasses LE: Oxidation of carbon-14-labelled endogenous lipids by isolated perfused rat heart. Science 143:371-373, 1964
109. Sobel BE, Shell WE, Klein MS: An isoenzyme of creatine phosphokinase associated with rabbit heart mitochondria. J Mol Cell Cardiol 4:367-380, 1972
110. Spector AA: Transport and utilization of free fatty acid. Ann NY Acad Sci 149:768-783, 1968
111. Spitzer JJ, Gold M: Free fatty acid metabolism by skeletal muscle. Am J Physiol 206:159-164, 1964
112. Stanisz J, Wice BM, Kennell DE: Comparative energy metabolism in cultured heart muscle and HeLa cells. J Cell Physiol 115:320-330, 1983
113. Stein O, Stein Y: Lipid synthesis, intracellular transport and storage. III. Electron microscopic radioautographic study of the rat heart perfused with tritiated oleic acid. J Cell Biol 36:63-77, 1968
114. Stewart HB, Tubbs PK, Stanley KK: Intermediates of fatty acid oxidation. Biochem J 132:61-76 1973
115. Takehori Y, Yang JJ, Ricchiuti NV, Seraydarian MW: Oxygen consumption of mammalian myocardial cells in culture: measurements in beating cells attached to the substrate in the culture dish. Anal Biochem 145:302-307, 1985
116. Taylor SH: Insulin and heart failure. Br Heart J 33:329-333, 1971

117. Thompson CI, Rubio R, Berne RM: Changes in adenosine and glycogen phosphorylase activity during the cardiac cycle. Am J Physiol 238:H389-H398, 1980
118. Van Brussel E, Yang JJ, Seraydarian MW: Isozymes of creatine kinase in mammalian cell cultures. J Cell Physiol 116:221-226, 1983
119. Van der Laarse A, Graf-Minor ML, Witteveen SAGJ: Release of hypoxanthine from and enzyme depletion in rat heart cultures deprived of oxygen and metabolic substrates. Clin Chim Acta 91:47-52, 1979
120. Van der Laarse A, Hollaar L, Kokshoorn LJM, Witteveen SAGJ: The activity of cardio-specific isoenzymes of creatine phosphokinase and lactate dehydrogenase in monolayer cultures of neonatal rat heart cells. J Mol Cell Cardiol 11:501-510, 1979
121. Van der Laarse A, Hollaar L, Van der Valk EJ, Witteveen SAGJ: Enzyme release from and enzyme depletion in rat heart cell cultures during anoxia. J Mol Med 3:123-131, 1978
122. Vemuri R: Biochemical alterations in cultured heart cells and their sarcolemma during ischaemia, hypoxia and anoxia. Acad thesis, Jerusalem, 1986
123. Vemuri R, De Jong JW, Huizer T, Hegge JAJ, Heller M, Pinson A: High energy phosphate metabolism in cultured rat heart cells during anoxia and volume restriction. 1988, submitted
124. Vemuri R, Heller M, Pinson A: Studies on oxygen and volume restriction in cultured cardiac cells. II. The glucose effect. Basic Res Cardiol 80:165-169, 1985
125. Vemuri R, Mersel M, Heller M, Pinson A: Studies on oxygen and volume restriction in cultured heart cells: Possible rearrangement of sarcolemmal lipid moieties during ischemia. Mol Cell Biochem 1988, in press
126. Vemuri R, Yagev S, Heller M, Pinson A: Studies on oxygen and volume restriction in cultured cardiac cells. I. A model for ischemia and anoxia with a new approach. In Vitro 21:521-525, 1985
127. Vial C, Godinot C, Gautheron DC: Creatine kinase in pig heart mitochondria. Properties and role in phosphate potential regulation. Biochimie 54:843-852, 1972
128. Wakil SJ, Mahler HR: Studies on the fatty acid oxidizing system of animal tissues. V. Unsaturated fatty acyl coenzyme A hydrolase. J Biol Chem 207:125-132, 1954
129. Warshaw JB, Rosenthal MD: Changes in glucose oxidation during growth of embryonic heart cells in culture. J Cell Biol 52:283-291, 1972
130. Webber RJ, Edmond J: Utilization of L(+)-3-hydroxy-butyrate, acetoacetate, and glucose for respiration and lipid synthesis in the 18-day-old rat. J Biol Chem 252:5222-5226, 1977
131. Weinman EO, Chaikoff IL, Dauben WG, Gee M, Entenman C: Relative rates of conversion of the various atoms of palmitic acid to carbon dioxide in the intact rat. J Biol Chem 184:735-744, 1950
132. Weisiger R, Golan I, Ockner R: Receptors for albumin on the liver cell surface may mediate uptake of fatty acid and other albumin-bound substances. Science 211:1048-1050, 1981
133. Williamson JR, Krebs HA: Acetoacetate as a fuel for respiration in the perfused rat heart. Biochem J 80:540-547, 1961
134. Wollenberger A, Babskii EB, Krause EG, Genz S, Blohm D, Bogdanova EV: Cyclic changes in the levels of cyclic AMP and cyclic GMP in frog myocardium during the cardiac cycle. Biochem Biophys Res Commun 55:446-452, 1973
135. Wosilait WD, Soler-Argilaga C, Nagy P: A theoretical analysis of the binding of palmitate by human serum albumin. Biochem Biophys Res Commun 71:419-426, 1976
136. Yagev S, Heller M, Pinson A: Changes in cytoplasmic and lysosomal enzyme activities in cultured rat heart cells: The relationship to cell differentiation and cell population in cultures. In Vitro 20:893-898, 1984
137. Yagi K, Mase R: Coupled reaction of creatine kinase and myosin A ATPase. J Biol Chem 237:397-403, 1962
138. Yagi K, Noda L: Phosphate transfer to myofibrils by creatine kinase. Biochim Biophys Acta 43:249-259, 1960
139. Yang WCT, Geiger PJ, Bessman SP: Formation of creatine phosphate from creatine on ^{32}P-labelled ATP by isolated rabbit heart mitochondria. Biochem Biophys Res Commun 76:882-887, 1977
140. Ziegler B, Lippmann HG, Mehling R, Jutzi E: Characterization of growth and glucose utilization of a cell culture of chicken embryo heart. Biol Med Germ 25:555-571, 1970
141. Zierler KL, Maseri A, Klassen G, Rabinowitz D, Burgess J: Muscle metabolism during exercise in man. Trans Assoc Am Physicians 81:266-273, 1968

Chapter 15

Energy Metabolism of Adult Cardiomyocytes

H.M. Piper, B. Siegmund and R. Spahr, Physiologisches
Institut I, Universität Düsseldorf, Düsseldorf, F.R.G.

By now the basic features of the energy metabolism of isolated cardiomyocytes have been described. Unstimulated cardiomyocytes are quiescent and exhibit the characteristics of cardiac metabolism at basal energic demand. Their energetic demand, however, can be stimulated to the maximal rates of the physiological range. As oxidative substrates fatty acids are preferred to lactate and glucose. The uptake kinetics for these three substrates have been characterized. For glucose and lactate defined transport mechanisms are identified, for fatty acids a purely physical diffusion process is also discussed. Under anoxia, energy production falls short of energy supply. Already during the reversible phase of hypoxic injury, cardiomyocytes have some cytosolic enzyme release. It is hypothesized that both release of enzymes and development of rigor are causally related to changes in the free energy change of cytosolic ATP hydrolysis rather than in absolute ATP concentrations. Loss of Ca^{2+} control follows energetic exhaustion. At cytosolic Ca^{2+} levels beyond 3 uM, it becomes irreversible. In detail the relation between energy metabolism and onset of irreversible hypoxic cell injury is not sufficiently understood.

I. Introduction

Isolated adult cardiomyocytes have become an established tool in heart research. A number of ways has been described to isolate muscle cells from myocardial tissue, and primary cultures of these non-dividing cells have also been developed (34). This review is designed to summarize the current knowledge about the basic features of their substrate and energy metabolism.

The results demonstrate that today cardiomyocyte preparations can be obtained that exhibit the typical metabolic characteristics of the myocardium. Studying metabolic behavior of a purified myocyte preparation allows to identify the contribution of the cardiac muscle cell type to the complex metabolic reaction of whole tissue. Furthermore, the properties of the cardiomyocyte can be investigated individually, i.e., free of the immediate influence of adjacent muscle cells. This latter possibility renders the use of cardiomyocytes particulary helpful in studies on hypoxic cell injury since in this model inhomogeneities in the exogenous factors, acting on the individual cell, can be avoided.

II. Energy Metabolism under Aerobic Conditions

1. Oxygen Consumption

In the absence of external stimulation, adult ventricular myocytes have a stable resting membrane potential and are mechanically quiescent. Therefore their oxygen demand is low. In newly isolated as well as in short-term cultured cardiomyocytes, this basal oxygen demand was found to be approximately 1 umol O_2/min per g wet wt. at 37°C (47,50,51,60). This value is close to the oxygen demand of a rat heart in cardioplegic arrest (3). With respiratory uncoupling, oxygen consumption of cardiomyocytes can be stimulated to a magnitude known for the heart under maximal catecholamine stimulation, i.e., 20 to 40 umol O_2/min per g wet wt. (60). This demonstrates that isolated cardiomyocytes possess the full physiological range for variations of energy demand. The Q_{10} value for lowering the temperature from 37 to 27°C was determined as 1.8 (60), closely resembling the corresponding Q_{10} values of the arrested rat heart (3).

Isolated cardiomyocytes have also been stimulated to contract by electrical field stimulation. The metabolic cost per contraction was found to be similar to that in myocardial tissue under zero load conditions (25). The increment in energetic demand caused by mechanical activity under such unloaded conditions is proportional to the square of length changes, indicating that it accounts primarily for the elastic compression of the myofilaments (54).

2. Uptake of Substrates

The uptake of glucose into the myocardial cell is mediated by a specific transport system of which the affinity and the capacity can vary. In isolated cardiomyocytes insulin accelerates initial uptake rates of 3-O-methylglucose, a nonmetabolized glucose analogue, into isolated cardiomyocytes by a factor of 2 (14) to 9 or 10 (19,24). Interestingly, the V_{max} values obtained with insulin were identical in these studies, whereas the basal rates varied. In the unstimulated state, the glucose transporter has a low affinity for glucose. K_m values range from 7-9 mM (19,52). Insulin leaves the transporter as one of low affinity. Thus, insulin acts mainly through enlarging the capacity of the transport system. The onset of insulin action on stimulating glucose uptake is observed in the first 30 s, the effect is maximal in 1-3 min (8,14). The promptness of the reaction indicates that it is not dependent on protein synthesis. This is consistent with the hypothesis that additional transporter molecules are only recruited from internal stores (38).

Under hypoxic and other forms of energetic stress the uptake of glucose by the myocardium is greatly enhanced (37,42,52). In contrast to the action of insulin, the affinity for glucose is fivefold increased, possibly helping the ischemic cell to continue aerobic glycolysis when interstitial concentrations of glucose decrease. The specific effect of insulin on glucose uptake under hypoxic conditions is much reduced. It seems that the maximal number of glucose transporters mobilized under the short-term action of insulin or anoxia is the same. When cellular energy reserves are largely exhausted glucose transport slows down again (21).

Increased pressure work of myocardium was also reported to enhance glucose transport with an increased V_{max} rate and decreased K_m value (44). In contrast, when unloaded isolated cardiomyocytes were stimulated to contractions (and this caused a fivefold increase in oxygen demand) V_{max} rates remained unchanged whereas the K_m value decreased by 50% (55). Thus, increased energy turnover per se cannot be the cause for a stimulated glucose uptake under pressure work.

Lactate is a major oxidative substrate for the myocardium and isolated cardiomyocytes. The transport of lactate into the isolated myocardial cell has been analyzed by Kammermeier et al. (36). They demonstrated that lactate uptake is characterized by a sigmoidal dose dependence, indicating a carrier-mediated process. The half-maximal transport rate was found in the range between 5-10 mM, i.e., the affinity of the transport system is low. At physiological concentrations of lactate, 2.3 mM pyruvate inhibited the uptake of lactate, suggesting a competitive mechanism. The uptake of lactate was enhanced when the pH was lowered. But the increment in transport rate exceeded the increase in the concentration of the protonated acid by about tenfold. From this the authors concluded that, instead of favoring nonionic diffusion, protons are more likely involved in the mechanism of a carrier-mediated transport process, e.g., in form of an H^+-lactate cotransport. Dennis et al. (13) have characterized the lactate transport system in the perfused rat heart, and came to similar conclusions.

Long-chain nonesterified fatty acids are major substrates for myocardial energy production. In the extracellular space, fatty acids are non-covalently bound to albumin. It has first been demonstrated by Hütter et al. (32) that the oxidation of fatty acids in the working heart is a function of the concentrations of the fatty acid/albumin complex, and not of the steady state concentration of unbound fatty acids. For constant fatty acid/albumin ratios, saturable Michaelis-Menten type kinetics were observed. In isolated cardiomyocytes, it was shown that transsarcolemmal transport obeys the same kind of kinetics (53). The V_{max} values obtained for various fatty acid/albumin ratios were virtually identical with those obtained for fatty acid oxidation in the whole heart. This indicates that the rate-limiting process in both experimental systems is the same, namely transsarcolemmal transport. Recently similar kinetics have also been reported for the translocation of long-chain fatty acids through the endothelium of the heart (63). But the maximal transport rate for fatty acids through the endothelium (160 nmol/min per g tissue, equivalent to approximately 5 umol/min per g endothelial mass) is far greater than that for transsarcolemmal transport (35 nmol/min per g cardiomyocyte mass), at a palmitate/albumin molar ratio of about 1 (53,63).

Because of the saturable kinetics for fatty acid uptake by various animal cells, some mechanism of facilitated diffusion is generally assumed. But the validity of this conception has been questioned since the purely physical partitioning of fatty acids between the complex with albumin and a phospholipid bilayer obeys the same type of saturable Michaelis-Menten-like kinetics (45). Model calculations indicate that the transport rates in the liver cell could be fully accounted for by this purely physical mechanism. Such a calculation can also be made for the myocardial cell. Similarly, DeGrella and Light (11,12) have argued that cardiomyocytes in the absence of albumin take up fatty acids purely by diffusion. The rates observed would fully account for transport rates observed in vivo.

3. Substrate Oxidation

In short-term cultured cardiomyocytes, oxidation rates for physiological substrates (glucose, lactate, pyruvate, palmitate) were found constant for several hours (50,51). At physiological concentrations, fatty acids are preferred to lactate and glucose. At the excessive concentration of 5 mM, pyruvate is oxidized at a rate similar to that for exogenous fatty acids. Thus, in cardiomyocytes, physiological substrates are used in a pattern characteristic for the myocardium. For cultured cardiomyocytes, however, these exogenous substrates represent only a minor energy source. After 4 h in serum-supplemented medium 199, the cells contain enlarged stores of glycogen and triacylglycerols (50). When incubated in a simple saline medium with selected exogenous carbon sources, they preferentially oxidize fatty acids from endogenous triacylglycerol.

It was shown that in quiescent short-term cardiomyocytes endogenous lipolysis accounts for 94% of the calculated oxygen demand, when exogenous substrates are absent (50). The remaining share is covered by glycogen degradation. Consistent with the pronounced preference for lipids, only 7% of pyruvate dehydrogenase is in its active form. The supply of exogenous glucose together with insulin stops glycogenolysis and increases glycolysis, but pyruvate dehydrogenase activity increases only slightly. Superphysiological concentrations of pyruvate (5 mM) displace glucose and glycogen as oxidative substrates and induce an activation of pyruvate dehydrogenase. The contribution of endogenous lipolysis to oxidative energy metabolism is conversely reduced. At equimolar concentration, lactate is less effective than pyruvate (50).

In detail, the metabolic behavior of quiescent short-term cultured cardiomyocytes reflects the characteristics of cardiac energy metabolism at low demand (29,33). In aerobic myocardium at low respiratory rates, concentrations of NADH and acetyl-CoA are elevated. This causes beta-oxidation rates to be low. Increased cytosolic long-chain acyl-CoA levels in the cytosol together with high concentrations of ATP and production of alpha-glycerophosphate favored by the reduced cytosolic state stimulate triacylglycerol synthesis. On the other hand, NADH and acetyl-CoA are known to promote conversion of pyruvate dehydrogenase into its inactive form (52). Another factor also reducing the activity of this enzyme consists in low mitochondrial Ca^{2+} levels in the quiescent cardiomyocytes (20). At low activity of this flux-regulating enzyme, fatty acids are by far the preferred substrate.

A reduced cytosolic state together with a low oxidation rate of pyruvate favors net lactate production in quiescent myocytes under fully oxygenated conditions (51). With glucose as sole exogenous substrate, they produce lactate in higher quantities than CO_2 from exogenous glucose. As expected, this relation is inverted when the energy demand is increased. It is well documented that lactate is produced also by the normoxic heart if perfused with glucose as the sole substrate, especially at a reduced work load (61,67). In extrapolating these results, a relatively large lactate production from aerobic, quiescent cardiomyocytes is indeed predicted. In contrast to the behavior of quiescent cardiomyocytes not long after isolation, cardiomyocytes apparently have a truly enlarged glycolytic capacity after several weeks in culture. After 25 days in culture, the energy obtained by degradation of glucose was 30% aerobic and 70% glycolytic (R. Spahr, H.M. Piper & S.L. Jacobson, unpublished observations).

At a normal energetic state, the phosphofructokinase reaction is the major rate-limiting step in the Embden-Meyerhof pathway. Its activity is regulated predominantly by the cytosolic concentration of citrate and of fructose 2,6-bisphosphate (31,52). The effect of both these regulators can be demonstrated also in quiescent cardiomyocytes. Citrate levels have been shown to correlate closely with glycolytic flux rates in experiments in which primarily the relative contribution of lipids to substrate oxidation is varied (51). Under insulin, the prevalent stimulus for phosphofructokinase activity is apparently fructose 2,6-bisphosphate (51).

III. Energy Metabolism and Cell Injury under Hypoxic Conditions

1. Energy Production and Demand

It has been demonstrated that the individual heart-muscle cell will not stop mitochondrial respiration and increase glycolytic flux unless exogenous pO_2 has dropped below 0.5 torr (68). This means that the cell as a whole is almost as sensitive to low oxygen tension as the isolated mitochondrion is. Conditions with an oxygen tension below this threshold level are usually referred to as "anoxia".

The hypoxic and the ischemic myocardium reduce their energy demand immediately after oxygen withdrawal (40). In the beating heart, the reduction in contractile activity, i.e., mechanical failure, accounts for most of this self-protective mechanism. This, however, is only part of the energy-saving mechanism, since myocardium, arrested prior to ischemia (5), and quiescent myocytes (48) also quickly reduce their energy demand under hypoxic conditions. In addition, the absolute saving in energy consumption depends on the prehypoxic energy expenditure (58). Except from the energy-conserving effect of early mechanical failure, little is known about the nature of these mechanisms.

In spite of this self-protective reduction of energy demand, anaerobic energy metabolism at the end always falls short of energy needs as indicated by the loss of high-energy phosphates (40). Cultured cardiomyocytes contain elevated amounts of glycogen, and they use glycogen for glycolytic energy production in substrate-free anoxia. But when ATP contents have fallen by about 70%, glycolytic energy production almost stops even though still half of this substrate is left (48). It has been suggested by Kübler and Spieckermann (40) that such inability to use a glycolytic substrate is due to the lack of phosphorylation energy at the phosphofructokinase level. In the hypoxic perfused heart (26) and in anoxic isolated cardiomyocytes (48), irreversible injury can be postponed considerably if glycolytic flux is stimulated already at the onset of oxygen deprivation. Since enhanced glycolysis delivers more ATP to the cell, cellular ATP contents decrease at a slower rate. In these systems the waste products of glycolysis, lactate and protons, do not accumulate to the same extent as in ischemic tissue.

A contribution of mitochondria to the loss of energy in the hypoxic cell is well documented (23,56). In the non-respiring state, the mitochondrial ATP-synthetase acts in a backward direction, i.e., as an ATP-hydrolase. Thus, one major effect by which mitochondria contribute to the aggravation of hypoxic cellular injury consists in the inversion of a normal physiological function. The functional injury of mitochondria isolated from ischemic and hypoxic myocardium has been the subject of many studies, but it has not been demonstrated that these impairments are responsible for the functional inability of the heart after reperfusion. In isolated cardiomyocytes impairment of mitochondrial metabolism was not found very pronounced, when compared with other aspects of metabolic disturbance (9,41). And even in cells becoming hypercontracted in anoxia, the changes in mitochondrial ultrastructure are relatively moderate (4,57).

Other results also indicate that a loss of mitochondrial function determines not the "point of no-return" of hypoxic cell injury. When cytosolic Ca^{2+} just starts to rise in a hypoxic cardiomyocyte, resupply of ATP by resumption of oxidative phosphorylation prevents the impending loss of Ca^{2+} control (2). But when cytosolic free Ca^{2+} concentrations have risen to 3 uM, loss of Ca^{2+} control has become irreversible (2). It is probably the inability of mitochondria to resume sufficient ATP production at this Ca^{2+} level, defining it as a crucial threshold. Since this behavior is based on a normal functional property of mitochondria (7), it does not need specific mitochondrial injury to bring about cell death. In summary, according to current evidence mitochondria are not a "limiting structure" for cell survival.

2. Enzyme Release

Release of cytosolic enzymes from the hypoxic perfused heart or from hypoxic isolated cardiomyocytes has been demonstrated to correlate with cellular ATP contents in certain experimental models (18,28,48). Already a small reduction in contents of energy-rich phosphates, as it may also occur in physiological stress, leads to a detectable release of enzymes (59). Studies on isolated cardiomyocytes in substrate-free anoxia have shown that this early enzyme release is due to protein loss from only the cytosolic compartment of cells which are still in the reversible phase of hypoxic injury (48,57). At this early stage, ultrastructural alterations are subtle (57); large holes in the sarcolemma can be ruled out as causing the protein leakage. The number of hypercontracted cells does not correlate with this early release of cytosolic enzymes.

There may be a causal relationship between protein release from hypoxic cells and a) increase in the number of subsarcolemmal vesicles (57); b) protrusion of cell surfaces into small sarcolemmal blebs (57); and c) rapid increase in fluidity of the hypoxic sarcolemma by an order of magnitude (16). Leakage of soluble proteins from reversibly energetically stressed cells has also been reported for other cells types (6,46,66,69). It has been hypothesized by Spieckermann et al. (59) that the heart cell releases cytosolic enzymes continuously, but at a low rate, and that the release through this unidentified channel is enhanced under conditions of energetic stress.

A comparison of energy depletion of isolated cardiomyocytes by anoxia and by glycolytic blockade, however, points to the free-energy change of cytosolic ATP hydrolysis (64) as a decisive factor for early enzyme release (49). This magnitude characterizes the thermodynamic availability of ATP and is determined by the ratio $(ATP/ADP \cdot P_i)$. Iodoacetate blocks glycolytic flux at the level of the glyceraldehyde-3-phosphate dehydrogenase reaction. It traps free cytosolic P_i by an enormous accumulation of phosphorylated intermediates. In comparison to anoxia, therefore, at a given ATP level in iodoacetate-treated myocytes, the $(ATP/ADP \cdot P_i)$ ratio will be higher, i.e., more phosphorylation energy will be available in the cytosol. After a loss of one third of initial ATP contents, release of lactate dehydrogenase from anoxic cells was indeed found three times higher then from iodoacetate-treated cells with the same ATP level (49).

In infarcted myocardium many cells are lysed. Therefore, a considerable part of the enzyme release from such tissue can be explained by mass loss from broken cells. Indeed, it has been shown that the size of the necrotic area correlates well with the amount of enzymes lost from infarcted myocardium (1). In the clinical setting, detection of cardiospecific enzymes in the circulation will generally indicate the occurrence of necrotic cells, because the moderate early release of cytosolic enzymes easily remains undetected due to the impeded washout from ischemic areas and the limited sensitivity of the methods in use.

3. Contracture Development

In the course of progressive energy breakdown, contracture development is observed in the globally-ischemic myocardium ("stone heart") and also in isolated, energy-depleted cardiomyocytes. A number of studies indicate that the development of tension in the myocardial cell under hypoxia and ischemia (2,30), and also under other conditions of energy depletion (10), is caused by the formation of rigor complexes, i.e., of fixed crossbridges between actin and myosin filaments. Thus it is not initiated by a rise in cytosolic Ca^{2+}, which would lead through fast crossbridge cycling to sustained contraction. Recently, experiments deciding between these alternative mechanisms became feasible. By simultaneous monitoring of the cytosolic free Ca^{2+} concentration and cell shortening in single isolated cardiomyocytes, it was demonstrated that cell shortening of hypoxic cardiomyocytes precedes a rise in cytosolic free Ca^{2+} (2).

The factors causing rigor bond formation in the oxygen-deficient heart cell are still a matter of debate. Based on the finding by Weber and Murray (65) that, in vitro, rigor bonds between actin and myosin are formed in a Ca^{2+}-independent manner if the ATP concentration is below 100 uM, it has been hypothesized that in the hypoxic cell, too, rigor is caused by a lowering of the cytosolic ATP concentration below this level (23,30,39). However, the average ATP concentration in hypoxic heart tissue in contracture can be equal to or greater than 1 mM. It was therefore assumed that the cytosol contains a subcompartment next to the myofibrils in which the ATP concentration is critically reduced. So far, no direct evidence has been given for the thruth of this assumption. In the following an alternative explanation will be proposed.

The crossbridge cycle is driven by the energy liberated during ATP hydrolysis. Thermodynamically, the energy available from ATP hydrolysis is characterized by the free-energy change ΔG (64), i.e., $\Delta G = \Delta G_o + RT \ln [ATP]/[ADP] \cdot [P_i]$. In this equation ΔG_o signifies the standard free-energy change; R, the universal gas constant; T, the absolute temperature; and [ ] refer to the free metabolite concentrations. The rigor state is supposed to occur transiently in a normal crossbridge cycle and represents the state lowest in potential energy (15). It is therefore conceivable that single crossbridges remain trapped in the rigor state, if the average ΔG value is too low to allow the cycling of all crossbridges to be completed. In fact, tissue tension in the ischemic rat heart is found to increase after about 10 min (27), at which time the cytosolic ΔG has been reduced from 60 to about 45 kJ/mol (17).

In the globally ischemic and hypoxic heart, rigor is accompanied by muscle shortening. This is a gradual process, leading to shortened sarcomeres throughout the tissue. A gradual shortening of the myofibrils is also observed in single isolated myocytes under anoxia, if these are attached to a substratum (57). At an average ATP content of about 2 umol/g wet wt., these cells are in a square, but still polygonal form. The sarcomere length is reduced from the initial 1.9 um (slack length of sarcomeres) to 1.4 um (57). At 1.4-um sarcomere length, myosin filaments are already distorted. This sarcomere length, however, is also observed during forceful contractions in cardiomyocytes and is fully reversible. If the cells in this shortened square form are reoxygenated, they relengthen (57). This behavior demonstrates the reversible nature of this rigor state. In contrast to attached cells, free-floating single cells do not shorten in a gradual fashion, but rapidly shrink to a minimal sarcomere length of about 1.1 um or 60% their initial length (2,23). This difference of behavior can probably be explained by the complete absence of any force opposing rigor forces in non-attached cells. Such extremely contracted cells are usually unable to relengthen upon reoxygenation, although they can regain control of their cytosolic free Ca^{2+} levels (2). Apparently, at 1.1-um sarcomere length, severe disruptions of the cytoskeleton have taken place, which normally disable the cells to reassume an elongated shape.

4. The "Point of No-Return"

From studies on short-term cultured surface-attached cardiomyocytes, which seem to form the most stable system currently available, it is now clear that, in the muscle cell itself, onset of enzyme release, irreversibility of structural cell injury, and cytolysis are separate events. In the heart, exposed to ischemia of hypoxia, these events occur in indistinguishably close temporal coincidence.

A sufficient criterion for impending cell death consists in the loss of cytosolic Ca^{2+} control. In isolated hypoxic cardiomyocytes, a rise in cytosolic Ca^{2+} is not the result of large membrane perforations. The Ca^{2+} accumulation in the cytosol enters the cell from outside, but the accumulation proceeds gradually and the disturbance of Ca^{2+} homeostasis can be reversed by early resumption of oxidative energy production. Therefore, it seems to be a shortage of energy which primes the heart cell for cell death. The sarcoplasmic reticulum plays a decisive role at the edge of irreversible loss of Ca^{2+} control: when its ability to sequester Ca^{2+} was blocked with caffeine, any recovery of Ca^{2+} control was abolished (2).

The exact role of energy metabolism in the process of progressive injury is still an open field for research. So far, in most studies average levels of high-energy phosphates have been determined and not the free-energy change of ATP hydrolysis. The latter is a parameter for the thermodynamically available energy. An apparent ATP threshold (1-2 umol/g wet wt.) for resuscitability has been described for ischemic dog myocardium (35,40), ischemic rat hearts (62), and anoxic cultured rat cardiomyocytes (48,57). But it is not definitely clear whether this threshold is a borderline under all circumstances. In energetically-poisoned cardiomyocytes, the ATP concentration may fall to much lower levels, before the cytosolic Ca^{2+} concentrations rises (22). This, however, must not mean that a true recovery from these ATP concentrations is possible before the loss of Ca^{2+} control. It has also been claimed that under certain conditions, ischemic hearts can recover from similar low ATP levels (43). But since in the latter study only some functional parameters for recovery were investigated, it does neither prove the possibility of true structural and metabolic recovery from very low energy levels.

Acknowledgement

This study has been supported by the Deutsche Forschungsgemeinschaft, grant Pi 162/2-1.

References

1. Ahmed SA, Williamson JR, Roberts E, Clark RE, Sobel BE: The association of increased plasma MB CPK activity and irreversible ischemic myocardial injury in the dog. Circulation 54:187-193, 1976
2. Allshire A, Piper HM, Cuthbertson KSR, Cobbold PH: Cytosolic free Ca^{2+} in single rat heart cells during anoxia and reoxygenation. Biochem J 244:381-385, 1987
3. Arnold G, Lochner W: Die Temperaturabhängigkeit des Sauerstoffverbrauchs stillgestellter künstlich perfundierter Warmblüterherzen zwischen 34°C und 4°C. Pflügers Arch 284:169-175, 1967
4. Borgers M, Piper HM: Ca^{2+}-shifts in anoxic cardiac myocytes. A cytochemical study. J Mol Cell Cardiol 18:439-448, 1986
5. Bretschneider HJ, Gebhard MM, Preusse CJ: Cardioplegia. Principles and problems. In: Sperelakis N, ed: Physiology and pathophysiology of the heart. The Hague: Nijhoff Publ, 1986:605-616
6. Bütikofer P, Ott P: The influence of cellular ATP levels on dimyristoylphosphatidylcholine-induced release of vesicles from human erythrocytes. Biochim Biophys Acta 821:91-96, 1985
7. Carafoli E: The homeostasis of calcium in heart cells. J Mol Cell Cardiol 17:203-212, 1985
8. Chen V, McDonough KH, Spitzer JJ: Effects of insulin on glucose metabolism in isolated heart myocytes from adult rats. Biochim Biophys Acta 846:398-404, 1985
9. Cheung JY, Leaf A, Bonventre JV: Mitochondrial function and intracellular calcium in anoxic cardiac myocytes. Am J Physiol 250:C18-C25, 1986
10. Cobbold PH, Bourne PK: Aequorin measurements of free calcium in single hearts cells. Nature 312:444-446, 1984
11. DeGrella RF, Light RJ: Uptake and metabolism of fatty acids by dispersed adult rat heart myocytes. I. Kinetics of homologous fatty acids. J Biol Chem 255:9731-9738, 1980
12. DeGrella RF, Light RJ: Uptake and metabolism of fatty acids by dispersed adult rat heart myocytes. II. Inhibition by albumin and fatty acid homologues, and the effect of temperature and metabolic reagents. J Biol Chem 255:9739-9745, 1980
13. Dennis SC, Kohn MC, Anderson GJ, Garfinkel D: Kinetic analysis of monocarboxylate uptake into perfused rat hearts. J Mol Cell Cardiol 17:987-995, 1985
14. Eckel J, Pandalis G, Reinauer H: Insulin action on the glucose transport system in isolated cardiocytes from adult rat. Biochem J 212:385-392, 1983
15. Eisenberg E, Hill TL: A cross-bridge model of muscle contraction. Progr Biophys Mol Biol 33:55-82, 1978
16. Finch SAE, Piper HM, Spieckermann PG, Stier A: Anoxia influences the lateral diffusion of a lipid probe in the plasma membrane of isolated cardiac myocytes. Basic Res Cardiol 80, Suppl 1:145-152, 1985
17. Fiolet JWT, Baartscheer A, Schumacher CA, Coronel R, Ter Welle HF: The change of the free energy of ATP-hydrolysis during global ischemia and anoxia in the rat heart. J Mol Cell Cardiol 16:1023-1036, 1985
18. Gebhard MM, Denkhaus H, Sakai K, Spieckermann PG: Energy metabolism and enzyme release. J Mol Med 2:271-283, 1977
19. Gerards P, Graf W, Kammermeier H: Glucose transfer studies in isolated cardiocytes of adult rats. J Mol Cell Cardiol 14:141-149, 1982
20. Hansford RG: Relation between cytosolic free Ca^{2+} concentration and the control of pyruvate dehydrogenase in isolated cardiac myocytes. Biochem J 241:145-151, 1987
21. Haworth RA, Berkhoff HA: The control of sugar uptake by metabolic demand in isolated adult rat heart cells. Circ Res 58:157-165, 1986
22. Haworth RA, Goknur AB, Hunter DR, Hegge JO, Berkhoff HA: Inhibition of calcium influx in isolated adult rat heart cells by ATP depletion. Circ Res 60:586-594, 1987
23. Haworth RA, Hunter DR, Berkhoff HA: Contracture in isolated adult rat heart cells: role of Ca^{2+}, ATP and compartmentation. Circ Res 49: 1119-1128, 1981

24. Haworth RA, Hunter DR, Berkhoff HA: Heterogenous response of isolated heart cells to insulin. Arch Biochem Biophys 233:106-114, 1984
25. Haworth RA, Hunter DR, Berkhoff HA, Moss RL: Metabolic cost of the stimulated beating of isolated adult heart cells in suspension. Circ Res 52:342-351, 1983
26. Hearse DJ, Chain EB: The role of glucose in the survival and recovery of the anoxic isolated perfused rat heart. Biochem J 128:1125-1133, 1972
27. Hearse DJ, Garlick PB, Humphrey SM: Ischemic contracture of the myocardium: Mechanisms and prevention. Am J Cardiol 39:986-993, 1977
28. Higgins TJC, Allsopp D, Bailey PJ, D'Souza EDA: The relationship between glycolysis, fatty acid metabolism and membrane integrity in neonatal myocytes. J Mol Cell Cardiol 13:599-615, 1981
29. Hiltunen JK, Hassinen IE: Energy-linked regulation of glucose and pyruvate oxidation in isolated perfused rat heart. Role of pyruvate dehydrogenase. Biochim Biophys Acta 440:377-390, 1976
30. Holubarsch C: Force generation in experimental tetanus, KCl contracture and oxygen and glucose deficiency contracture in mammalian myocardium. Pflügers Arch 396:277-284, 1983
31. Hue L, Rider M: Role of fructose 2,6-bisphosphate in the control of glycolysis in mammalian tissues. Biochem J 245:313-324, 1987
32. Hütter JF, Piper HM, Spieckermann PG: Kinetic analysis of myocardial fatty acid oxidation suggesting an albumin receptor mediated uptake process. J Mol Cell Cardiol 16:219-226, 1984
33. Idell-Wenger JA, Neely JR: Regulation of uptake and metabolism of fatty acids by muscle. In: Dietschy L, Gotto AM, Ontko JA, eds: Disturbances in lipid and lipoprotein metabolism. Bethesda: American Physiological Society, 1978:269-284
34. Jacobson SL, Piper HM: Cell cultures of adult cardiomyocytes as models of the myocardium. J Mol Cell Cardiol 18:439-448, 1986
35. Jennings RB, Hawkins HK, Lowe JE, Hill ML, Klotman S, Reimer KA: Relation between high energy phosphate and lethal injury in myocardial injury in the dog. Am J Pathol 92:187-241, 1978
36. Kammermeier H, Wein B, Graf W: Characteristics of lactate transfer in isolated cardiac myocytes. Basic Res Cardiol 80, Suppl 1:57-60, 1985
37. Kao RL, Christman EW, Luh SL, Krauhs JM, Tyers GF, Williams EH: The effect of insulin and anoxia on the metabolism of isolated mature rat cardiac myocytes. Arch Biochem Biophys 203:587-599, 1980
38. Karnieli E, Zarnowski MJ, Hissin PJ, Simpson IA, Salans LB, Cushman SW: Insulin stimulated translocation of glucose transport systems in the isolated rat adipose cell. J Biol Chem 256:4772-4777, 1981
39. Katz AM, Tada M: The "stone heart" and other challenges to the biochemist. Am J Cardiol 39:1073-1077, 1977
40. Kübler W, Spieckermann PG: Regulation of glycolysis in the ischemic and the anoxic myocardium. J Mol Cell Cardiol 1:351-377, 1970
41. McDonough KH, Spitzer JJ: Effects of hypoxia and reoxygenation on adult rat heart cell metabolism. Proc Soc Exp Biol Med 173:519-526, 1983
42. Morgan HE, Henderson MJ, Regen DM, Park CR: Regulation of glucose uptake in muscle. I. The effects of insulin and anoxia on glucose transport and phosphorylation in the isolated, perfused heart of normal rats. J Biol Chem 236:253-261, 1961
43. Neely JR, Grotyohann LW: Role of glycolytic products in damage to ischemic myocardium. Circ Res 55:816-824, 1984
44. Neely JR, Liebermeister H, Morgan HE: Effect of pressure development on membrane transport of glucose in isolated rat heart. Am J Physiol 212:815-822, 1967
45. Noy N, Donelly TM, Zakim D: Physical-chemical model for the entry of water-insoluble compounds into cells. Studies of fatty acid uptake by the liver. Biochemistry 25:2013-2021, 1986
46. Orrenius S, Thor H, Rajs J, Berggren M: Isolated rat hepatocytes as an experimental tool in the study of cell injury. Effect of anoxia. Forensic Sci 8:255-263, 1976
47. Piper HM, Probst I, Schwartz P, Hütter JF, Spieckermann PG: Culturing of calcium stable adult cardiac myocytes. J Mol Cell Cardiol 14:397-412, 1982
48. Piper HM, Schwarz P, Hütter JF, Spieckermann PG: Energy metabolism and enzyme release of cultured adult rat heart muscle cells during anoxia. J Mol Cell Cardiol 16:995-1007, 1984
49. Piper HM, Schwartz P, Siegmund B: Energy metabolism and hypoxic injury in cardiomyocytes. In: Piper HM, Isenberg G, eds: Isolated adult cardiomyocytes. Boca Raton: CRC Press, in press
50. Piper HM, Spahr R, Schweickhardt C, Hunneman D: Importance of endogenous substrates for cultured adult cardiac myocytes. Biochim Biophys Acta 883:531-541, 1986

180   H. M. Piper et al.

51. Probst I, Spahr R, Piper HM: Carbohydrate and fatty acid metabolism of adult cardiac myocytes maintained in short-term culture. Am J Physiol 250:H853-860, 1986
52. Randle PJ, Tubbs PK: Carbohydrate and fatty acid metabolism. In: Berne RM, Sperelakis N, Geiger SR, eds: Handbook of physiology. Section 2, Vol I. Bethesda: American Physiological Society, 1979:805-844
53. Rauch B, Bode C, Piper HM, Hütter JF, Zimmerman R, Braunwell E, Hasselbach W, Kübler W: Palmitate uptake in calcium tolerant, adult rat myocardial single cells - evidence for an albumin mediated transport across sarcolemma. J Mol Cell Cardiol 19:159-166, 1987
54. Rose H, Kammermeier H: Contraction and metabolic activity of electrically stimulated cardiac myocytes from adult rats. Pflügers Arch 407:116-118, 1986
55. Rose H, Schnitzler N, Kammermeier H: Influence of metabolic rate and electrical stimulation resp. on the affinity of the glucose transporter of isolated cardiac myocytes. J Mol Cell Cardiol 19, Suppl III:80, 1987 (Abstr)
56. Rouslin W, Erickson JL, Solaro RJ: Effects of oligomycin and acidosis on rates of ATP depletion in ischemic heart muscle. Am J Physiol 250:H503-508, 1986
57. Schwartz P, Piper HM, Spahr R, Spieckermann PG: Ultrastructure of adult myocardial cells during anoxia and reoxygenation. Am J Pathol 115:349-361, 1984
58. Spieckermann PG, Brückner J, Kübler W, Lohr B, Bretschneider HJ: Präischämische Belastung und Wiederbelebungszeit des Herzens. Verh Dtsch Ges Kreislauf-Forsch 35:358-364, 1968
59. Spieckermann PG, Norbeck H, Preusse CJ: From heart to plasma. In: Hearse DJ, de Leiris J, eds: Enzymes in cardiology: Diagnosis and research, New York: Wiley, 1979:81-95
60. Spieckermann PG, Piper HM: Oxygen demand of calcium-tolerant adult cardiac myocytes, Basic Res Cardiol 80, Suppl 2:71-74, 1985
61. Taegtmeyer H, Hems R, Krebs HA: Utilization of energy-providing substrates in the isolated working rat heart. Biochem J 186:701-711, 1980
62. Taegtmeyer H, Roberts AFC, Raine AEG: Energy metabolism in reperfused heart muscle: metabolic correlates to return of function. J Am Coll Cardiol 6:864-870, 1985
63. Van der Vusse GJ, Little SE, Bassingthwaighte JB: Transendothelial transport of arachidonic acid and palmitic acid in the isolated rabbit heart. J Mol Cell Cardiol 19, Suppl III:100, 1987 (Abstr)
64. Veech RL, Lawson JWR, Cornell NW, Krebs HA: Cytosolic phosphorylation potential. J Biol Chem 254:551-561, 1979
65. Weber A, Murray JM: Molecular control mechanism in muscle contraction. Physiol Rev 53:613-673, 1974
66. Wilkinson JH, Roinson JM: Effect of ATP on release of intracellular enzymes from damaged cells. Nature 249:662-663, 1974
67. Williamson JR, Ford C, Illingworth J, Safer B: Coordination of citric acid cycle activity with electron transport flux. Circ Res 38, Suppl 1:39-51, 1976
68. Wittenberg BA, Wittenberg JB: Oxygen pressure gradient in isolated cardiac myocytes. J Biol Chem 260:6548-6554, 1985
69. Zierler KL: Muscle membranes as a dynamic structure and its permeability to aldolase. Ann NY Acad Sci 75:227-234, 1958

Chapter 16

Ischemia, High-Energy Phosphate Metabolism and Sarcomere Dynamics in Myocardium

H.E.D.J. ter Keurs, J.J.J. Bucx, E. Harmsen, P.P. de Tombe and
W.J. Leijendekker, Department of Medicine, Health
Sciences Center, Calgary, Alberta, Canada

1. Effects of ischemia or anoxia on the dynamics of sarcomere contraction of myocardium are described on the basis of a model of excitation-contraction coupling in the cardiac cell.
2. A model of regional ATP supply to the components of the excitation-contraction coupling apparatus is proposed. This model emphasizes that ischemia/anoxia may affect various cell organelles to a different degree and with a different time-course.
3. The role of lowered ATP levels and accumulation of breakdown products of high-energy phosphates, and of glycogenolysis, in the development of abnormalities of force development and of sarcomere dynamics during ischemia/anoxia is discussed. It is proposed that early during ischemia an important effect of the rise of inorganic phosphate ions is to be anticipated; this effect is enhanced by intracellular acidosis and possibly magnesium accumulation. A striking series of recent findings suggests that the cell may be able to reduce calcium ion influx into the cytosol at low ATP concentrations.
4. The time-course of active force development and of force development between stimuli - the so-called unstimulated force during ischemia - and the mechanisms that may be responsible for the decrease of active force and for the development of unstimulated force during ischemia are discussed.
5. The role of accumulation of ions in the cell in abnormalities of excitation-contraction coupling is reviewed. It is proposed that the complex of ion-exchanging proteins at the surface membrane of the cell plays an important role in accumulation of sodium ions during ischemia which may underlie functional abnormalities that occur during reperfusion or re-oxygenation.

Introduction

The effects of ischemia on contraction and relaxation of cardiac muscle function are still largely an enigma, understandably so because of the complexity of the effects of interruption of the flow through the coronary bed. The effects of stasis of coronary flow are evidently both to block the supply of nutrients and oxygen to the myocardium and to prevent washout of metabolites. Hence the ischemic insult is at least composed of the effects of impeded regulation of the extracellular milieu and of the derailment of the intracellular milieu as a consequence of the blockade of oxidative phosphorylation. Anoxia, which is often used as a model of ischemia, differs from ischemia itself because membrane-permeable metabolites are washed out of the extracellular space and hence also accumulate less inside the cell.

We wish to dissect the effects of components of this insult on the contractile process in this chapter. In the course of this "dissection", we will emphasize the effects of ischemia or anoxia or metabolic blockade on excitation-contraction coupling and on contraction. In particular the rapid initial changes of high-energy phosphate levels and of their breakdown products will be emphasized. We will then

J.W. de Jong (Ed.), Myocardial Energy Metabolism, Martinus Nijhoff Publishers, Dordrecht/Boston/Lancaster, 1988

discuss the possible contribution of these concentration changes to the rapid loss of "contractility", to the failure of relaxation and to the mechanism that underlies the development of force in the unstimulated myocardium, i.e., the diastolic contracture.

The discussion will be based on the effect of anoxia on isolated trabeculae from the right ventricle of rat heart. These trabeculae were studied with the aid of laser diffraction techniques (61) to allow analysis of sarcomere behavior as a tool to evaluate both contractile processes and to investigate properties of the excitation-contraction coupling apparatus.

ATP-synthesis and Excitation-Contraction Coupling

Normoxia

ATP is predominantly generated by oxidative phosphorylation in the mitochondria of normoxic working myocardium. The average oxygen consumption of this process in a heart operating under normal conditions is 150 umol/s/L (57). Incorporation of phosphate ions (P_i) in ADP yields 1 mmol ATP/s/L ($\pm$ 10%) depending on the main substrate used (57,70). Feedback mechanisms serve to maintain the levels of ATP, creatine phosphate (CP), ADP, AMP and protons as well as those of the breakdown products of AMP nearly constant in the working cell. Consequently, one may expect that the energy-requiring processes, i.e., excitation-contraction (EC) coupling, contraction and relaxation occur while the concentrations of ions and non-ionic metabolites fluctuate only modestly. Both the rate and magnitude of energy turnover by the EC-coupling apparatus and by contraction is thus expected to be independent of fluctuations of energy metabolism during normoxia.

Figure 1 depicts pathways of ATP supply to the most important components of excitation-contraction coupling and to the contractile filaments, that have emerged over the past decades. ATP is generated in the mitochondria by oxidative phosphorylation and exchanged for ADP by translocase at the inner mitochondrial membrane. Mitochondrial creatine kinase converts creatine in the intermembranous space into CP at the cost of ATP, and the released ADP returns from there to the mitochondrial core to be rephosphorylated. CP diffuses from the intermembranous space to the sites of ATP hydrolysis on the contractile filaments, the sarcoplasmic reticulum and the sarcolemma. Each of these structures is occupied with a creatine kinase that generates ATP close to the ATPases in the structure (5). The creatine formed in the process diffuses back toward the mitochondria.

It has been shown that this ATP supply to the ATPases at the cell's effector sites is complemented by localized glycolytic units. This is of importance because breakdown of glycogen is the main pathway for the generation of ATP during ischemia. Glycogen has been shown by ultrastructural techniques to be present in particles at specific strategic locations in the sarcoplasm, i.e., near the sarcolemmal Na^+/K^+ ATPase, near the contractile filaments (the myosin ATPase and near the sarcoplasmic reticulum Ca^{++} ATPase). Heilmeyer et al. (33,51) described in detail a particle isolated from skeletal muscle which contained glycogen, glycogenolytic enzymes, phosphorylase phosphatase and Ca^{++} ATPase. Entmann et al. (17,18,23) described an SR-glycogen complex isolated from cardiac muscle, which also contains all the enzymes for glycogenolysis. These particles appear to be associated with the sarcoplasmic reticulum in situ as well as in the isolated microsomal fraction and seem to allow a specific functional coupling between ATP synthesis from glycogen and the SR. More recently, Weiss and Lamp have shown that local glycolysis may be a preferential source for ATP for membrane function (74).

The evidence that the protein-glycogen complexes are located near the ATP-converting enzymes could imply that during ischemia the rate of breakdown, and therefore the rate of depletion of glycogen, varies with location in the cell. Depletion of glycogen at a specific location could possibly have a large effect on the nearby ATPases while at other sites ATP supply and conversion will continue. This phenomenon may contribute to metabolic nonuniformity in the cell.

Another intriguing concept, emerging over recent years, is that ATP not only is a source of energy, but also regulates the function of membrane channels and ion exchangers (see Fig. 1). For instance the K^+-conducting channels mentioned above

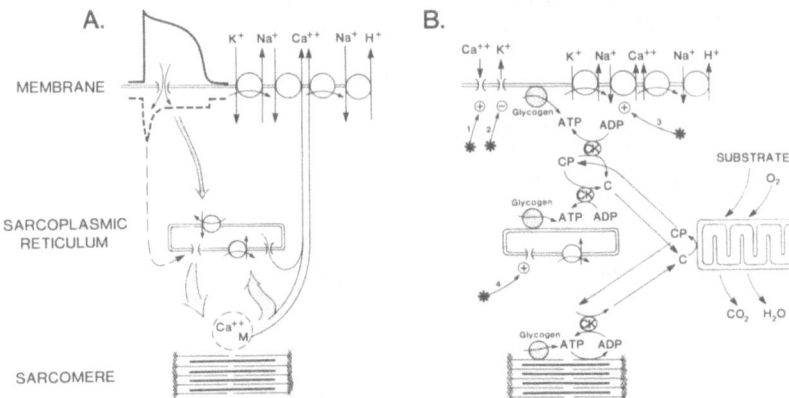

Fig. 1. This figure shows a diagram of the excitation-contraction coupling system in the cardiac cell and a model of the structures involved in excitation-contraction coupling with ATP derived from oxidative phosphorylation and from regional glycogenolysis. Panel A illustrates that during the action potential, calcium enters the cell as a rapid flux followed by a maintained component of the slow inward current. The calcium inward current (indicated by the dashed line) does not lead directly to force development as the calcium ions that enter the cell are bound rapidly to the sarcoplasmic reticulum (SR) that envelops the myofibrils. The rapid influx of calcium is thought to trigger calcium-induced release of calcium by release sites of the sarcoplasmic reticulum. The released calcium ions activate the contractile filaments, and contraction ensues; relaxation follows because the cytosolic calcium is sequestered again by the sarcoplasmic reticulum and partly extruded through the cell membrane by the Na^+/Ca^{++} exchanger and by a low-capacity high-affinity Ca^{++} pump. The force of a contraction is thus determined by the circulation of calcium from the SR to the myofilaments and back to the SR, and by the amount of calcium that entered the cell during the preceding action potential. The relaxation of the twitch depends on the rate of calcium dissociation from the contractile filaments and on the rates of calcium sequestration and extrusion. It is important to note that the Na^+/Ca^{++} exchanger is electrogenic such that calcium extrusion through the exchanger leads to depolarization of the cell. Panel B. Oxidative phosphorylation by the mitochondria yields ATP that is used to convert creatine (C) into creatine phosphate (CP) at the mitochondrial inner membrane. Both C and CP diffuse relatively freely through the outer membrane of the mitochondria and diffuse toward the myofilaments, the SR and to the sarcolemma. There, CP is converted into C while ADP is converted into ATP by creatine kinase (CK) thus supplying the enzymes involved in the excitation-concentration coupling process with energy. Enzymes involved in glycogenolysis are present in glycogen granules found at each of the structures that are relevant to excitation-contraction coupling. This warrants local supply of ATP irrespective of the ATP flux at other sites in the cell. Another important aspect of the role of ATP in the cell is that ATP (indicated by asterisks) controls the properties of the Ca^{++} channel at the surface membrane (indicated by 1); the K^+ channels (indicated by 2); the Na^+/Ca^{++} exchanger (indicated by 3); and Ca^{++} channels in the SR (indicated by 4); in such a way that, if the ATP level drops below 0.5 mM, K^+ channels open and "clamp" the membrane potential at the K^+ equilibrium potential. Then, Ca^{++} fluxes through the sarcolemmal and SR channels are blocked while the Na^+/Ca^{++} exchanger is inhibited. Hence Ca^{++} entry and efflux in a cell, in which the metabolism is blocked, is diminished and Ca^{++}-induced cell damage is less likely to occur. For further explanation, see text.

are regulated by ATP (58,75). Similarly, ATP dependence of the Ca^{++} current in myocardium has been described (37). Another example with implications for ischemia is the observation that Na^+/Ca^{++} exchange in the squid axon is inhibited after cyanide blockade of oxidative metabolism; the exchange is restored by intracellular ATP (4). Regulation of SR calcium channels by ATP in skeletal muscle has been shown by Smith et al. (67). Recently Meissner's group also observed that the threshold at which Ca^{++}-sensitive Ca^{++} channels in cardiac SR open, increases if the [ATP] is lowered (62). The effect of reduction of the [ATP] on the K^+- and Ca^{++}-transport systems seems to be that the cell membrane potential is "clamped" near the K^+ equilibrium potential and that both calcium entry into the cell and Ca^{++} release by the SR are suppressed.

Ischemia

The situation described above that during normoxia EC-coupling and contraction are independent of fluctuations of the energy metabolism is suddenly lost when acute ischemia develops. An interruption of coronary flow exhausts the extracellular fluids acutely of oxygen. Oxygen flow into the oxidative phosphorylation stops. When oxidative metabolism stops, CP which normally is an intermediate of the high-energy phosphate (HEP) traffic, becomes the prime source of energy. The [CP] in the normoxic cells amounts to about 15 mM. It should therefore suffice to supply the cell following interruption of coronary flow for a period of 15 s, provided that the rate of consumption of ATP is initially unchanged.

The extremely high rate of conversion of ADP to ATP through hydrolysis of CP by the creatine kinases initially maintains the high ATP levels (5,38), but evidently at the expense of liberating P_i ions and creatine, and uptake of protons. The [P_i] and the pH rise. Because protons are strongly buffered in the cell (buffer capacity 70-80 mEq H^+/pH unit; ref. 16), the pH rises only modestly.

The [P_i] rises rapidly following onset of ischemia as these ions are unbuffered. The initial rate of rise of the P_i level should logically equal the rate of ATP consumption in the cell with fully operative oxidative phosphorylation. As we have outlined above, the rate of ATP consumption and thus of P_i generation during normoxia, and hence immediately after onset of ischemia, equal about 1 mmol/s/L. Because [P_i] ions hardly permeate through the cell membrane, this implies that the intracellular [P_i] is expected to rise with 10 mM in 10 s.

As will be discussed below, the change of metabolite levels in itself has drastic effects on EC-coupling and on the contractile process. Consequently, the ATP expenditure by these processes will change rapidly after onset of ischemia as a result of changes in the metabolite levels. The decrease of the ATP consumption then evidently decreases in its turn the rate of generation of metabolites.

ATP levels fall when CP hydrolysis cannot keep up with ATP breakdown, and ATP metabolites are formed. The rise of the concentration of ADP, AMP and P_i accelerates anaerobic glycolysis rapidly after onset of ischemia (57), with a modest ATP yield, but at the expense both of a large increase in the [lactate] and of a large release of protons. The latter evidently causes a sharp decrease of the pH despite the ample capacity of the intracellular buffers (39). Moreover, the lactate production leads to a proportional and substantial increase in the osmolarity.

ATP levels decline rapidly during ischemia indicating that the maximal rate of ATP synthesis by anaerobic glycolysis (which amounts to about 0.3 mmol/s/L; ref. 63) fails to meet the ATP hydrolysis. ADP is partly converted to AMP, which allows a small amount of ATP to be regenerated through the adenylate kinase equilibrium (73). Most of the AMP, however, is degraded eventually into uric acid through adenosine, inosine, hypoxanthine and xanthine (28-30). The breakdown of the nucleotides in the cell is also accompanied by an increase in the cellular osmolarity.

Effects of the HEP-Metabolite Levels on the Function of the Energy-Requiring Processes

The ATPases of the mechanisms that serve excitation-contraction coupling and contraction bind and hydrolyze ATP. They convert the available energy from these reactions into chemical or mechanical work and heat, while the products P_i and ADP are released. The rates of binding ATP and of release of the products are functions of the concentrations of ATP, ADP and P_i. The predictable effect of a decrease of the [ATP] and a rise of the [ADP] and [P_i] is thus to inhibit the energy-converting ATPases as is well-known for the contractile system (59,75). We will put relatively little emphasis on the concept that the phosphorylation potential (11,22,43) determines force development by ischemic cardiac muscle, as various authors have shown that the concentration of the breakdown products of hydrolysis is more important than the ATP/ADP·P_i ratio (2,44).

The Km's of the ATP-hydrolyzing systems involved in EC-coupling are in the range of 1 mM (2). However, the depression of cardiac function occurs already at ATP levels that are substantially above the Km's of these enzymes (11). ATP shortage would, therefore, only influence these enzymes and the ATP-sensitive channels and ATP-sensitive exchangers, mentioned above, if spatial non-uniformity of the ATP distribution occurs in the ischemic cells, e.g., as a result of the high energy turnover in the contractile system. However, myocardium seems to be protected to some extent against such failure of its membrane functions by the localized ATP supply partially through glycolytic units (74).

The accumulation of ATP-breakdown products (e.g., adenosine), on the other hand, is known to affect sarcolemmal ion channels, and therefore to change the currents that flow through the cell membrane at rest and during the action potential. Hence, the initiating event of EC-coupling is depressed shortly after onset of ischemia. Because the action potential not only triggers contraction in the myocardium but also modulates the strength of contraction through variation of the influx of calcium during the plateau phase (53), a concomitant change in the strength of the contraction will ensue (see Fig. 1).

Many of the processes involved in EC-coupling and contraction are activated by the entry of Ca^{++} in the cell and subsequent release of Ca^{++} by the SR. Thus abnormalities of Ca^{++} handling by the cell are expected to affect cardiac function during ischemia and hypoxia (65). Activation of proteins by Ca^{++} invariably requires binding of Ca^{++} to anionic amino acid residues, such as carboxyl groups of tyrosine, as is the case with troponin C (48). It is well-known that protons compete with Ca^{++} for these binding sites but do not activate the protein to which they bind themselves (19). Similary Mg^{++} released during ATP breakdown competes for binding sites for Ca^{++} (20).

The rate and the efficiency at which the ATPases convert energy of ATP into other forms of energy, is further determined by the ionic conditions to which they are exposed. These effects are well-known for the contractile process, which, for example, is strongly depressed when the contractile filaments are exposed to high ionic strength (14). Similarly, Ca^{++} release by the SR is depressed in hypertonic solutions (24,36).

The Effects of Ischemia or Hypoxia on Force Development by Trabeculae

Figure 2 shows that the typical response of a cardiac trabecula to hypoxia consists of a rapid decline of twitch-force development to low levels. The decline may be monophasic or may follow a more complex pattern with even an initial increase in force depending on the stimulus rate and on the loading conditions (Bucx et al., in preparation). The decline in active force development is accompanied by a slight increase in the unstimulated force. While unstimulated force develops occasionally, but not always, slight spontaneous activity of the sarcomeres may be observed microscopically or by laser diffraction methods (9). After 5-15 min of anoxia, unstimulated force rises steeply to levels that ultimately may exceed maximal twitch force by 50% (see Fig. 2).

This pattern is typical both for studies in which myocardium of isolated perfused hearts or of isolated superfused muscles is made ischemic or anoxic with maintained perfusion or in which oxidative phosphorylation has been stopped by CN- ions, with or without suppression of anaerobic glycolysis (1). The rate at which the twitch force decays and unstimulated force develops, however, depends on the experimental paradigm that has been used.

The initial rise of twitch force has been attributed to a transient increase in the intracellular pH (1) due to breakdown of CP. The mechanism underlying the subsequent suppression of force development has not completely been elucidated yet. Both reduction of Ca^{++} release by the SR and depression of the contractile response to Ca^{++} are considered responsible. Decrease of Ca^{++} loading of the SR probably results from a diminished calcium influx into the cell (55). Two components of Ca^{++} entry into the cell have been studied particularly; both were depressed during ischemia. First, the amount of Ca^{++} entering during the action potential is reduced with the decrease of the duration of the plateau of the action potential. This occurs within a few min when hypoxia is combined with inhibition of glycolysis (49).

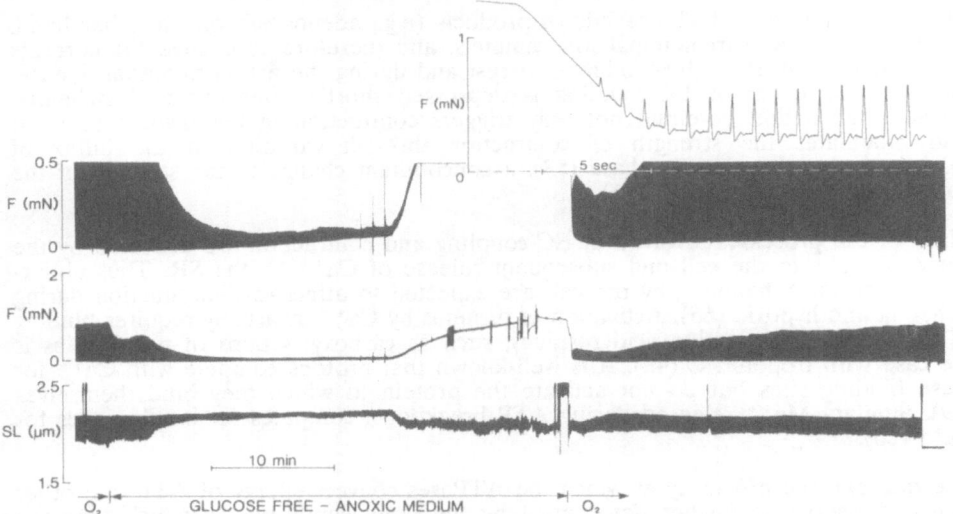

Fig. 2. This figure illustrates the time-course of active force (at 1 Hz stimulus rate, middle panel), unstimulated force (top panel) and sarcomere length (SL) in a cardiac trabecula of a rat during perfusion with a hypoxic Krebs-Henseleit solution without glucose at a temperature of 30°C. Force development during the twitches rapidly declined following onset of hypoxia followed by transient partial recovery and a further decline to zero. Unstimulated force initially rose slightly, which was accompanied by slight spontaneous activity of the sarcomeres; after 15 min, unstimulated force rose rapidly to a level that exceeded active force development in the control period by 50%. This force disappeared within 15 s after reoxygenation of the muscle (see inset); a few seconds later twitch force was apparent again. Spontaneous contractions immediately followed the electrically-induced twitches during the first 10 min after reoxygenation, and were accompanied by a Ca-dependent contracture that caused an increase of the unstimulated force. The signs of calcium overload subsequently disappeared and twitch-force recovery was complete after 30 min.

One mechanism for the shortening of the action potential may be that adenosine, AMP or ADP activate a purinergic (P1) receptor and thereby increase the current through a K^+ channel (45). Weiss et al. (74) provided another mechanism for this phenomenon by evidence that glycolysis at the sarcolemma in a metabolically active cell is essential to keep the local ATP-sensitive K^+ channels (58) closed even during normoxia. Opening of these channels during hypoxia and interrupted glycolysis would cause a repolarizing current that shortens the action potential (49) and extracellular K^+ accumulation (66).

A reduction of the Ca^{++} current during the action potential as a result of the effect of lowered ATP levels such as described by Horie et al. (37) further decreases Ca^{++} influx during the action potential. This results in decreased loading of the SR and thus decreased Ca^{++} release. In a later stage of hypoxia, accumulation of Mg^{++}, released as a result of ATP hydrolysis, further decreases SR loading (20). Eventually, at low ATP level, the Ca^{++} channels in the SR are less responsive (62), so that Ca^{++}-induced Ca^{++} release (53) is suppressed (see Fig. 1).

Secondly, Ca^{++} influx into the hypoxic cell in the period between action potentials is reduced. This has convincingly been shown in cardiac cells (31); it may result from a reduction of Na^+/Ca^{++} exchange at the lowered pH (60). Moreover, a reduction of the ATP concentration has been postulated to reduce Na^+/Ca^{++} exchange (4). The mechanism of this effect is unknown, but binding of ATP to the exchanger has been assumed to modify the affinity of the exchanger for Na^+ (54). Although these mechanisms are likely to play a role during ischemia, the channels and the exchanger are sensitive to ATP at levels around 0.2 mM. They therefore do not explain the rapid decrease of twitch force following onset of events mimicking ischemia.

The decrease of twitch force occurs too rapidly to be explained by a decrease of the free [ATP] (2,44). Both the rise in $[P_i]$ and the fall of intracellular pH have been shown to be important (44). The effects of P_i and of protons occur so rapidly that they probably reflect a strong negative inotropic action on the contractile filaments (44). Protons and Mg^{++} act by their competition with Ca^{++} for binding to troponin C on the actin filaments (20), and P_i probably through its inhibition of P_i release from the crossbridge (44,75), which allows the development of the force-generating state of the bridge. Other possibilities that have been investigated are the accumulation of lactate and the concomitant rise in osmolarity; blockade of the lactate transport over the cell membrane does not mimic the effect of ischemia, however (15).

The Effects of Ischemia or Hypoxia on Relaxation

The rate of relaxation of the twitch of cardiac muscle decreases shortly after onset of ischemia, as is illustrated in Fig. 3. The fact that this has not universally been found, though, may be related to the simultaneous decrease of the amplitude of the twitch which itself determines the rate of relaxation, as was found in rat cardiac trabeculae (Bucx et al., in preparation). The mechanism underlying the decrease of the rate of relaxation is located in a decreased rate of sequestration and exhaustion of Ca^{++} from the cytosol. Ca^{++} sequestration by the SR slows down (19), probably as a result of protons for the SR Ca^{++}-ATPase. The effect of lowered pH alone is to decrease the rate of uptake of calcium ions into the sarcoplasmic reticulum (19).

Furthermore, acidosis causes a substantial Na^+ influx during ischemia and anoxia as a result of pH-dependent Na^+/H^+ exchange over the cell membrane (6,41), while the resultant Na^+ accumulation is incompletely corrected as a result of inhibition of the Na^+/K^+ pump (46). Intracellular Na^+ accumulation causes a diminished efflux of Ca^{++} during the relaxation phase and hence is expected to further delay relaxation (64).

The diminished efflux of Ca^{++} is not completely explained by the rise in Na_i (26), suggesting dissociation of H^+, Na^+ and Ca^{++} transport over the cell membrane, as might be expected if the Na^+/Ca^{++} exchanger is inhibited at low ATP levels (4,54).

An increase of the heart rate causes an increased Na^+ and Ca^{++} influx over time and therefore induces an increase in the intracellular Na^+ level. This again leads to slowing of relaxation and induces on the other hand more Ca^{++} uptake by the sarcoplasmic reticulum. Ineffective extrusion from the cytosol at a high heart rate causes again the cytosolic calcium concentration to rise. It is also to be expected that, at a high heart rate, the amount of Ca^{++} in the SR is increased. It is known that, under these conditions, spontaneous calcium release may occur (69). If such calcium release ensues during the relaxation phase, further prolongation of relaxation would occur, as has been observed in studies of sarcomere dynamics in hypoxic rat cardiac trabeculae (9).

An increase in the heart rate, e.g., by pacing, further delays relaxation and causes an increase in the diastolic pressure as has been shown by Apstein and Grossman (3). This phenomenon, which has been coined "demand ischemia", is consistent with the mechanism outlined above.

Diastolic Compliance and Ischemia

Shortly after onset of anoxia in isolated heart-muscle preparations, a conspicuous rise in the unstimulated force follows, as was illustrated in Fig. 2. Many studies have been performed to clarify the source of this force, which also increases diastolic compliance. Two possible sources have been addressed in particular. First, it is conceivable that the diastolic force increases due to activation of the contractile filaments by a rise in the free intracellular $[Ca^{++}]$, as was discussed in the preceding paragraph. The second possibility is that the HEP levels and the levels of HEP metabolites have changed sufficiently to induce rigor.

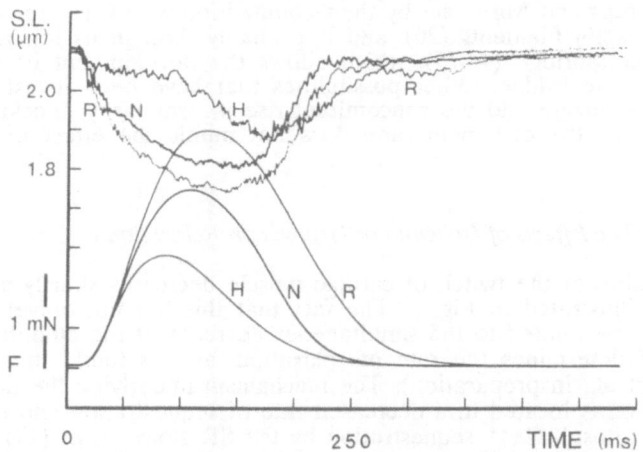

Fig. 3. This figure illustrates the effect of 5 min of hypoxia and reoxygenation on the time-course of the twitch. The top panel of the figure shows the tracings of sarcomere length (S.L.) during control (N), during hypoxia (H) and during reoxygenation (R). The bottom panel shows the concurrent force transients during the twitches at N, H, and R. Note the occurrence of spontaneous late contractions manifest as sarcomere shortening transients during H and R. It is clear that the relaxation phase is protracted during H and R. For further explanation, see text. (From Bucx, Sethi, Ter Keurs; unpublished observations.)

The first possibility has been investigated in detail in isolated myocytes (12,31) and in trabeculae (2) with the use of calcium-sensitive luminescent or fluorescent substances. These studies have shown that the unstimulated force increased or the cells shortened clearly before an increase in the free Ca^{++} level in the cell (2,31). It is therefore unlikely that the force is a result of the Ca^{++}-mediated activation of the contractile filaments, especially because simultaneous acidosis of the cell reduces the sensitivity of the filaments to Ca^{++}. The conclusion of these studies and of energetic studies was that the mechanical response to persistent anoxia is due to the development of rigor bonds between actin and myosin (35).

The hypothesis that rigor bonds exist between actin and myosin is supported by mechanical studies of the dynamic properties of the sarcomeres, as is illustrated in Fig. 4. The figure shows that the dynamic mechanical response of a rigor bond can be best compared to that of a simple stiff spring located in the S2 element of the crossbridge between actin and myosin. The spring should exhibit a stiffness which is independent of the rate (47,72) or frequency with which the length of the spring is altered. Such frequency-independent stiffness was indeed observed in our analysis of anoxic cardiac trabeculae. Figure 4 also shows that the stiffness of sarcomeres which develop a sustained calcium-dependent force - as in a caffeine-mediated tetanus - is clearly frequency dependent (10). Such studies are further required to establish the time-course of change of the elastic properties of myocardium during ischemia. This will allow us to unravel the contribution of calcium-dependent active crossbridge cycling and of development of rigor bridges during the rise in unstimulated force.

The question: "Why does rigor develop?" is still unsettled. Many studies have shown considerable rigor force development, e.g., 150% of maximal active force, while the ATP levels were still in excess of 0.5 mM. This contrasts the observations on skinned fibers both in skeletal muscle and in heart that rigor only develops, if the ATP concentration is lower than 50 uM (20). Three possible explanations are presently under investigation.

First, the rigor force may result from a complete rigor in a fraction (e.g., 90%) of the cells of the preparation, while the remainder of the cells is still relaxed. This would require severe non-uniformity of the cell properties throughout myocardium. Non-uniformity of the response of cardiac tissue during ischemia or anoxia (Bucx et al., in preparation) and of cardiac cells (68) to anoxia has been observed. The hypothesis requires that the cells in rigor are indeed devoid of ATP (<50 uM), while the remaining cells still contain a normal concentration of ATP. With a suitable ratio of cells with normal ATP levels to those without ATP, one would indeed expect that the residual ATP level in the bulk of the tissue could be 0.5 mM, while the myocardium behaves as in rigor.

Alternatively, rigor may have developed at much higher ATP levels than in isolated skinned muscle studies, because of the simultaneous elevation of HEP-metabolite levels. Prime candidates are P_i and ADP, as experiments (52) have shown that the ATP threshold for rigor force development is indeed elevated in the presence of increased concentrations of these metabolites. Cooperative interaction between Ca^{++} binding at elevated Ca^{++} levels and rigor-bond formation would further enhance the development of rigor bridges, as has been shown by Bremel and Weber (7). This question awaits further study of skinned cells at varied levels of ATP, ADP, P_i and Ca^{++}.

Another hypothesis to explain the rigor under these conditions has been put forward by Gudbjarnason et al. (27) and has later been supported by the work of Bricknell et al. (8). The hypothesis states that during anoxia a block of transport of ATP may occur from the mitochondria to the cytosol. ATP would thus stagnate in the mitochondria, while ATP shortage would manifest itself at the sites of consumption, either as rigor or as dysfunction of the other ATP-consuming enzyme systems. Some evidence for the latter hypothesis has been obtained in studies of the ATP content of the mitochondrial and cytosolic compartments of cardiac (and other) cells. These studies suggested compartmentation of ATP in the mitochondria during hypoxia (21,34).

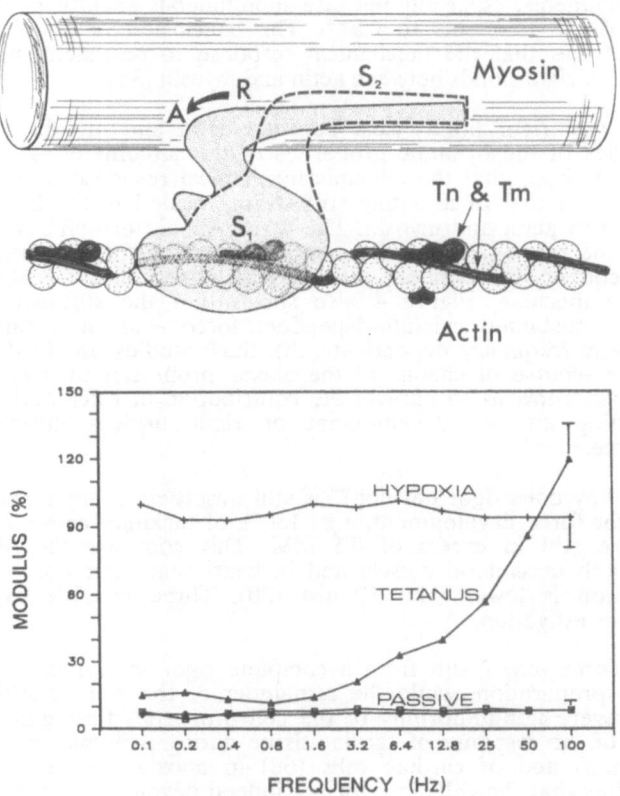

Fig. 4. The top panel illustrates a model of the structure of the crossbridge of myosin attached to actin prior to and during the power stroke (R and A). The deformation of the crossbridge that leads to force development is assumed to take place in the S1 head of the bridge; as a result of the power stroke the S2 link between myosin and the S1 head is stretched. Stretch of the elastic S2 link causes a force that pulls actin along myosin. The bottom panel shows the response of muscle force as a function of the frequency of sinusoidal length changes. Cyclic attachment and deformation of the bridge is assumed to occur during the calcium- and ATP-dependent contraction, while the bridge attaches and stays in state R during rigor. The muscle in the rigor state, therefore, behaves as a simple stretched spring, and is stiff at all velocities at which the spring is stretched. The cyclic behavior of the crossbridges manifests itself in a time-dependence of the force response upon stretch. The time-dependence is easily appreciated if the muscle is stretched with sinusoidal movements of small amplitude and the force response is expressed as a function of the frequency of the length changes as in the bottom panel. These properties are represented for the unstimulated muscle under control conditions, at maximal force development during hypoxia and during a caffeine-mediated tetanus. It is clear from the illustration that the properties of the unstimulated muscle during control and hypoxia only indicate a difference in stiffness, while both are frequency independent, in contrast with the stiffness of the muscle during the caffeine tetanus which strongly depends on the frequency of the length perturbations. These observations confirm that the hypoxic muscle is in rigor, while crossbridge cycling takes place in the muscle during the tetanus. Tm = tropomyosin; Tn = troponin.

Arrhythmias During Ischemia or Hypoxia

Accumulation of Na^+ causes, as we have seen, a diminished efflux of calcium through the Na^+/Ca^{++} exchanger. This effect is enhanced by the development of acidosis as a result of anaerobic glycolysis, which causes an increase in the Na^+ level in the cell (6,41,60). Hence, phenomena known to occur during calcium overload of the sarcoplasmic reticulum could develop (64). They include spontaneous contractions with concomitant transient depolarizations and consequently triggered arrhythmias. But they are depressed by other components of ischemia (13), e.g., inhibition by low ATP levels of Na^+/Ca^{++} exchange mediated currents that may underlie the transient depolarization (64).

Opie and his collaborators have shown (13) that under conditions of ischemia, several factors reduce the amplitude of the spontaneous after-depolarizations such as: lactate accumulation, acidosis and K^+ accumulation in the extracellular space. These effects have been interpreted to mean that triggered arrhythmias which are presumably based upon the occurrence of spontaneous after-depolarizations are unlikely events in a severely ischemic region. However, in a region in which ischemia is less severe and where the inhibitory effects are less strong, this mechanism may induce arrhythmias.

Washout of the above metabolites during reperfusion of the ischemic tissue rapidly allows the spontaneous after-depolarizations to occur again. Moreover, the increased levels of intracellular Na^+ persist during reperfusion (25), which enhances the occurrence of spontaneous contractions and triggered arrhythmias (32), even if during ischemia no signs of Ca^{++} overload were present. Hence, further Ca^{++} entry may occur immediately following reperfusion or reoxygenation and this may lead to development of spontaneous contractions, transient depolarizations, arrhythmias, and ultimately cell death.

Acknowledgements

The authors would like to thank Ms. Lenore Doell for skilled secretarial assistance. This work was supported by a grant from the Alberta Heritage Foundation for Medical Research.

References

1. Allen DG, Morris PG, Orchard CH, Pirolo, JS: A nuclear magnetic resonance study of metabolism in the ferret heart during hypoxia and inhibition of glycolysis. J Physiol (Lond) 361:185-204, 1985
2. Allen DG, Orchard CH: Myocardial contractile function during ischemia and hypoxia. Circ Res 60:153-168, 1987
3. Apstein CS, Grossman D: Opposite initial effects of supply and demand ischemia on left ventricular compliance. J Mol Cell Cardiol 19:119-128, 1987
4. Baker PF, McNaughton PA: Kinetics and energetics of calcium efflux from intact squid giant axons. J Physiol (Lond) 259:103-144, 1976
5. Bessman SP, Carpenter CL: The creatine-creatine phosphate energy shuttle. Annu Rev Biochem 54:831-862, 1985
6. Bielin FV, Bosteels S, Verdonck F: Effect of intracellular pH on sodium in rabbit cardiac Purkinje fibers. J Physiol (Lond) 390:55 (Abstr)
7. Bremel RD, Weber A: Cooperation within actin filaments in vertebrate skeletal muscle. Nature 238:97-101, 1972
8. Bricknell OL, Daries PS, Opie LH: A relationship between adenosine triphosphate, glycolysis and ischaemic contracture in the isolated rat heart. J Mol Cell Cardiol 13:941-945, 1981
9. Bucx JJJ, Sethi S, Ter Keurs HEDJ: Sarcomere dynamics and relaxation during hypoxia and reoxygenation in rat myocardium. Circulation 76, Suppl IV:IV-57, 1987 (Abstr)

10. Bucx JJ, Backx P, De Tombe PP, Ter Keurs HE: Rigor development in hypoxic rat cardiac trabeculae. Circulation 74, Suppl II:II-165, 1986 (Abstr)
11. Clark K, O'Connor AJ, Willis RJ: Temporal relation between energy metabolism and myocardial function during ischemia and reperfusion. Am J Physiol 253:H412-H421, 1987
12. Cobbold PH, Bourne PK: Aequorin measurements of free calcium in single heart cells. Nature 312:444-446, 1984
13. Coetzee WA, Opie LH: Effects of components of ischemia and metabolic inhibition on delayed afterdepolatization in guinea pig papillary muscle. Circ Res 61:157-165, 1987
14. Edman KAP, Hwang JC: The force-velocity relationship in vertebrate muscle fibres at varied tonicity of the extracellular medium. J Physiol (Lond) 269:255-272, 1977
15. Eisner DA, Elliot AC, Smith GL: Why does ischemia decrease developed pressure in isolated ferret hearts? J Physiol (Lond) 390:57, 1987 (Abstr)
16. Ellis D, Thomas RC: Direct measurement of the intracellular pH of mammalian cardiac musle. J Physiol (Lond) 262:755-771, 1976
17. Entman ML, Bornet EP, van Winkle WB, Goldstein MA, Schwarz A: Association of glycogenolysis with cardiac sarcoplasmic reticulum: II. Effect of glycogen depletion, deoxycholate solubilization and cardiac ischemia: Evidence for a phosphorylase kinase membrane complex. J Mol Cell Cardiol 9:515-528, 1977
18. Entman ML, Kanike K, Goldstein MA, Nelson TE, Bornet EP, Futch TW, Schwartz A: Association of glycogenolysis with cardiac sarcoplasmic reticulum. J Biol Chem 251:3140-3146, 1976
19. Fabiato A, Fabiato F: Effect of pH on myofilaments and the sarcoplasmic reticulum of skinned cells of cardiac and skeletal muscle. J Physiol (Lond) 276:233-255, 1978
20. Fabiato A, Fabiato F: Effects of magnesium on contractile activation of skinned cardiac cells. J Physiol (Lond) 249:497-517, 1975
21. Geisbuhler T, Altschuld RA, Trewyn RW, Ansel AZ, Lamka K, Brierley GP: Adenine nucleotide metabolism and compartmentalization in isolated adult rat heart cells. Circ Res 54:536-546, 1984
22. Gibbs C: The cytoplasmic phosphorylation potential: its possible role in the control of myocardial respiration and cardiac contractillity. J Mol Cell Cardiol 17:727-731, 1985
23. Goldstein MA, Murphy DL, van Winkle WB, Entman ML: Cytochemical studies of a glycogen-sarcoplasmic reticulum complex. J Muscle Res Cell Motil 6:177-187, 1985
24. Gordon AM, Godt, RE, Donaldson SKB, Harris CE: Tension in skinned frog muscle fibers in solutions of varying ionic strength and neutral salt composition. J Gen Physiol 62:550-574, 1973
25. Grinvald PM: Calcium uptake during post-ischemic reperfusion in the isolated rat heart: Influence of extracellular sodium. J Mol Cell Cardiol 14:359-365, 1982
26. Guarnieri T: Intracellular sodium-calcium dissociation in early contractile failure in hypoxic ferret papillary muscles. J Physiol (Lond) 388:449-465, 1987
27. Gudbjarnason S, Mathes P, Ravens KG: Functional compartmentation of ATP and creatine phosphate in heart muscle. J Mol Cell Cardiol 1:325-339, 1970
28. Harmsen E: Myocardial purine synthesis. Acad Thesis, Rotterdam, 1984
29. Harmsen E, Seymour A-MS, Hogan G, Radda GK: The correlation between myocardial high-energy phosphates and purine release during restricted flow. A ^{31}P-NMR study. Proc Int Soc NMR, London, 1985 (Abstr)
30. Harmsen E, Seymour A-MS, Hogan G, Radda GK: The correlation between myocardial high-energy phosphates and purine release during restricted flow. A ^{31}P-NMR study. Am J Physiol (submitted)
31. Haworth, RA, Goknur AB, Hunter DR, Hegge JO, Berkott HA: Inhibition of calcium influx in isolated adult rat cells by ATP depletion. Circ Res 60:586-594, 1987
32. Hayashi H, Ponnambalam C, McDonald TF: Arrhythmic activity in reoxygenated guinea pig papillary muscle and ventricular cells. Circ Res 61:124-133, 1987
33. Heilmeyer LMG, Meyer JF, Haschke RH, Fisher EH: Control of phosphorylase activity in a muscle glycogen particle. II. Activation by calcium. J Biol Chem 245:6649-6656, 1970
34. Hohl C, Ansel A, Altschuld R, Brierley GP: Contracture of isolated rat heart cells on anaerobic to aerobic transition. Am J Physiol 242:H1022-H1030, 1982
35. Holubarsch C, Alpert NR, Goulette R, Mulieri LA: Heat production during hypoxic contracture of rat myocardium. Circ Res 51:777-786, 1982
36. Homsher E, Briggs FN, Wise RM: The effects of hypertonicity on resting and contracting frog skeletal muscles. Am J Physiol 226:855-863, 1974
37. Horie M, Irisawa H, Noma A: Voltage dependent magnesium block of adenosine-triphosphate-

sensitive potassium channel in guinea pig ventricular cells. J Physiol (Lond) 383:000-000, 1987 (in press)

38 Jacobus WE: Respiratory control and the integration of heart high-energy phosphate metabolism by mitochondrial creatine kinase. Annu Rev Physiol 47:707-725, 1985

39. Jacobus WE, Poreas IH, Lucas SK, Weisfeldt ML, Flaherty JT: Intracellular acidosis and contractility in the normal and ischemic heart as examined by ^{31}P-NMR. J Mol Cell Cardiol 14:13-20, 1982

41. Kaila K, Vaughan-Jones RD: Influence of sodium hydrogen exchange on intracellular pH, sodium and tension in sheep cardiac Purkinje fibres. J Physiol (Lond) 390:93-118, 1987

42. Kakei M, Noma A, Shibasaki T: Properties of adenosine triphosphate regulated potassium channels in guinea-pig ventricular cells. J Physiol (Lond) 363:441-463, 1985

43. Kammermeier H, Schmidt P, Jungling E: Free energy change of ATP-hydrolysis: a causal factor of early hypoxic failure of the myocardium. J Mol Cell Cardiol 14:267-277, 1982

44. Kentish JC: The effects of inorganic phosphate and creatine phosphate on force production in skinned muscles from rat ventricle. J Physiol (Lond) 370:585-604, 1986

45. Kurachi Y, Nakajima T: Induction of K current by adenosine and adenine nucleotides in isolated atrial cells. Circulation 76:IV-112, 1987 (Abstr)

46. Lamers JMJ, Hülsmann WC: Inhibition of Na^+/K^+- stimulated ATPase of hearts by fatty acids. J Mol Cell Cardiol 9:343-346, 1977

47. Lewis MJ, Housmans PR, Claes VA, Brutsaert DL, Henderson AH: Myocardial stiffness during hypoxic and reoxygenation contraction. Cardiovasc Res 14:339-344, 1980

48. McCubin WD, Beyers DM, Kay CM: Regulation of the actin-myosin interaction by calcium; the troponin tropomyosin complex. In: Ter Keurs HEDJ, Tyberg JV, eds: Mechanics of the circulation. Dordrecht: Nijhoff Publ, 1987:113-130

49. McDonald TF, MacLeod DP: Metabolism and the electrical activity of anoxic ventricular muscle. J Physiol (Lond) 229:559-582, 1973

50. McDonald RH, Taylor RR, Cingolani HE. Measurement of myocardial developed tension and its relation to oxygen consumption. Am J Physiol 211:667-673, 1966

51. Meyer F, Heilmeyer LMG Jr, Haschke RH, Fisher EH: Control of phosphorylase activity in a muscle glycogen particle. I. Isolation and characterization of the protein-glycogen complex. J Biol Chem 245:6642-6648, 1970

52. Miller AJ, Smith GL: The contractile behavior of EGTA- and detergent-treated heart muscle. J Muscle Res Cell Motil 6:541-567, 1985

53. Morad M, Cleeman L: Role of Ca^{2+} in development of tension in heart muscle. J Mol Cell Cardiol 19:527-553, 1987

54. Mullins LJ: A mechanism for Na^+/Ca^{2+} transport. J Gen Physiol 70:681-695, 1977

55. Nayler WG, Poole-Wilson PA, Williams A: Hypoxia and calcium. J Mol Cell Cardiol 11:683-706, 1979

56. Neely JR, Denton RM, England PJ, Randle PJ: The effects of increased heart work on the tricarboxylate cycle and its interactions with glycolysis in the perfused heart. Biochem J 128:147-159, 1972

57. Neely JR, Liebermeister H, Battersby EJ, Morgan HE: Effect of pressure development on oxygen consumption by isolated heart. Am J Physiol 212:804-814, 1967

58. Noma A: ATP regulated K channels in cardiac muscle. Nature 305:147-148, 1983

59. Nosek TM, Fender KY, Godt RE. It is diprotonated inorganic phosphate that depresses force in skinned skeletal muscle fibers. Science 236: 191-193, 1987

60. Philipson KD, Bersohn MM, Nishimoto AY: Effects of pH on Na-Ca exchange in canine cardiac sarcolemmal vesicles. Circ Res 50:287-293, 1982

61. Ricciardi L, Bucx JJJ, Ter Keurs HEDJ: Effects of acidosis on force-sarcomere length and force-velocity relations of rat cardiac muscle. Cardiovasc Res 20:117-123, 1986

62. Rousseau E, Meissner G: Ca^{2+} release channel from cardiac sarcoplasmic reticulum: Conductance and activation by Ca^{2+} and ATP. Biophys J 51:197a, 1987 (Abstr)

63. Rovetto MJ, Whitmer JT, Neely JR: Comparison of the effects of anoxia and whole heart ischemia on carbohydrate utilization in isolated working rat hearts. Circ Res 32:699-711, 1973

64. Schouten VJA, Ter Keurs HEDJ; The slow repolarization phase of the action potential in rat heart. J Physiol (Lond) 360:13-25, 1984

65. Shen AC, Jennings RB: Myocardial calcium and magnesium in acute ischemic injury. Am J Pathol 67:417-440, 1972

66. Shine KI, Douglas AM, Ricchiuti N: Ischemia in isolated interventricular septa: Mechanical events. Am J Physiol 231:1221-1232, 1977
67. Smith JS, Coronado R, Meissner G: Sarcoplasmic reticulum contains adenine-nucleotide activated calcium channels. Nature 316:446-449, 1985
68. Stern MD, Chien AM, Capogrossi MC, Pelto DJ, Lakatta EG: Direct observation of the "oxygen paradox" in single rat ventricular myocytes. Circ Res 56:899-903, 1985
69. Stern MD, Kort AA, Bathnagar GM, Lakatta EG: Scattered light intensity fluctuations in diastolic rat cardiac muscle caused by spontaneous Ca^{2+} dependent cellular mechanical oscillations. J Gen Physiol 82:119-153, 1983
70. Taegtmeyer H, Hemp R, Krebs HA: Utilization of energy-providing substrates in the isolated working rat heart. Biochem J 186:701-711, 1980
71. Ter Keurs HEDJ, Schouten VJA, Bucx JJJ, Mulder BM, De Tombe PP: Excitation-contraction coupling in myocardium: Implications of calcium release and Na^+/Ca^{2+} exchange. Can J Physiol Pharmacol 65:619-626, 1987
72. Ventura-Clapier R, Vassort G: Rigor tension in metabolic blockade and ionic rises in resting tension in rat heart. J Mol Cell Cardiol 13:551-561, 1981
73. Walker EJ, Dow JW: Localization and properties of two isoenzymes of cardiac adenylate kinase. Biochem J 203:361-369, 1982
74. Weiss JN, Lamp ST: Glycolysis preferentially inhibits ATP-sensitive K^+ channels in isolated guinea pig cardiac myocytes. Science 238:67-69, 1987
75. Woledge RC, Curtin NA, Homsher E: Energetic aspects of muscle contraction. Monographs of the Physiological Society, No 41. London: Acad Press, 1985

Chapter 17

Myocardial Function and Metabolism during Ischemia and Reperfusion in Anesthetized Pigs

P.D. Verdouw[1], W.J. van der Giessen[1] and J.M.J. Lamers[2], [1]Laboratory for Experimental Cardiology, Thoraxcenter, and [2]Department of Biochemistry I, Erasmus University Rotterdam, The Netherlands

Reduction of coronary blood flow in anesthetized pigs offers an attractive model of myocardial ischemia. In addition to data obtained in canine models, the study of myocardial ischemia in pigs yields important information for situations in which an extensive collateral coronary circulation is absent. In this chapter the effects of the most widely used anesthetic agents during normal and reduced porcine myocardial perfusion are reviewed. Recent data on myocardial function and metabolism during partial and complete coronary occlusion, and after reperfusion, are critically examined. Emphasis is given to the relationship between metabolic markers and recovery of myocardial function after reperfusion.

Introduction

A large number of animal models is available to study myocardial ischemia and interventions which aim to protect jeopardized tissue or limit irreversible injury. The dog has usually been the animal of choice in in-vivo studies, but the use of domestic pigs has increased considerably during recent years (39,46). The pig is a suitable animal for cardiovascular research, in particular on myocardial ischemia and atherosclerosis, because its distribution of the coronary arteries (47) and its serum lipoprotein profile (21) are comparable to that of man. Moreover, at variance with dogs, in pigs atherosclerosis can not only be easily induced, but also develops spontaneously. Location and nature of these atherosclerotic plaques are comparable to those observed in man (35).

For the study of acute myocardial ischemia, it is important to know that hearts of domestic pigs, in contrast to those of dogs, lack an extensive collateral circulation. In pigs the residual flow is usually less than 5% after complete occlusion of a coronary artery (22). Thus the choice of the pig as an experimental animal in the study of acute myocardial ischemia may only be relevant for those clinical situations in which we are dealing with human hearts with low collateral flow as in previously undiseased hearts (32,47). Collateral growth can be induced in young growing pigs by placement of a plastic cylinder around a major coronary artery, which becomes flow-limiting as the animals grow older. With this technique collateral blood flow may become as high as 40% of the flow through the coronary artery before placement of the constrictor (23). The advantage of now having a preparation which has gone through episodes of myocardial ischemia for some length of time may be outweighed by the considerable loss of animals due to sudden cardiac death during this process of collateral induction. In our laboratory this amounted to 30% of the animals and occurred frequently in connection with excitement.

Occurrence of Ventricular Fibrillation and the Effects on Myocardial Metabolism

In anesthetized pigs a complete occlusion of the left anterior descending coronary artery at its origin invariably leads to ventricular fibrillation in all animals within

30 min after the artery has been occluded (42). Ligation at a more distal site only slightly reduces the incidence of this arrhythmia (41). Using a flow reduction rather than a complete occlusion will decrease the incidence of ventricular fibrillation as a reduction of coronary blood flow to 25% of baseline causes ventricular fibrillation in only 30% of the animals. A still smaller decrease (to 50% of baseline) almost completely abolishes serious arrhythmias during the first 30 min (41-43).

The loss of animals can be avoided by defibrillation, but the metabolic changes that occur during the periods of ventricular fibrillation could invalidate the results obtained after such an event. However, when in pigs on cardiopulmonary bypass the myocardium was reperfused and sinus rhythm restored after ischemic ventricular fibrillation lasting 10 min, there was almost complete and immediate reversal of the changes observed in mitochondrial preparations isolated from the ischemic segments. These included partial uncoupling of oxidative phosphorylation, decreased oxygen uptake and loss of cytochrome oxidase activity (34). However, myocardial creatine phosphate (CP), which had decreased from 6 umol/g wet wt. to almost zero, and AMP and ADP, both of which doubled during ventricular fibrillation, recovered only partially (to 60%) after defibrillation. ATP (40% decrease during ventricular fibrillation) did not recover during the first 10 min after restoration of flow and sinus rhythm. Myocardial lactate levels (1 umol/g wet wt.) increased 20- and 30-fold during 5 and 10 min of ventricular fibrillation, respectively. They remained higher (5 and 7 umol/g wet wt., respectively) after 10 min of reperfusion and restoration of sinus rhythm. Because in most studies animals encountering a ventricular fibrillation during acute myocardial ischemia are only included for further study when sinus rhythm has been restored within 60 s, it is unlikely that, in particular in the ischemic tissue, significant damage may be caused during this short period. A critical evaluation of the effect of fibrillation on the metabolism of the adjacent non-ischemic myocardium has not been carried out.

Anesthetics and the Effects on Cardiac Metabolism

The use of anesthetics is unavoidable in open-chest preparations. It is therefore important to realize that these agents may also affect cardiovascular responses to pharmacologic agents and modify the effects of myocardial ischemia and reperfusion. So far no evidence has been presented that anesthetics disturb the balance between oxygen supply and demand in normal perfused hearts in spite of effects on its most prominent variables. In pigs the most commonly used anesthetic regimens include pentobarbital, butyrophenones (azaperone, droperidol), alpha-chloralose, ketamine and inhalation anesthetics, such as halothane, enflurane and nitrous oxide. Neuromuscular blocking agents (pancuronium bromide and vecuronium bromide) are frequently combined with these drugs (37).

Pentobarbital reduces stroke volume by a direct cardiac depressant action, while heart rate and systemic vascular resistance increase. The drug, however, depresses the baroreceptor-induced reflex tachycardia as mean arterial pressure falls. Since the latter is a common feature during myocardial ischemia, the ischemia-induced responses may be modified accordingly. The butyrophenones, on the other hand, cause vasodilatation and thereby increase heart rate. A rather surprising finding has been that the combination of azaperone-metomidate anesthesia yields rather high arterial lactate concentrations (4 mM), possibly due to inhibition of cellular respiration which may account for the negative inotropic effects. The high arterial lactate levels make lactate the preferred substrate for the myocardium (60 - 70% extraction). Free fatty acid and glucose extraction are almost non-existent with this anesthetic regimen (5,43).

When two end-tidal concentrations of halothane (0.46 and 1.04 vol%) were tested in pigs, myocardial oxygen consumption decreased dose-dependently to 40% of the value obtained at 0 vol% due to the effects on myocardial contractility and mean arterial blood pressure. Myocardial lactate uptake was not affected (19). Arterial lactate levels increased up to 40%, possibly due to the inhibitory effect of halothane on hepatic lactate metabolism. Myocardial concentrations of ATP, CP and glycogen were not affected by halothane, suggesting that there was no interference with the energy balance. Inhibition of mitochondrial oxidative phosphorylation by halothane has been suggested to be responsible for its negative inotropic effects. However, on the basis of the data described above, it may very well be that the drug effects on cardiac work are predominantly caused by interference with myocardial cellular calcium fluxes. Halothane is one of the few anesthetic agents of which the effects on ischemic porcine myocardium have been studied. In one such study (20), it was demonstrated that halothane (0.32 vol% end-tidal) and fentanyl (50 ug·kg^{-1} as bolus followed by an infusion of 100 ug·kg^{-1}·h^{-1}) supplemented with nitrous oxide and pancuronium bromide had similar effects on arterial coronary venous differences of metabolic markers of myocardial ischemia (ATP catabolites and lactate production), cardiac output, myocardial contractility and left ventricular filling pressure after coronary artery blood flow was reduced by 60%. As halothane may induce malignant hyperthermia in a number of breeds, the use of enflurane, which has also less cardiodepressant actions, is preferable.

Pancuronium bromide is the most widely used neuromuscular blocking agent, but its monoquaternary analog vecuronium bromide exerts less cardiovascular actions. Neither compound has any effect on myocardial substrate utilization (6).

The Effect of Restricted Coronary Blood Supply on Myocardial Function and Metabolism

In open-chest anesthetized pigs with unimpeded coronary blood supply almost all oxygen is extracted by the heart; oxygen saturation values of less than 15% are not unusual in blood sampled from the great cardiac vein (in pigs sampling of the coronary sinus does not render representative values because this vein receives also blood from the hemiazygous vein). In view of this high myocardial oxygen extraction, it is not surprising that a relatively small (25 - 30%) reduction in coronary blood flow already results in some loss of regional myocardial function (44). A reduction in flow by 60 - 70% results in a complete loss of function.

A relation, although relatively weak, has been established between the loss of contractile function and the appearance of the ATP-catabolite inosine in the coronary venous effluent during the first minutes of ischemia (4). The importance of this relationship between function and metabolism is weakened by the observation that the coronary venous concentrations of not only inosine, but also of most other metabolic markers of myocardial ischemia, remain not constant during prolonged periods of flow impairment (5,43). Thus after a reduction in coronary artery flow by 75% of baseline, the inosine concentration in the great cardiac vein increases rapidly from 20 to 120 uM during the first 20 min of ischemia. It then decreases gradually to 60 uM during the next 40 min (5). A similar time course, although less pronounced, has also been described for other nucleosides and for glycolytic breakdown products (5,43). Extreme care must therefore be taken if one wants to correlate functional changes, which are usually persistent, with those in appearance of breakdown products of ATP and glycolysis. Not only the rate of oxidative phosphorylation is markedly reduced at the onset of anaerobic conditions, but also the rate of ATP utilization is profoundly affected. Sufficient ATP may be present in severely ischemic myocytes to maintain certain functions (such as active ion transport) for a significant period of time. A more reliable parameter is therefore the tissue's energy charge, (ATP + 1/2 ADP)/(ATP + ADP + AMP), which reflects the imbalance between ATP utilization and ADP phosphorylation.

As stated before, presumably due to its high arterial concentrations, lactate is the preferred substrate during azaperone/metomidate anesthesia and both glucose and free fatty acids are not taken up (43). With this anesthetic regimen, myocardial lactate uptake (31 ± 8 umol/min) changes into release with a maximum value of 48 ± 9 umol/min after coronary blood flow in open-chest pigs was reduced by 60%. Furthermore, now a slight uptake of free fatty acids (1.8 ± 0.5 umol/min) was found. Using an in-situ working swine heart, Liedtke et al. (16) found that under pentobarbital anesthesia a 50% flow reduction yielded a lower glucose consumption and a lower free fatty acid extraction. Tissue levels of CP, ATP and lactate were not affected after 30 min. A further reduction to 40% of baseline was required before CP (by 69%) and ATP (by 25%) started to decrease.

Janse et al. characterized the "border zone" during myocardial ischemia in isolated blood perfused pig hearts (9,10). Histochemistry in biopsies revealed that glycogen-depleted cells interdigitated with cells rich in glycogen. Because of the small size of the samples (3 mm in diameter), one must conclude that the demarcation between normal and ischemic myocardium is sharp. A similar conclusion was drawn from electrophysiological mapping. Of interest is also the observation that no transmural electrical and metabolic gradients were present within the ischemic myocardium. This is in agreement with the spatial development of irreversible tissue damage following coronary artery occlusion (36). On the other hand, Fujiwara et al. found a spatial wave front phenomenon of irreversible damage spreading from the subendocardium to the subepicardium during the first 40 min of ischemia which had disappeared after 2 h (8). The wave front phenomenon was independent of collateral blood flow. It may be related to the higher wall stress for the subendocardial layers which poses a higher oxygen-demand on these myocardial segments. When occlusions, complete or partially, lasted longer than 24 h an excellent relation existed between regional residual blood flow and regional function (7,31).

In contrast to the metabolic effects on the jeopardized myocardium, those on the adjacent non-ischemic myocardium have not very well been studied. Savage et al. (31) showed that in conscious pigs function of the non-ischemic myocardium increased considerably during the first 15 min after a coronary artery occlusion and persisted during the first 48 h. Since these changes in regional function were accompanied by a 20% rise in heart rate, myocardial oxygen-demand must also have increased. On the other hand, myocardial perfusion (42%) was only significantly elevated after 24 h. Liedtke et al. demonstrated that a moderate level of ischemia (52% flow reduction) caused not only a decrease in oxygen consumption of the underperfused myocardium, but increased oxygen consumption and free fatty acid uptake of adjacent non-ischemic myocardium (17). There was no commensurate increase in CO_2 production from labeled palmitate, suggesting that fatty acids were utilized for acylation of glycerol-3-phosphate (ultimately leading to the formation of triglycerides). Tissue levels of long-chain acyl-CoA and acylcarnitine appeared to be increased. We also observed elevated tissue levels of long-chain acylcarnitine in adjacent non-ischemic myocardium after 60 min of complete coronary artery occlusion in pigs (12). This may be related to increased utilization of fatty acids for triglyceride formation in that region (12). In agreement with this finding is the observation by Liedtke et al. that during the last part of the 40-min period of ischemia, the energy balance of the adjacent myocardium was maintained by a two-fold increase in glucose uptake (17).

Recovery of Function and Metabolism of Ischemic Myocardium after Reperfusion

Depending on the duration and severity of flow reduction, the size of the coronary collateral circulation and the oxygen demand at the onset of ischemia, myocardial tissue may be reversibly or irreversibly damaged (33). The ischemic injury is reversible if restoration of coronary blood flow allows myocytes to survive and resume their function. Recovery of contractile function upon reinstitution of flow is usually not prompt and may last from several days to weeks. Thus the study of the recovery of function requires a preparation which will not deteriorate for prolonged periods of time. This excludes isolated heart preparations and the open-chest anesthetized animal for such studies unless early markers of recovery of function become available.

If in pigs a coronary artery is occluded for less than 2 min, recovery of function is complete within 5 min (30). In contrast, when the preceding period of ischemia exceeds 20 min, no sign of functional recovery is observed after 2 h of reperfusion (28,40). After 30 min of ischemic flow reduction to 40% of baseline, recovery of function is only 50% during the first hours (45). Lactate production occurs during early reperfusion after a partial occlusion, but does not persist with the postischemic reduction in function and myocardial oxygen consumption (18). With the longer periods of severe ischemia, there was not only no sign of any recovery of systolic wall function, but end-diastolic wall thickness was increased after two hours of reperfusion (24). The latter was accompanied by a large accumulation of calcium in the tissue and in isolated mitochondria, while the ability of the isolated mitochondria to produce ATP was impaired. The authors concluded from these data that 30 min of ischemia in pigs leads to irreversible damage. Observations in our own laboratory undermine this conclusion. We found that animals, subjected to the same period of coronary artery occlusion, also exhibited no sign or recovery of function after 2 h of reperfusion. However, they demonstrated almost complete recovery of regional systolic wall thickening assessed by 2-dimensional echocardiography two weeks later (15,28).

Murphy et al. also observed that after 2 h of reperfusion following 30 min of ischemia, oxygen consumption, the ADP/O ratio and respiratory control index of isolated mitochondria, and myocardial ATP content, were all approximately 50% of that of the control segment. These were all less than 20% of control parameters in myocardium, which had been reperfused for 2 h after 60 min of ischemia (24). In view of these data, it is of interest that the mitochondrial and total tissue calcium contents increased to a similar extent suggesting that myocardial Ca^{2+} accumulation is mainly due to active Ca^{2+} transport into the mitochondria energized by substrate oxidation.

It is questionable whether the ATP concentration in myocardium is a limiting factor in the recovery of function. Van der Giessen et al. showed that pretreatment with the stable prostacyclin analog Iloprost enhanced recovery of function following 20 min of ischemia independent of the ATP content of the hypoperfused tissue at the end of ischemia and after 2 h of reperfusion (40). Peng et al. also showed that the ATP content of myocardium, which had been ischemic for 30 min, was 45% of control after 10 min of reperfusion and 50% of control almost 2 h later. This suggests that these levels do not increase gradually during early reperfusion (24,27). The time-course of ATP levels during ischemia and reperfusion should be known before definite conclusions regarding the value of ATP concentration measurements as a critical parameter, determining cell survival, can be drawn.

Reperfusion of ischemic myocardium has been reported to exacerbate damage, (often) leading to an acceleration of necrosis of potentially reversibly injured myocardium. The mechanisms involved are, despite strenuous efforts, still not fully understood (29). The swelling of cells, which occurs when normotonic fluids are used for reperfusion, may be one of the factors involved. In isolated porcine hearts, swelling may be so severe that capillaries will be obstructed and results in the no-reflow phenomenon (11,38). With normotonic fluids, lactate levels remain high, and CP and ATP levels low during reperfusion, but with hypertonic fluids CP and ATP levels are partially restored, and lactate levels are lowered. In pigs the role of hydrolysis of membrane phospholipids and formation of oxygen-derived free radicals in development of reperfusion damage has also been extensively studied (2,3,26). In a porcine preparation on an extracorporeal circulation, tissue levels of CP and the energy charge decreased during ischemia. They recovered only partially during reperfusion, but completely when the free radical scavengers superoxide dismutase or catalase were present in the perfusion medium (2). In this study the ATP values did again not reflect recovery of function. The energy charge, which was lowered by only 5% during ischemia, recovered promptly with reperfusion when the free radical scavengers were present. In this respect it is important to note that xanthine oxidase possesses almost no activity in porcine and human myocardium and is therefore an unlikely candidate for generation of oxygen radicals during ischemia and reperfusion in these species (25).

Recently we have reported that the loss of integrity and function of the sarcolemmal and sarcoplasmic reticulum membrane, detected by protein phosphorylation assays and freeze-fracture electron microscopy, may represent one of the earliest derangements causing irreversible myocardial damage (1,12,14,15,28). Several processes may be involved in the morphological changes: stretching of the bilayer by severe contractions and osmotic pressure changes, breakdown of membrane protein and lipids by proteases, phospholipases and free radical attack, rearrangement of phospholipid and protein molecules in the bilayer by lipid blebbing and fusion processes (12-15,28). The lipid blebbing and membrane protein aggregation, induced by solidification of negatively charged phospholipids, hexagonal II phase separation of phospholipids and neutralization of the charge of membrane proteins, may all be initiated by the increases in cytosolic H^+ and Ca^{2+} concentrations. The membrane disruption that follows may cause the leakage of cytosolic proteins and bring the cellular Ca^{2+} movement further out of control. Spoiling of ATP by Ca^{2+}-activated myofibrillar ATPase, excessive mitochondrial Ca^{2+} accumulation and Ca^{2+} activated pumping ATPases in sarcolemma and sarcoplasmic reticulum are induced by the Ca^{2+} overload in the reperfusion phase. These data lend further support to the hypothesis that parameters other than the disturbance of energy metabolism should be taken into account in determining the extent of recovery after prolonged reperfusion.

References

1. Blom J, Verdouw PD, Lamers JMJ: Sarcoplasmatic reticulum function in the ischemic myocardium. Biomed Biochim Acta 46, Suppl 8/9:589-592, 1987
2. Das DK, Engelman RM, Rousou JA, Breyer RH, Otani H, Lemeshow S: Pathophysiology of superoxide radical as potential mediator of reperfusion injury in pig heart. Basic Res Cardiol 81:155-166, 1986
3. Das DK, Engelman RM, Rousou JA, Breyer RH, Otani H, Lemeshow S: Role of membrane phospholipids in myocardial injury induced by ischemia and reperfusion. Am J Physiol 251:H71-H79, 1986
4. De Jong JW, Goldstein S: Changes in coronary venous inosine concentration and myocardial wall thickening during regional ischemia in the pig. Circ Res 35:111-116, 1974
5. De Jong JW, Verdouw PD, Remme WJ: Myocardial nucleoside and carbohydrate metabolism and hemodynamics during partial occlusion and reperfusion of pig coronary artery. J Mol Cell Cardiol 9:297-312, 1977
6. Dhasmana KM, Verdouw PD, Prakash O, Saxena PR: Effect of neuromuscular blocking agents pancuronium bromide and its monoquaternary analogue ORG NC 5 on regional blood flow and cardiac metabolism in pigs. Drug Dev Res 2:393-402, 1982
7. Duncker DJ, Heiligers JPC, Saxena PR, Verdouw PD: Nisoldipine and perfusion of post-stenotic myocardium in conscious pigs with different degrees of concentric stenosis. Br J Pharmacol 1988, in press
8. Fujiwara H, Ashraf M, Sato S, Millard RW: Transmural cellular damage and blood flow distribution in early ischemia in pig hearts. Circ Res 51:683-693, 1982
9. Janse MJ, Cinca J, Moréna H, Fiolet JWT, Kléber AG, De Vries GP, Becker AE, Durrer D: The "border zone" in myocardial ischemia, an electrophysiological, metabolic, and histochemical correlation in the pig heart. Circ Res 44:576-588, 1979

10. Janse MJ, Moréna H, Cinca J, Fiolet JWT, Krieger WJ, Durrer D: Electrophysiological, metabolic and morphological aspects of acute myocardial ischemia in the isolated porcine heart. Characterization of the "border zone". J Physiol (Lond) 76:785-790, 1980

11. Krieger WJG, Ter Welle HF, Fiolet JWT, Janse MJ: Tissue osmolality, metabolic response, and reperfusion in myocardial ischemia. Basic Res Cardiol 79:562-571, 1984

12. Lamers JMJ, De Jonge-Stinis JT, Hülsmann WC, Verdouw PD: Reduced in vitro ^{32}P incorporation into phospholamban-like protein of sarcolemma due to myocardial ischaemia in anaesthetized pigs. J Mol Cell Cardiol 18:115-125, 1986

13. Lamers JMJ, De Jonge-Stinis JT, Verdouw PD, Hülsmann WC: On the possible role of long chain fatty acylcarnitine accumulation in producing functional and calcium permeability changes in membranes during myocardial ischaemia. Cardiovasc Res 21:313-322, 1987

14. Lamers JMJ, Essed CE, Pourquie MEM, Hugenholtz PG, Verdouw PD: The effect of nifedipine on ischemia-induced changes in the biochemical properties of isolated sarcolemmal vesicles and the ultrastructure of myocardium. Can J Cardiol 3:39-50, 1987

15. Lamers JMJ, Post JA, Verkleij AJ, Ten Cate FJ, Van der Giessen WJ, Verdouw PD: Loss of functional and structural integrity of the sarcolemma: an early indicator of irreversible injury of myocardium? Biomed Biochim Acta 46, Suppl 8/9:517-521, 1987

16. Liedtke AJ, Hughes HC, Neely JR: Metabolic responses to varying restrictions of coronary blood flow in swine. Am J Physiol 228:655-662, 1975

17. Liedtke AJ, Nellis S, Neely JR: Effects of excess free fatty acids on mechanical and metabolic function in normal and ischemic myocardium. Circ Res 43:652-661, 1978

18. McFalls EO, Pantely GA, Ophuis TO, Anselone CH, Bristow JD: Relation of lactate production to postischaemic reduction in function and myocardial oxygen consumption after partial coronary occlusion in swine. Cardiovasc Res 21:856-862, 1987

19. Merin RG, Verdouw PD, De Jong JW: Dose-dependent depression of cardiac function and metabolism by halothane in swine (*Sus scrofa*). Anesthesiology 46:417-423, 1977

20. Merin RG, Verdouw PD, De Jong JW: Myocardial functional and metabolic responses to ischemia in swine during halothane and fentanyl anesthesia. Anesthesiology 56:84-92, 1982

21. Mersmann HJ: Lipid metabolism in swine. In: Stanton HD, Mersmann HJ, eds: Swine in cardiovascular research. Vol I. Boca Raton: CRC Press, 1986:75-104

22. Millard RW: Changes in cardiac mechanics and coronary blood flow of regionally ischemic porcine myocardium induced by diltiazem. Chest 78, Suppl:193-199, 1980

23. Millard RW: Induction of functional collaterals in swine heart. Basic Res Cardiol 76:468-473, 1981

24. Murphy ML, Peng CF, Kane JJ Jr, Straub KD: Ventricular performance and biochemical alteration of regional ischemic myocardium after reperfusion in the pig. Am J Cardiol 50:821-828, 1982

25. Muxfeldt M, Schaper W: The activity of xanthine oxidase in heart of pigs, guinea pigs, rabbits, rats, and humans. Basic Res Cardiol 82:486-492, 1987

26. Otani H, Engelman RM, Rousou JA, Breyer RH, Das DK: Enhanced prostaglandin synthesis due to phospholipid breakdown in ischemic-reperfused myocardium. J Mol Cell Cardiol 18:953-961, 1986

27. Peng CF, Davis JL, Murphy ML, Straub KD: Effects of reperfusion on myocardial wall thickness, oxidative phosphorylation, and Ca^{2+} metabolism following total and partial myocardial ischemia. Am Heart J 112:1238-1244, 1986

28. Post JA, Lamers JMJ, Verdouw PD, Ten Cate FJ, Van der Giessen WJ, Verkleij AJ: Sarcolemmal destabilization and destruction after ischaemia and reperfusion and its relation with long-term recovery of regional left ventricular function in pigs. Eur Heart J 8:423-430, 1987

29. Reimer KA, Jennings RB: Myocardial ischemia, hypoxia, and infarction. In: Fozzard HA, Haber E, Jennings RB, Katz AM, eds: The heart and cardiovascular system. New York: Raven Press, 1986: 1133-1201

30. Roelandt J, Ten Cate FJ, Verdouw PD, Bom AH, Vogel JA: Effects of coronary artery occlusion and reperfusion on the time course of myocardial contraction. In: Bleifeld W, Effert S, Hanrath P, Mathey D, eds: Evaluation of cardiac function by echocardiography. Berlin: Springer Verlag, 1980:36-43

31. Savage RM, Guth B, White FC, Hagan AD, Bloor CM: Correlation of regional myocardial blood flow and function with myocardial infarct size during acute myocardial ischemia in the conscious pig. Circulation 64:699-707, 1981

32. Schaper W: Experimental infarcts and the microcirculation. In: Hearse DJ, Yellon DM, eds: Therapeutic approaches to myocardial infarct size limitation. New York: Raven Press, 1984:79-90

33. Schaper W, Binz K, Sass S, Winkler B: Influence of collateral blood flow and of variations in MVO$_2$ on tissue-ATP content in ischemic and infarcted myocardium. J Mol Cell Cardiol 19:19-37, 1987
34. Skinner FP, Levitsky MG, Scott RF, Frick J: Restoration of myocardial bioenergetic metabolism in swine after periods of ischemic ventricular fibrillation. J Thorac Cardiovasc Surg 69:729-735, 1975
35. St Clair RW: Atherosclerosis regression in animal models: Current concepts of cellular and biochemical mechanisms. Progr Cardiovasc Dis 26:109-132, 1983
36. Taylor IM, Shaikh NA, Downar E: Ultrastructural changes of ischemic injury due to coronary artery occlusion in the porcine heart. J Mol Cell Cardiol 16:79-94, 1981
37. Thurmon JC, Tranquilli WJ: Anesthesia for cardiovascular research. In: Stanton HC, Mersmann HJ, eds: Swine in cardiovascular research. Vol I. Boca Raton: CRC Press, 1986:40-59
38. Tranum-Jensen J, Janse MJ, Fiolet JWT, Krieger WJG, Naumann d'Alnoncourt C, Durrer D: Tissue osmolality, cell swelling, and reperfusion in acute regional myocardial ischemia in the isolated porcine heart. Circ Res 49:364-381, 1981
39. Van der Giessen WJ: Experimental models of myocardial ischemia. Progr Pharmacol 5/4:47-59, 1985
40. Van der Giessen WJ, Schoutsen B, Tijssen JGP, Verdouw PD: Iloprost (ZK 36374) enhances recovery of regional myocardial function during reperfusion after coronary artery occlusion in the pig. Br J Pharmacol 87:23-27, 1986
41. Verdouw PD, Deckers JW, Conrad GJ: Antiarrhythmic and hemodynamic actions of flecainide acetate (R-818) in the ischemic porcine heart. J Cardiovasc Pharmacol 1:473-486, 1979
42. Verdouw PD, Hartog JM: Provocation and suppression of ventricular arrhythmias in domestic swine. In: Stanton HC, Mersmann HJ, eds: Swine in cardiovascular research. Vol II. Boca Raton: CRC Press, 1986:121-156
43. Verdouw PD, Remme WJ, De Jong JW, Breeman WAP: Myocardial substrate utilization and hemodynamics following repeated coronary flow reduction in pigs. Basic Res Cardiol 74:477-493, 1979
44. Verdouw PD, Ten Cate FJ, Schamhardt HC, Van der Hoek TM, Bastiaans OL: Segmental myocardial function during progressive coronary flow reduction and its modification by pharmacologic intervention. In: Weiss HW, Witzstrock G, eds: Advances in clinical cardiology. New York: Publishing House, 1980:270-283
45. Verdouw PD, Wolffenbuttel BHR, Ten Cate FJ: Nifedipine with and without propranolol in the treatment of myocardial ischemia: effect on ventricular arrhythmias and recovery of regional wall function. Eur Heart J 4, Suppl C:101-108, 1983
46. Verdouw PD, Wolffenbuttel BHR, Van der Giessen WJ: Domestic pigs in the study of myocardial ischemia. Eur Heart J 4, Suppl C:61-67, 1983
47. Weaver ME, Pantely GA, Bristow JD, Ladley HD: A quantitative study of the anatomy and distribution of coronary arteries in swine in comparison with other animals and man. Cardiovasc Res 20:907-917, 1986

Chapter 18

The Energetics of "Stunned" Myocardium

W. Schaper and B.R. Ito, Department of Experimental Cardiology,
Max-Planck-Institute, Bad Nauheim, F.R.G.

*We review and report on experiments that address the mechanisms underlying
"stunning" and "conditioning" of postischemic myocardium. Short coronary occlusions
cause a long-lasting reduction of regional contractile function in the absence of cell
necrosis. Regional oxygen-demand is lower as a consequence of reduced myofibrillar
ATP turnover which leads to higher-than-normal phosphocreatine-levels. This
sequence of events, coined "stunning", conditions the myocardium in such a way
that a subsequent coronary occlusion is tolerated much better. Several arguments
are provided that "stunning" is probably not caused by the only moderate reduction
of the tissue ATP concentration. The hypothesis is put forward that the molecular
mechanism of "stunning" is associated with a change in the sarcolemmal calcium ion
homeostasis which reduces the calcium availability of the myofibrils. We hypothesize
further that "stunning" may not be a special form of reversible ischemic damage but
that it may be the unmasking of a pathophysiologic mechanism to protect against
cardiac overload.*

A New Concept

In 1975 Heyndrickx et al. (7) described an observation, the significance of which
became more important with time after the discovery: a short lasting coronary
occlusion that did not lead to cell death produced a regional contractile dysfunction
that did not recover for a surprisingly long time (days), whereas the electro-
cardiogram normalized within minutes. It was soon noted that the tissue
concentration of ATP, that was moderately depressed by the short coronary
occlusion, stayed depressed for a similarly long time (28). The intimate association
between contraction and ATP was so suggestive of a causal relationship that a false
sense of understanding prevailed, which inhibited awhile a true insight into the
nature of "stunning". Recently, however, the buzz-word has entered the clinical
laboratory because clinicians become aware of the fact that regional contractile
function recovers slowly after successful thrombolysis and may therefore not be a
good early end point to judge the success of the procedure. Clinicians have also
reported cases of coronary occlusion with greatly impaired regional function, where
revascularization procedures had revived tissue that conventional wisdom would have
expected to be dead (26). These cases were classified as "hibernating" myocardium
by which is meant that the afflicted myocardium ceases to function and uses the
little available energy to maintain its structure. These observations lead to the
concept *that the heart may possess and utilize mechanisms that enable it to down-
regulate its regional energy expenditure in the presence of reduced supply (37).*

This is a relatively new concept which may soften the hard dogma of the supply-
demand concept that dominated the pathophysiology of coronary artery disease for a
long time. It would also be prudent not to overemphasize this new concept because
the hard reality is that many patient's myocardium cannot be salvaged by
reperfusion because it failed to enter the "stunning" or "hibernating" state within
the allotted time window of reversible damage. However, *it appears to be of great*

scientific and practical interest to study the processes that enable the heart to cut its contractile engine in times of reduced supply in order to maintain structural integrity. Evidence that such mechanisms exist came from experiments where coronary occlusions of short duration were repeated. These experiments showed that the first occlusion left a trace, induced a memory and set the stage for the following occlusions whose impact on the myocardium was then much less injurious. One of the most convincing experiments was reported by Jennings' and Reimer's group (19,29), where a coronary occlusion of 40-min duration that always produced large infarcts, did not produce sizeable infarcts when the 40-min occlusion was preceded by a 10-min occlusion. A longer (total) ischemia time had in essence greatly reduced the ischemic damage, because the first occlusion had "conditioned" the myocardium in a way that is not mechanistically clear. The experiment does not say that infarcts are avoidable, if myocardium is "conditioned". However, the ultimate cell death is significantly delayed. Delayed-but-not-prevented is the important message (35,36) that we have learned from 10 years of hard, but not always successful, labor to reduce infarct size (36). The observation that a few drugs can delay, but not prevent, infarcts hints at these mechanisms that the heart can utilize to defend itself against undue demand.

All experiments described in this review point toward a change that occurs in the perception that we have of the role of the heart in general circulatory physiology: classical dogma has it that the heart under physiological conditions is always the obedient slave to the needs of other organs, ready to accept short-term energetic deficits, unable to refuse a demand even if selfinterest (in pathologic states) would dictate a slow-down. The classical observation for the physiologic energy deficit is the reaction of the coronary vasculature to a postextrasystolic beat: the potentiated beat produces a larger power output, but coronary flow increases only with the next beat, i.e., the deficit is first, payment later. This dogma holds also for pathological states: if a coronary occlusion occurs in the "unconditioned" state the afflicted myocardium does not down-regulate its needs. It continues to spend more (in futile attempts of contraction) than it receives. It beats itself, literally, to death.

The research of the last 10 years has modified this dogma in that we became more aware of situations where the heart shows the limited ability to defend itself against injury, or to protect itself against excessive loads. Besides stunning, other possible mechanisms of defense are the down-regulation of beta-receptors in congestive heart failure (17), the myocyte atrophy in chronic ischemia (33), the release of natriuretic factor at volume overload (3), and the development of a collateral circulation in chronic ischemia (34,35).

In the following paragraphs, we will describe the experiments that have led to this concept that cedes the myocardium a limited escape from the rigors of the control system.

The Role of Adenine Nucleotides in "Stunned" Myocardium

As said above: the relationships between contractile function and ATP are so close that a situation where both are decreased is almost selfexplanatory. However, it has been pointed out that there is no close relationship between the tissue [ATP] and contraction (6,20,23,27) other than a very general one in the sense that a very low ATP content does not enable a high ATP-turnover. It is the turnover of ATP rather than its tissue concentration that is closely related to contraction. There is, however, also good evidence that the tissue [ATP] may control its turnover, as shown by Kammermeier et al. (12), because ATP-utilizing enzymes require a high energy level. We became interested in the problem some years ago, when we studied the myocardial synthesis of ATP from salvaged precursors. The question was: can tissue-ATP levels that had been decreased by reversible ischemia be restored by selective administration of nuclosides? And: does the restoration of ATP improve the function of "stunned" myocardium? The background of this problem was the observation that the de-novo synthesis of ATP is very slow (about 3 nmol/h per g heart muscle; ref. 40), requiring about 4 days for nearly complete replenishment. We infused adenosine, 5-amino-4-imidazolecarboxamide riboside (AICA-riboside) and ribose into the left coronary artery of in situ beating dog hearts (anesthetized, open chest; ref. 18) after a period of 45 min of ischemia, and obtained needle biopsies at regular intervals after the beginning of infusion. We measured the tissue concentration of adenine nucleotides and nucleosides with HPLC and we measured the specific radioactivity of the nucleotide pool after infusion of radioactive adenosine, ^{3}H-AICA-riboside and of ^{14}C-glycine (for de-novo synthesis).

Stimulation of ATP-synthesis by ribose was described by Zimmer in the rat heart (20). Phosphoribosylpyrophosphate is in short supply in the postischemic heart and the exogenous supply of ribose increases the ATP-synthesis. AICA-riboside is the immediate precursor of IMP. It was also shown to increase the ATP pool size and to accelerate the postischemic repletion of ATP (18,38). Adenosine is phosphorylated to AMP, which is converted to ATP by the adenylate-kinase reaction. As Table 1 shows, adenosine infusion is the most rapid means to replenish postischemic ATP. It increases the rate of ATP-synthesis over that of the de-novo rate by a factor of 200. It was the only substrate that increased the tissue [ATP] significantly during the first 3 h of reperfusion. Ribose and AICA-riboside increased the rate of synthesis, but not to an extent that would have resulted in increased steady-state levels.

We then tested the hypothesis whether repletion of ATP resulted in improved regional postischemic contractile function (8). We tested this by implanting two pairs of sonomicrometer crystals into the subendocardium, one into a potentially ischemic, the other into a control region. We then studied the return of function after a period of 45 min of coronary occlusion in the presence or absence of exogenous (intracoronary infusion) adenosine and AICA-riboside. Although adenosine (but not AICA-riboside) caused significant increases in the tissue [ATP], postischemic regional function was not improved. AICA-riboside significantly depressed regional function via the synthesis of AICA-triphosphate (41). We concluded from this experiment that the dysfunction of "stunned" myocardium may have little to do with the tissue [ATP].

As we know now, other variables may enter the equation. Our own experiments (5) have shown that the coronary endothelium rapidly deaminates adenosine. Others have shown that endothelium predominantly phosphorylates adenosine (21). The true [adenosine] at the myocyte level following exogenous infusion is therefore not known. Although it is feasible that a part of the synthesized ATP resides also in the endothelium, it is not very probable that all of it is concentrated in endothelial cells. Given a volume fraction of endothelium of 3%, the final "endothelial" [ATP] after adenosine infusion would reach over 60 mM, i.e., clearly not compatible with solubility and osmolality. Another complicating factor is the inhibition of adenosine

Table 1. Purine nucleotide de novo synthesis and accelerated synthesis rates after 3 hours of intracoronary substrate infusion in previously ischemic and nonischemic myocardium

	De novo	Ribose	AICA-riboside	Adenosine
Postischemic	3.1 ± 0.3	15.1 ± 0.4	27 ± 3	268 ± 45
Nonischemic	1.5 ± 0.3	9.8 ± 0.5	28 ± 4	219 ± 36

Rate of incorporation of precursors into the canine myocardial ATP-pool during precursor-infusion at reperfusion following a 45-min occlusion period. Adenosine is the most potent precursor. The rate of ATP synthesis from adenosine is 180-times faster than the de-novo synthesis without exogenous infusion of precursors. Data are given in nmol/h per g wet wt. AICA-riboside = 5-amino-4-imidazolecarboxamide riboside.

kinase by one of its co-substrates, i.e., adenosine itself (22). The infusion of relatively large amounts of adenosine into a coronary artery, which resulted after 3 h in a significant increase of ATP levels in the dog heart, may have been the fortuitous outcome of activation of adenosine deaminase in the endothelium that reduced the actual [adenosine] at the myocytes to levels that favor phosphorylation. Own experiments in isolated buffer-perfused rat hearts (Görge and Schaper, unpublished) that may not be entirely comparable with in situ beating blood-perfused dog hearts, showed that adenosine was phosphorylated only at effective concentrations below $5 \cdot 10^{-5}$ M.

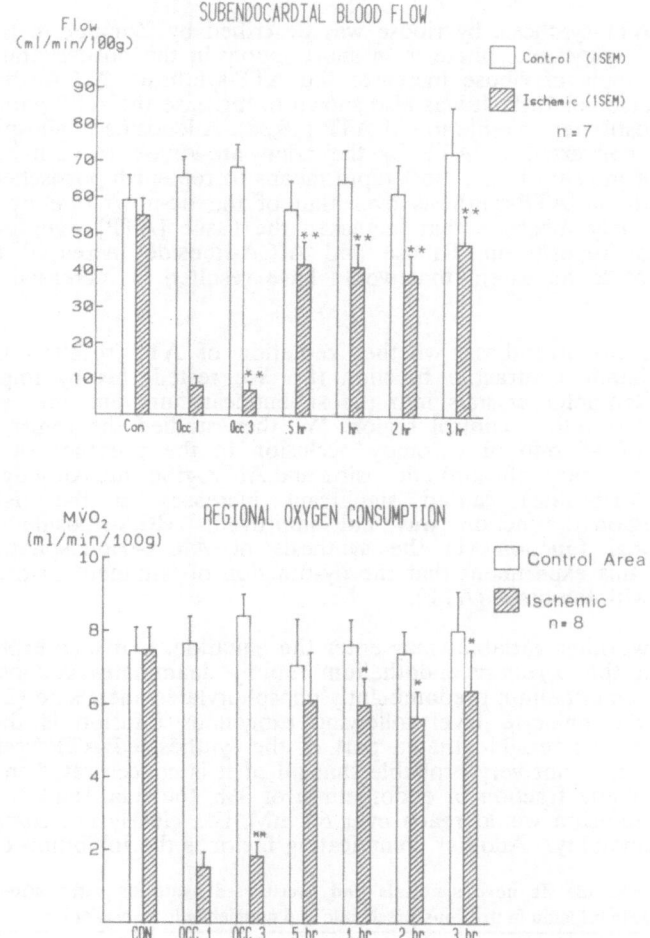

Fig. 1. Reduced regional O_2-consumption leads to down-regulation of blood flow. Subendocardial blood flow (upper panel, radioactive tracer microspheres) was measured in a control region (Con) and in an ischemic region (Occ 1, Occ 3) that was reperfused after three occlusions lasting 10 min each. Subendocardial blood flow was also measured at 0.5, 1, 2, and 3 h after reperfusion. Subendocardial blood flow falls to about 60% of that in a control region. Regional transmural oxygen consumption was measured from the transmural flows multiplied with the regional (arterio-)venous oxygen saturations. The figure shows that "stunning" was accompanied by reduced regional MVO$_2$ that led to a down-regulation of blood flow. * P<0.05 vs. Control; ** P<0.001 vs. Control.

In conclusion, we have provided evidence that intracoronary infusion of adenosine into a region of "stunning" results in an ATP increase in that region, which is much more than can be accounted for by an increase only of the endothelial pool of ATP. On the other hand, the increase in myocyte ATP may have been too small to definitively refute the hypothesis that changes in the ATP level are causative in "stunning".

Defective Energy Utilization vs. Energy Supply as a Cause of "Stunning"

One of the symptoms of "stunned" myocardium is the reduction of subendocardial blood flow after the hyperemic phase of reperfusion (see Fig. 1). We investigated the question of a microcirculatory disorder as a cause of dysfunction. We studied the distribution of left ventricular blood flow with radioactive tracer microspheres and the saturation of coronary venous blood with a differential oximeter. The oxygen saturation in two epicardial veins was studied, i.e., one draining from a previously ischemic region, the other from a normal region. About 1 h after onset of reflow following 10 min of occlusion, no differences in oxygen saturation between the two sampling sites were observed and it was assumed that these values were representative for the regions drained. The subendocardial blood flow determined at this moment had fallen by 25% below the value of normal neighbor-territories of the same heart. This leads to the conclusion that the "stunned" subendocardium consumed less oxygen, i.e., the general relationship between work output and oxygen consumption was maintained and blood flow was down-regulated to meet the lower demand. In order to exclude a primary microcirculatory disorder, we infused adenosine and we were able to show that dilatory reserve was normal, which makes "plugging" of microvessels as a cause of subendocardial flow reduction very unlikely. A microcirculatory disorder should also maintain the ischemic state well into the reperfusion period and it should further deplete ATP, which in fact does not occur.

Our experiments lead us to believe that the reduced O_2-uptake of "stunned" myocardium is not the expression of an energetic deficit, but rather a disturbance of energy utilization. In another series of experiments, we intracoronarily infused $CaCl_2$ into a "stunned" region. We could demonstrate that the region contracted normally again, even regained its maximal level of systolic shortening (10,37). Function returned to normal levels after stopping calcium infusion and the responses did not fade after several cycles. In yet another experiment, postextrasystolic potentiation showed almost complete restoration of function in a segment of "stunned" myocardium (see Fig. 2). These experiments demonstrate clearly that energy supply is not yet limited and that the contractile machinery is not damaged: vascular adaptation to increased demand is intact, mitochondria convert enough energy, myofibrils contract to the same maximal level if stimulated. We conclude therefore that "stunning" may be an expression of altered electromechanical coupling and that the most likely organelle, that underwent the change, is the sarcolemmal membrane. We tend to believe that "stunning" is not a defect or a damage, but rather a functional response to stress, perhaps related to down-regulation of voltage-dependent calcium-channels. We look upon "stunning" as an unmasking of a physiological response that is better developed in other organs (i.e., skeletal muscle), because the defense against myocardial ischemia was never put under the pressure of natural selection, because coronary heart disease develops relatively late in life.

Repeated Coronary Occlusions do not Produce Additive Effects

Several groups (9,16,19,29,38) have studied repeated short coronary occlusions for a variety of reasons: to study the influence of several drugs on ischemia in the same heart (2), to study the behavior of the nucleotide pool, and to study the

development of infarcts (19). We have tried repeated ischemia to test a hypothesis
that eventually proved to be wrong. We and others (9,38) have shown that only one
short occlusion, lasting between 3 and 5 min, is sufficient to produce a measurable
fall of myocardial ATP. Given the known long time requirements for ATP de-novo
synthesis, we argued that repeated occlusions of relatively short duration must
therefore lead to cumulative ATP depletions that may cause irreversible tissue
damage. This appeared to be an attractive clinical hypothesis, because repetitive
anginal attacks (such as in unstable angina) may lead to infarction even without
complete coronary artery occlusion. Cases were reported (30) where patent coronary
arteries existed shortly after a myocardial infarction. Alternative explanations are of
course possible in these cases. Our experiments soon showed that repeated
occlusions did not produce cumulative losses of ATP. The relatively largest loss
occurred after the first occlusion, when that lasted for 10 min, or after the first 5
occlusions, when these lasted for 3 min. More occlusions had an increasingly
diminishing effect, even 40 occlusions of 3 min interspersed by 3 min of reperfusion
did not produce irreversible damage. From previous experiments it was quite clear
that the first occlusion(s) had reduced the myocardial O_2-consumption, that
"stunning" had occurred which changed the energetic situation at the onset of the
next occlusion. The myocardium at risk will hence become less ischemic with the
next occlusion. The analysis of nucleotide breakdown products confirmed this: less
ADP, AMP, and relatively much less adenosine were produced compared with the
first occlusion. We hypothesized that in repeated occlusion the myocardium "learns"
how to stop the contractile process quicker in order to conserve energy and we
expected an earlier and more complete "bulging" with repeated occlusions. This was
difficult to show, but in these experiments systolic shortening was close to zero
during the first occlusion whereas bulging was -20% in repeated occlusions.

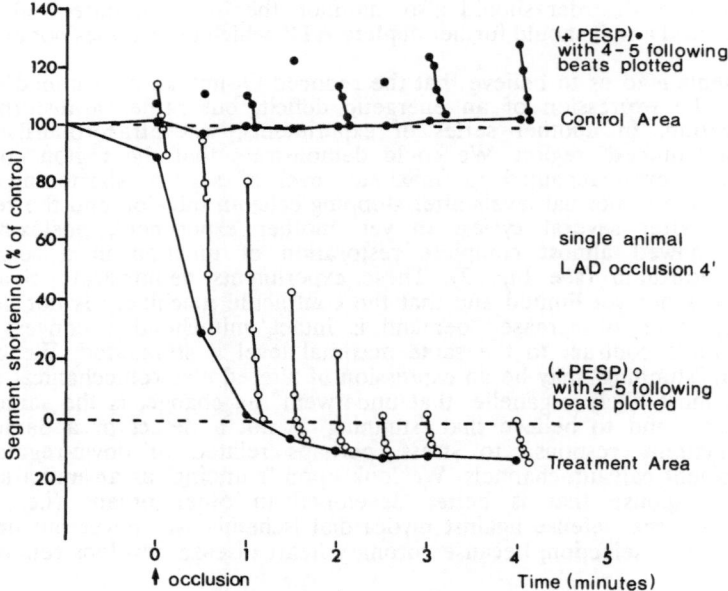

Fig. 2. Effect of postextrasystolic potentiation (PESP) on segment function of dog heart. Systolic
segment shortening was measured with subendocardial sonomicrometer crystals. Following occlusion,
systolic shortening falls quickly below zero within 1 min. A properly spaced extrasystole leads to
PESP of contraction that elicits a systolic shortening up to 2 min of coronary occlusion. At 1 min
into the occlusion, an almost normal contraction can be elicited (open circles). Solid circles: effect
of PESP on non-ischemic region contraction. LAD = left anterior descending coronary artery.

The Energy Balance of Ischemic Myocardium

When a coronary artery is suddenly occluded, the contractile function becomes weaker after 6 to 8 beats, thereafter contraction ceases and the affected ventricular wall shows systolic outward movement known as "bulging". Although it is understandable that the risk region ceases to contract when the substrate supply is abruptly cut, the still unanswered question is why the tissue [ATP] stays constant for a few minutes. Several hypotheses have been forwarded:

1. The decline in contraction following occlusion follows roughly the time course of fall of phosphocreatine (PC). This would support the hypothesis that only ATP synthesized from PC in close proximity to the myofibrils is the substrate for contraction. It is the "phosphocreatine shuttle" (11,32) that is responsible for the energy transfer between mitochondria and myofibrils. The fall in pO_2 in the mitochondria produces fewer molecules of PC, on-site phosphorylation of ADP at the myofibrils falls and contraction stops. The high cytosolic [ATP] remains untouched for several minutes, because its accessability for the myofibrils is low (1).

2. The inorganic phosphate that is produced by the cleavage of high-energy bonds may combine with cytosolic calcium, which reduces the availability of calcium ions at the myofibrils, thereby reducing the strength of contraction (15).

Whatever the precise mechanism for the cutting of the contractile engine is, two processes occur at the same time but in different directions: one that continues to respond with contraction to the stimulus of depolarization; and the other tries to reduce the electromechanical coupling.

Figure 2 shows an unpublished experiment where extrasystolic depolarizations caused potentiated beats during the course of declining contractility following coronary occlusions. The record shows that potentiated beats can be elicited up to 3 min following coronary occlusion, i.e., absolute lack of energy cannot have been the only cause of contractile failure. It is feasible that the amount of calcium ions available to the myofibrils had also decreased. The extrasystolic beats had provided an extra amount of calcium ions, which in turn had produced the potentiated contraction. This experiment hints at the existence of regulation of contraction in ischemia that is not exclusively dictated by the energy needs.

On the other hand, ischemic myocardium continues to respond to the stimulus of depolarization. However weak and inefficient, it accumulates a debt and proceeds toward irreversible injury.

That ischemic myocardium continues to beat during reversible ischemia is not obvious from sonomicrometer records. The nature of this method is, however, that small differences in regional contraction are amplified. It is intuitively clear that the two segments studied (an ischemic and a control segment) may differ only a little in the strength of contraction in order to obtain a large effect in the ischemic segment, because both regions are connected to each other in a tug-of-war situation.

We approached this problem by comparing the systolic bulging with the bulging caused by intracoronary infusion of procaine amide (see Fig. 3). It is apparent that the procaine hearts bulged more, indicating that hidden in the systemic bulge weak contractions are "buried".

The importance of this discussion about the mechanisms of "stunning" and "conditioning" lies in the fact that, if the heart would better utilize the mechanisms of ischemia-related-electromechanical uncoupling, infarcts can almost be completely avoided.

Our experiments in the canine heart showed (36) with the exception of the sub-endocardium of the posterior papillary muscle, that collateral flow is perfectly capable to maintain the structure of the myocardium, if all attempts at mechanical activity were reduced to zero. Our experiments also showed that the heart has the ability to adjust to the lower level of oxygen supply as provided by collateral flow. Jennings' and Reimer's group (19,29) has shown that "conditioning" significantly delays infarction. Our experiments with intracoronary verapamil, that did not produce infarctions following 3 h of left anterior descending coronary artery occlusion, support our hypothesis that a partial electromechanical uncoupling with reduced O_2-uptake is the immediate mechanism of protection. The following energetic balance shows that ischemic myocardium can reduce, however imperfect, its metabolic-electromechanical coupling, but that adverse stimuli can override this protective effect.

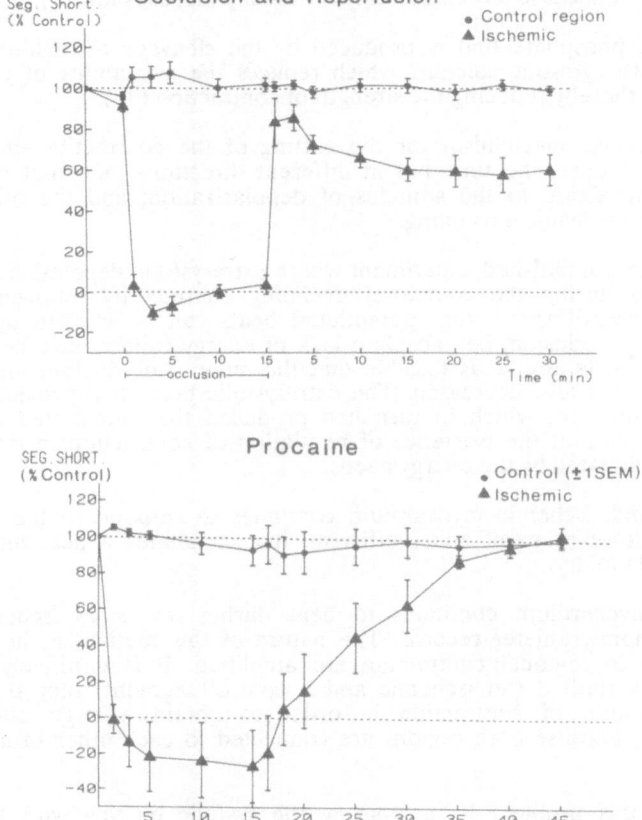

Fig. 3. Systolic shortening following coronary occlusion and reperfusion (upper panel), and following intracoronary procaine infusion in the same protocol (lower panel). "Bulging" in ischemia is much less pronounced than with procaine infusion, indicating some active force production in the ischemic region. Function completely returns to normal after stopping of procaine and does not produce "stunning" which indicates that "stunning" is not caused by mechanical factors. Seg. short. = segment shortening.

Under normal basal conditions (anesthesia with physiologically low heart rates), coronary flow in a canine heart is about 50 ml/min/100 g. This represents 8 ml of extractable O_2/min which is converted into a rate of 22 umol ATP/min per g wet wt. Shortly after coronary occlusion, tissue perfusion by collateral flow is maintained at a rate of 25 ml/min/100 g for the subepicardium and of 8 ml/min/100 g for the subendocardium. These represent an aerobic ATP production of 9 vs. 3.5 umol/min per g wet wt., respectively. These rates of aerobic ATP production are supplemented with glycolytic ATP production in the order of 0.2 umol/min per g of high-energy phosphate bonds. The rate of metabolic spending in ischemia is, however, faster. When we contrast the rate of fall of the tissue [ATP] with the rate of ATP production, we can define the degree of "overspending". This rate of overspending is surprisingly small in the subepicardium (which usually survives the occlusion), but fairly substantial in the subendocardium (3.5% more energy is spent than received *per minute*).

Repetitive occlusions (which unmask the existence of energy-saving mechanisms) reduce the amount of overspending by mechanisms that are as yet not well understood. The indicator for reduced overspending is the tissue [ATP] which is the result of ATP production and ATP utilization. The influence of repeated occlusions on tissue ATP is greatly blunted, which is most probably due to the O_2-sparing effect of "stunning" that results from changes in Ca^{2+} homeostasis. Another possible mechanism would be a decline in Ca^{2+} sensitivity of the sarcomeres (31).

Controversies

Not all workers in this field believe that "stunned" myocardium is a desirable, useful and probably physiological mechanism that one should amplify and utilize in order to delay infarction or to reduce infarct size. Several reports (4,25) show evidence of a cellular damage by oxygen free radicals because radical scavengers can prevent "stunning". However, beta-adrenoreceptor blockade was also shown to prevent "stunning" (14) as well as treatment with adenosine and with calcium channel antagonists (13,24). The unifying theory would be that calcium homeostasis is involved in the phenomenon, because all mechanisms mentioned utilize calcium as a final common pathway including adenosine which is a calcium chelator.

References

1. Bessman SP, Geiger PJ: Transport of energy in muscle: the phosphoryl-creatine shuttle. Science 211:448-452, 1981
2. Braunwald E, Maroko PR: Protection of the ischemic myocardium. In: Schaper W (ed): The pathophysiology of myocardial perfusion. Amsterdam: Elsevier/North-Holland Biomed Press, 1979:379-413
3. Cantin M, Genest J: The heart and the atrial natriuretic factor. Endocrinol Rev 6:107-127, 1985
4. Engler R, Covell JW: Granulocytes cause reperfusion ventricular dysfunction after 15 minute ischemia in the dog. Circ Res 61:20-28, 1986
5. Fliedner R, Podzuweit T, Schaper W: Katabolismus von Adenosin bei isolierten Koronar- und Aortenendothelzellen. Z Kardiol 75:23, 1986 (Abstr)
6. Hearse DJ: Oxygen deprivation and early myocardial contractile failure. A reassessment of the possible role of adenosine triphosphate. Am J Cardiol 44:1115-1121, 1979
7. Heyndrickx GR, Millard RW, McRitchie RJ, Maroko PR, Vatner SF: Regional myocardial, functional, and electrophysiological alterations after brief coronary occlusion in conscious dogs. J Clin Invest 56:978-985, 1975
8. Hoffmeister HM, Mauser M, Schaper W: Effect of adenosine and AICAR on ATP content and regional contractile function in reperfused canine myocardium. Basic Res Cardiol 80:445-458, 1985
9. Hoffmeister HM, Mauser M, Schaper W: Repeated short periods of regional myocardial ischemia: effect on local function and high energy phosphate levels. Basic Res Cardiol 81:361-372, 1986

10. Ito B, Tate H, Kobayashi M, Schaper W: Reversibly injured, post-ischemic canine myocardium retains normal contractile reserve. Circ Res 61:834-846, 1987
11. Jacobus WE: Myocardial energy transport: current concepts of the problem. In: Jacobus WE, Ingwall JS, eds: Heart creatine kinase. Baltimore: Williams & Wilkins, 1980
12. Kammermeier H, Schmidt P, Juengling E: Free energy change of ATP-hydrolysis: a causal factor of early hypoxic failure of the myocardium. J Mol Cell Cardiol 14:267-277, 1982
13. Klein HH: Acad thesis, Göttingen, 1987
14. Kloner RA, Ellis SG, Lange R, Braunwald E: Studies of experimental coronary artery reperfusion: effects on infarct size, myocardial function, biochemistry, ultrastructure and microvascular damage. Circulation 68:I8-I15, 1983
15. Kübler W, Katz AM: Mechanism of early "pump" failure of the ischemic heart: possible role of adenosine triphosphate depletion and inorganic phosphate accumulation. Am J Cardiol 40:467-471, 1977
16. Lange R, Ingwall J, Hale SL, Alker KJ, Kloner RA: Effects of recurrent ischemia on myocardial high energy phosphate content in canine hearts. Basic Res Cardiol 79:469-478, 1984
17. Lefkowitz RJ, Caron MG: Regulation of adrenergic receptor function by phosphorylation. J Mol Cell Cardiol 18:885-895, 1986
18. Mauser M, Hoffmeister HM, Nienaber C, Schaper W: Influence of ribose, adenosine, and "AICAR" on the rate of myocardial adenosine triphosphate synthesis during reperfusion after coronary artery occlusion in the dog. Circ Res 56:220-230, 1985
19. Murry CE, Reimer KA, Long JB, Jennings RB: Preconditioning with ischemia protects ischemic myocardium. Circulation 72, Suppl III:475, 1985 (Abstr)
20. Neely JR, Rovetto MJ, Whitmer JT, Morgan HE: Effect of ischemia on function and metabolism of the isolated working rat heart. Am J Physiol 225:651-658, 1973
21. Nees S, Gerlach E: Adenine nucleotide and adenosine metabolism in cultured coronary endothelial cells: formation and release of adenine compounds and possible functional implications. In: Berne RM, Rall TW, Rubio R, eds: Regulatory function of adenosine. The Hague: Nijhoff Publ, 347-360, 1983
22. Newsholme EA, Fisher MN: Adenosine kinase and the control of adenosine concentration in the heart. Biochem J 226:344, 1985 (Abstr)
23. Nishioka K, Jarmakani JM: Effect of ischemia on mechanical function and high energy phosphates in rabbit myocardium. Am J Physiol 242:H1077-H1083, 1982
24. Olafsson B, Formann MB, Puett DW, Pou A, Cates CU, Friesinger GC, Virmani R: Reduction of reperfusion injury in the canine preparation by intracoronary adenosine: importance of the endothelium and the no-reflow phenomenon. Circulation 76:1135-1145, 1987
25. Przyklenk K, Kloner RA: Superoxide dismutase plus catalase improve contractile function in the canine model of the "stunned myocardium". Circ Res 58:148-156, 1986
26. Puett DW, Forman MB, Cates CU, Wilson BH, Hande KR, Friesinger GC, Virami R: Oxypurinol limits myocardial stunning but does not reduce infarct size after reperfusion. Circulation 76:678-686, 1987
27. Reibel DK, Rovetto MJ: Myocardial ATP synthesis and mechanical function following oxygen deficiency. Am J Physiol 234:H620-H624, 1978
28. Reimer KA, Hill ML, Jennings RB: Prolonged depletion of ATP and of the adenine nucleotide pool due to delayed resynthesis of adenine nucleotides following reversible myocardial ischemic injury in dogs. J Mol Cell Cardiol 13:229-239, 1981
29. Reimer KA, Murry CE, Yamasawa I, Hill ML, Jennings RB: Four brief periods of myocardial ischemia cause no cumulative ATP loss or necrosis. Am J Physiol 251:H1306-H1315, 1986
30. Roberts WC, Buja LM: The frequency and significance of coronary arterial thrombi and other observations in fatal acute myocardial infarction. A study of 107 necropsy patients. Am J Med 52:425-443, 1972
31. Rüegg JC, ed: Calcium in muscle activation. A comparative approach. Zoophysiology. Berlin: Springer Verlag, 1986
32. Saks VA, Kupriyanov VV, Elizarova GV, Jacobus WE: Studies of energy transport in heart cells. The importance of creatine kinase localization for the coupling of mitochondrial phosphorylcreatine production to oxidative phosphorylation. J Biol Chem 255:755-763, 1980
33. Schaper J, Schaper W: Morphological changes in myocardium from patients with coronary heart disease and cardiac hypertrophy. Adv Cardiol 34:16-24, 1986
34. Schaper W: The collateral circulation of the heart. Amsterdam: North-Holland Publ, 1971

35. Schaper W, ed: The pathophysiology of myocardial perfusion. Amsterdam: Elsevier/North-Holland Biomed Press, 1979
36. Schaper W, Binz K, Sass S, Winkler B: Influence of collateral blood flow and of variations in MVO_2 on tissue-ATP content in ischemic and infarcted myocardium. J Mol Cell Cardiol 19:19-37, 1987
37. Schaper W, Ito B, Buchwald A, Tate H, Schaper J: Molecular and ultrastructural basis of left ventricular reperfusion dysfunction. Adv Cardiol 34:1-15, 1986
38. Swain JL, Sabina RL, Hines JJ, Greenfield Jr JC, Holmes EW: Repetitive episodes of brief ischaemia (12 min) do not produce a cumulative depletion of high energy phosphate compounds. Cardiovasc Res 18:264-269, 1984
39. Zimmer H-G, Gerlach E: Stimulation of myocardial adenine nucleotide biosynthesis by pentoses and pentitols. Pflügers Arch 376:223-227, 1978
40. Zimmer H-G, Trendelenburg C, Kammermeier H, Gerlach E: De novo synthesis of myocardial adenine nucleotides in the rat. Acceleration during recovery from oxygen deficiency. Circ Res 32:635-642, 1973
41. Zimmermann PT, Deeprose RD: Metabolism of 5-amino-l-beta-D-ribofuranosyl imidazole-4-carboxamide and related five-membered heterocycles to 5-triphosphates in human blood and L 5178 Y cells. Biochem Pharmacol 27:709-716, 1978

Chapter 19

Energy Depletion due to the Calcium Paradox

T.J.C. Ruigrok[1,2], J.H. Kirkels[2] and C.J.A. van Echteld[2],
[1]Department of Cardiology, University Hospital, Utrecht,
The Netherlands, and [2]Interuniversity Cardiology Institute
of the Netherlands

Reperfusion of an isolated mammalian heart with a calcium-containing solution after a brief calcium-free perfusion results in an excessive influx of calcium into the cells and irreversible cell damage: the calcium paradox. A rapid depletion of myocardial high-energy phosphate stores is one of the characteristics of the calcium paradox. During the first 30 seconds of calcium repletion in isolated rat heart, intracellular creatine phosphate and ATP levels decreased by 65% and 45%, respectively, and creatine, ADP and AMP levels increased by 15%, 85% and 2800%, respectively. The effect of different energy states of the heart on the capacity to develop the calcium paradox was studied in the anoxically perfused (with and without substrate) and reoxygenated rat heart. The presence of either ATP or electron transport appeared to be a prerequisite for the occurrence of the calcium paradox. It has been suggested that acidification of the cytosol, as a result of hydrolysis of ATP and accumulation of calcium by mitochondria, is an important factor in the development of the calcium paradox. By using phosphorus nuclear magnetic resonance spectroscopy, the time course of intracellular pH, measured from the chemical shift of the intracellular inorganic phosphate peak, was investigated in isolated rabbit heart. Intracellular pH during the calcium paradox varied from 7.1 to 7.0, indicating that acidification of the cytosol does not play a causal role in the development of the calcium paradox.

Introduction

Ringer demonstrated that contraction of the heart rapidly ceases when calcium is removed from the extracellular fluid (22). Zimmerman and Hülsmann reported that reintroduction of calcium into the extracellular fluid does not result in recovery of contraction of the heart, but in irreversible cell damage (28,29). They introduced the term "calcium paradox" to describe this phenomenon. The calcium paradox is characterized by an excessive influx of calcium into the cells (1), depletion of endogenous high-energy phosphate stores (2), severe contracture of the myofibrils (8), and rapid loss of intracellular constituents (29).

The influx of calcium into the myocardial cells may be divided into an early, relatively small gain in cytosolic calcium, and a subsequent, massive influx (18). Evidence has been provided that entry through the slow channels (17-19) and via the sodium-calcium exchange mechanism (3,10,18,23,26,27) are involved in the primary gain in calcium. Additional factors that may be responsible for the early gain in cytosolic calcium are the calcium pumps of the sarcolemma and the sarcoplasmic reticulum, whose activities are decreased by calcium-free perfusion (15,21). The raised cytosolic calcium may then trigger a number of events, including energy-dependent mitochondrial calcium accumulation (24) and activation of various ATPases (11), contracture-mediated disruption of intercalated discs (8), disruption of the sarcolemma (20), loss of intracellular constituents (29), and a secondary uncontrolled entry of calcium (18). Neither the sequence of these events nor their relative importance is known at present.

Energy Depletion during the Calcium Paradox

A rapid decline of myocardial high-energy phosphate stores is one of the characteristics of the calcium paradox. This has been demonstrated by the freeze-clamp method (2) and by using phosphorus nuclear magnetic resonance (^{31}P NMR) spectroscopy (5).

Table 1 shows data, obtained with the freeze-clamp method, on creatine phosphate (CP), creatine and adenine nucleotide (ATP, ADP and AMP) levels in isolated rat heart during calcium-free perfusion and the calcium paradox (2). During calcium-free perfusion, when no energy is required for mechanical activity, CP and ATP levels increased by 20% and 10%, respectively. During the first 30 seconds of calcium repletion, myocardial CP and ATP levels decreased by 65% and 45%, respectively. In the same period there was an increase in creatine (15%), ADP (85%), and AMP (2800%). During continued reperfusion with calcium, the concentration of all compounds gradually decreased. The effluent fluid contained large amounts of creatine and AMP, and relatively small amounts of CP, ATP and ADP.

The finding that myocardial CP and ATP levels were decreased after 30 seconds of reperfusion with calcium, whereas creatine, ADP and AMP levels increased during this interval, points to a sudden consumption of high-energy phosphate stores. When the sum totals of ATP + ADP + AMP (Ad Nu) and of CP + creatine are considered, it is clear that there was hardly any loss of these intracellular compounds up to the first 30 seconds of reperfusion. After this period a decrease of these sum totals can be explained by loss into the extracellular space and by further degradation of AMP.

Energy Dependence of the Calcium Paradox

By perfusing isolated rat hearts at 37°C under anoxic conditions with and without substrate, and by reoxygenating hearts after an anoxic period, the effect of different energy states of the heart on the capacity to develop the calcium paradox has been studied (24).

After 35 min of anoxic perfusion (30 min with calcium; 5 min without calcium) in the *presence* of glucose (Fig. 1a), the hearts still contained 15 ± 3 umol ATP per g dry heart tissue (mean ± S.D.; n = 4), which is 80% of the amount of ATP present

Table 1. Effect of calcium-free perfusion (-Ca) and reperfusion with calcium-containing solution (+Ca) at 37°C on creatine phosphate (CP), creatine and adenine nucleotide levels in isolated rat heart

Perfusion sequence		CP		Creatine	ATP		ADP		AMP		Ad Nu
+Ca	15 min	30	± 2	32 ± 5	18.8	± 0.8	3.2	± 0.4	0.21	± 0.04	22.2
-Ca	4 min	36	± 3	32 ± 3	20.8	± 1.4	3.3	± 0.3	0.18	± 0.06	24.3
+Ca	0.5 min	12	± 5	37 ± 8	11	± 2	6.1	± 0.6	5	± 3	22.1
	1 min	5	± 3	33 ± 5	6	± 3	4.7	± 0.5	5	± 2	15.7
	2 min	4	± 2	23 ± 5	5	± 2	3.2	± 0.4	1.8	± 0.5	10.0
	4 min	1.6	± 1.2	10 ± 5	2.9	± 1.0	2.4	± 0.8	1.1	± 0.5	6.4

Results are expressed as umol/g dry heart tissue (mean ± S.D.; n = 4). Ad Nu = Sum of adenine nucleotides.

at the end of the control perfusion period (see Table 1). No measurable amount of creatine kinase (CK) was released from the hearts during this period of anoxic perfusion. Reintroduction of calcium to the anoxic glucose-containing perfusate resulted in a massive release of CK from the hearts.

After 35 min of anoxic perfusion (30 min with calcium; 5 min without calcium) in the *absence* of glucose (Fig. 1b), the myocardial ATP content was only 2% (0.39 ± 0.13 umol/g dry heart tissue) of that at the end of the control perfusion period. The absence of glucose and calcium in the perfusate caused a slow release of CK. After reintroducing calcium to the anoxic glucose-free perfusate, enzyme release increased slightly, indicating that the calcium paradox did not occur in hearts subjected to anoxic substrate-free perfusion. However, reoxygenation resulted in a prompt and massive release of CK. This oxygen-induced calcium paradox was completely inhibited when KCN was present during the last 15 min of anoxic perfusion and during reoxygenation. This indicates that reactivation of the electron transport

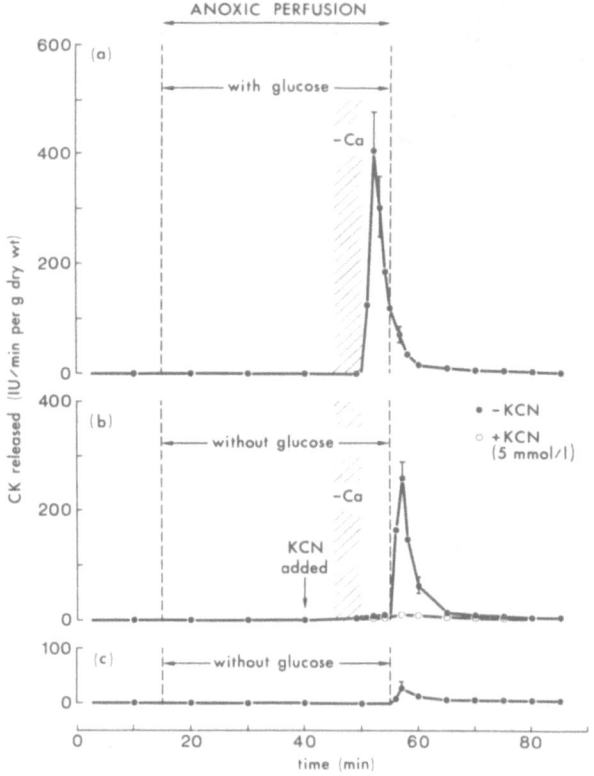

Fig. 1. Effect of reperfusion with calcium after a calcium-free period (shaded bar), in the anoxic rat heart (a: in the presence of glucose; b: in the absence of glucose), on the release of CK. a) Reperfusion with calcium resulted in an immediate and massive release of CK. b) Massive release of CK did not occur on reperfusion with calcium, but only on reoxygenation of the heart. In the presence of KCN oxygen-induced CK release was completely inhibited. c) Effect of reoxygenation after anoxic, glucose-free perfusion on the release of CK from isolated rat heart: Reoxygenation resulted in a relatively slight release of CK. Values are given as mean ± S.E.M. (n = 8). Reproduced with permission from ref. 24.

system is responsible for the sudden cell damage. CK release on reoxygenation after a merely anoxic substrate-free perfusion is much less pronounced than with a preceding calcium-free period (Fig. 1c).

These data show that the calcium paradox fails to occur in the absence of energy, but occurs as soon as the electron transport system is reactivated (Fig. 1b). Electron transport is not a prerequisite for the calcium paradox; in the absence of electron transport the paradox occurs, provided that ATP is present (Fig. 1a).

Intracellular pH during the Calcium Paradox

Hydrolysis of ATP and accumulation of calcium by mitochondria (supported by electron transport or ATP) with deposition of insoluble calcium phosphate, are accompanied by a release of protons (4,9). Hence, it has been suggested that acidification of the cytosol, with a consequent stimulation of cytosolic and lysosomal (phospho)-lipases and proteases, plays an important role in the development of the calcium paradox damage (11,24). ^{31}P NMR spectroscopy was used to investigate the time course of intracellular pH during the calcium paradox in isolated rabbit heart at 37°C (25). Intracellular pH was measured from the chemical shift of the intracellular inorganic phosphate (P_i) peak.

Until the end of the calcium-free period (10 min), intracellular pH amounted to 7.1 in all hearts (n = 6). Figure 2 shows ^{31}P NMR spectra of one of the hearts, taken

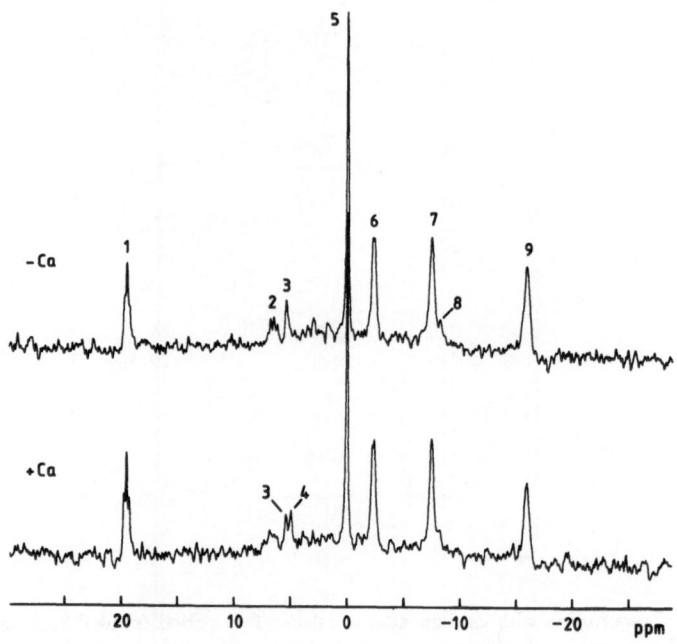

Fig. 2. ^{31}P NMR spectra of a rabbit heart obtained from 128 scans and taken between 5 and 10 min of control perfusion (+Ca), and between 5 and 10 min of the subsequent calcium-free perfusion (-Ca). Resonances are assigned as follows: 1, methylene diphosphonate (external reference compound); 2, sugar phosphates; 3, extracellular P_i; 4, intracellular P_i; 5, creatine phosphate; 6, gamma phosphate of ATP; 7, alpha phosphate of ATP; 8, NAD(H); 9, beta phosphate of ATP.

during control perfusion (+Ca), and during the subsequent calcium-free perfusion (-Ca). From the position of the intracellular P_i peak, intracellular pH during control perfusion was calculated to be 7.1. In this experiment the intracellular P_i peak was absent in the spectrum that was taken during calcium-free perfusion, most likely as a result of incorporation of P_i into high-energy phosphates, since myocardial CP and ATP levels increase during calcium-free perfusion (2).

Reperfusion with the calcium-containing solution resulted in a rapid decline of intracellular CP and ATP levels. During the first 100 seconds of reperfusion, intracellular pH in the heart used for Fig. 3 varied from 7.1 to 6.9. In the subsequent spectra the intracellular P_i peak, and also the CP and ATP peaks, were no longer perceptible. Intracellular pH values (mean ± S.D.) of all six hearts amounted to 7.1 ± 0.1 (0 to 20 s); 7.1 ± 0.1 (40 to 60 s); 7.0 ± 0.1 (80 to 100 s).

These results demonstrate that there was no appreciable fall of intracellular pH during reperfusion with calcium-containing solution. It may by argued that pH data were obtained only during the first 100 s of reperfusion, i.e., the period that the intracellular P_i peak was perceptible. Figure 4, however, shows that the calcium paradox damage develops so rapidly that the release of enzymes, which is one of

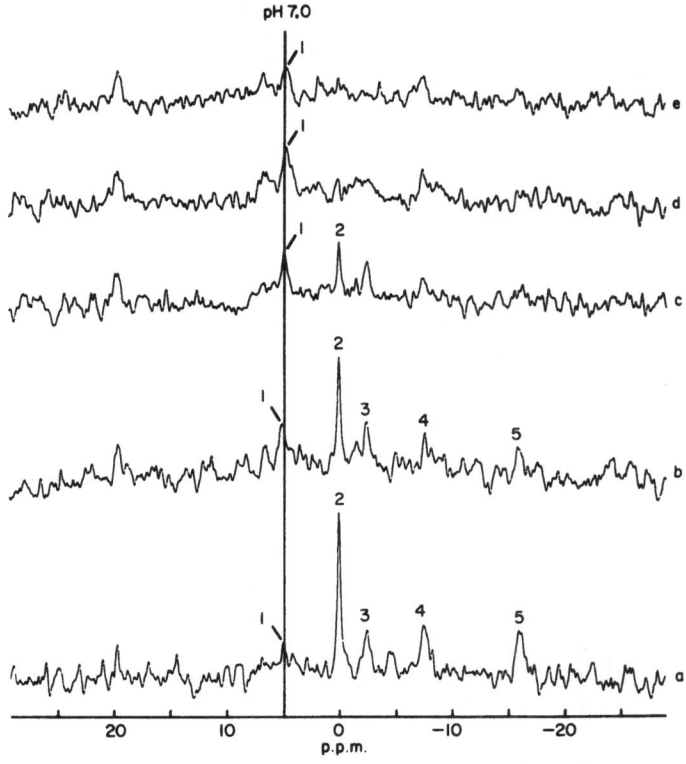

Fig. 3. ^{31}P NMR spectra of a rabbit heart obtained from 6 scans and taken between 0 and 20 s (a); 20 to 40 s (b); 40 to 60 s (c); 60 to 80 s (d) and 80 to 100 s (e) of reperfusion with calcium containing solution. Resonances are assigned as follows: 1, intracellular P_i; 2, creatine phosphate; 3, 4 and 5, the gamma, alpha and beta phosphate groups of ATP, respectively. Intracellular pH was measured from the chemical shift of the intracellular P_i peak, and amounted to 7.1 (a); 7.2 (b); 7.0 (c); 6.9 (d) and 6.9 (e). The vertical line indicates the position of the P_i peak at pH 7.0. Reproduced with permission from ref. 25.

the characteristics of the calcium paradox, is maximal between 40 and 80 s of reperfusion. The rapid disappearence of P_i from the cytosol during the calcium paradox may be the result of leakage into the extracellular space and uptake by mitochondria together with calcium.

As mentioned before, both hydrolysis of ATP and accumulation of calcium by mitochondria are accompanied by a release of protons. A third process, however, that has to be taken into account, is the breakdown of CP, which is a proton-consuming reaction (9,11). These opposite effects, in combination with the buffering capacity of the cell, may explain why the calcium paradox is accompanied by little or no net change in intracellular pH.

Concluding Remarks

At present we do not know which energy-dependent processes are responsible for the occurrence of the calcium paradox. The recent observation by Haworth et al. (12), however, that ATP depletion inhibits calcium influx in isolated myocardial cells, may explain why the calcium paradox does not occur in hearts subjected to anoxic substrate-free perfusion (see Fig. 1b). Reoxygenation and reactivation of the electron transport system could then promote calcium entry through the slow channels and in exchange for intracellular sodium, which increases during calcium-free perfusion as a result of a decrease in sodium/potassium-ATPase activity of the sar-

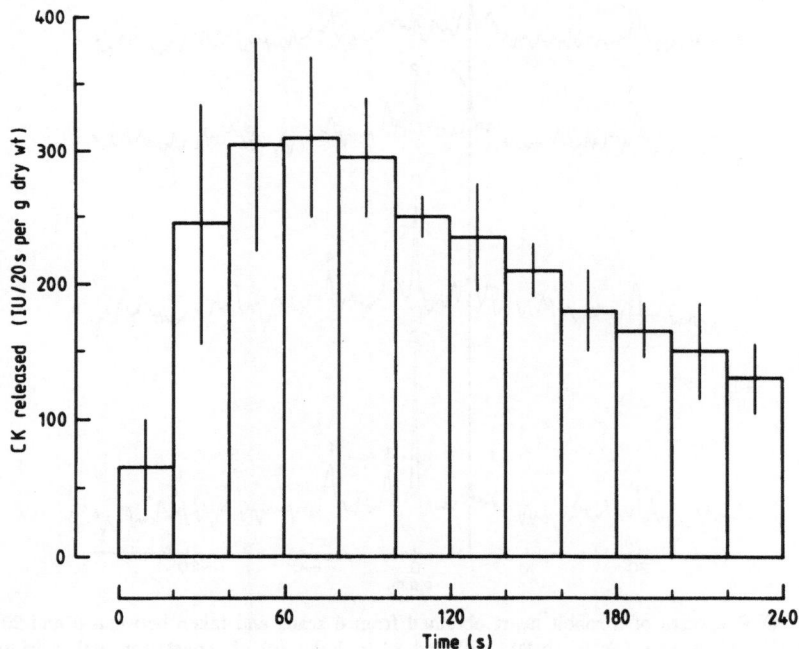

Fig. 4. CK release from isolated rabbit heart during the first 4 min of reperfusion with calcium-containing solution, after 10 min of calcium-free perfusion (37°C). CK activity was analyzed as described before (25). Results are expressed as mean ± S.D. (n = 5).

colemma (16). There is strong evidence that the slow channels (14) and the sodium-calcium exchange mechanism (6,7) are modulated by ATP. This hypothesis does not explain, however, why the calcium paradox is much more pronounced than the damage that occurs upon reoxygenation after a merely anoxic substrate-free perfusion (ref. 13, cf. Fig. 1b and c).

References

1. Alto LE, Dhalla NS: Myocardial cation contents during induction of calcium paradox. Am J-Physiol 237:H713-H719, 1979
2. Boink ABTJ, Ruigrok TJC, Maas AHJ, Zimmerman ANE: Changes in high-energy phosphate compounds of isolated rat hearts during Ca^{2+}-free perfusion and reperfusion with Ca^{2+}. J Mol Cell Cardiol 8:973-979, 1976
3. Bonvallet R, Rougier O, Tourneur Y: Role of the Na^+-Ca^{2+} exchange in the calcium paradox in frog auricular trabeculae. J Mol Cell Cardiol 16:623-632, 1984
4. Brierley GP, Murer E, Bachmann E: Studies on ion transport. III. The accumulation of calcium and inorganic phosphate by heart mitochondria. Arch Biochem Biophys 105:89-102, 1964
5. Bulkley BH, Nunnally RL, Hollis DP: "Calcium paradox" and the effect of varied temperature on its development; a phosphorus nuclear magnetic resonance and morphologic study. Lab Invest 39:133-140, 1978
6. Caroni P, Carafoli E: The regulation of the Na^+-Ca^{2+} exchange of heart sarcolemma. Eur J Biochem 132:451-460, 1983
7. DiPolo R, Beaugé L: In squid axons, ATP modulates Na^+-Ca^{2+} exchange by a Ca_i^{2+}-dependent phosphorylation. Biochim Biophys Acta 897:347-354, 1987
8. Ganote CE, Nayler WG: Contracture and the calcium paradox. J Mol Cell Cardiol 17:733-745, 1985
9. Gevers W: Generation of protons by metabolic processes in heart cells. J Mol Cell Cardiol 9:867-874, 1977
10. Goshima K, Wakabayashi S, Masuda A: Ionic mechanism of morphological changes of cultured myocardial cells on successive incubation in media without and with Ca^{2+}. J Mol Cell Cardiol 12:1135-1157, 1980
11. Grinwald PM, Nayler WG: Calcium entry in the calcium paradox. J Mol Cell Cardiol 13:867-880, 1981
12. Haworth RA, Goknur AB, Hunter DR, Hegge JO, Berkoff HA: Inhibition of calcium influx in isolated adult rat heart cells by ATP depletion. Circ Res 60:586-594, 1987
13. Hearse DJ, Humphrey SM, Bullock GR: The oxygen paradox and the calcium paradox: Two facets of the same problem? J Mol Cell Cardiol 10:641-668,1978
14. Irisawa H, Kokubun S: Modulation by intracellular ATP and cyclic AMP of the slow inward-current in isolated single ventricular cells of the guinea-pig. J Physiol (Lond) 338:321-337, 1983
15. Lamers JMJ, Ruigrok TJC: Diminished Na^+/K^+ and Ca^{2+} pump activities in the Ca^{2+} depleted heart: possible role in the development of Ca^{2+} overload during the Ca^{2+} paradox. Eur Heart J 4, Suppl H:73-79, 1983
16. Lamers JMJ, Stinis JT, Ruigrok TJC: Biochemical properties of membranes isolated from calcium-depleted rabbit hearts. Circ Res 54:217-226, 1984
17. Meno H, Kanaide H, Nakamura M: Effects of diltiazem on the calcium paradox in isolated rat hearts. J Pharmacol Exp Ther 228:220-224, 1984
18. Nayler WG, Perry SE, Elz JS, Daly MJ: Calcium, sodium, and the calcium paradox. Circ Res 55:227-237, 1984
19. Ohhara H, Kanaide H, Nakamura M: A protective effect of verapamil on the calcium paradox in the isolated perfused rat heart. J Mol Cell Cardiol 14:13-20, 1982
20. Post JA, Nievelstein PFEM, Leunissen-Bijvelt J, Verkleij AJ, Ruigrok TJC: Sarcolemmal disruption during the calcium paradox. J Mol Cell Cardiol 17:265-273, 1985
21. Preuner J: Functional alterations of the sarcolemma in Ca^{2+}-free perfused hearts. Basic Res Cardiol 80, Suppl 2:19-24, 1985
22. Ringer S: A further contribution regarding the influence of the different constituents of the blood on the contraction of the heart. J Physiol (Lond) 4:29-42, 1883
23. Ruano-Arroyo G, Gerstenblith G, Lakatta EG: 'Calcium paradox' in the heart is modulated by cell sodium during the calcium-free period. J Mol Cell Cardiol 16:783-793, 1984

24. Ruigrok TJC, Boink ABTJ, Spies F, Blok FJ, Maas AHJ, Zimmerman ANE: Energy dependence of the calcium paradox. J Mol Cell Cardiol 10:991-1002, 1978
25. Ruigrok TJC, Kirkels JH, Van Echteld CJA, Borst C, Meijler FL: ^{31}P NMR study of intracellular pH during the calcium paradox. J Mol Cell Cardiol 19:135-139, 1987
26. Tunstall J, Busselen P, Rodrigo GC, Chapman RA: Pathways for the movements of ions during calcium-free perfusion and the induction of the 'calcium paradox'. J Mol Cell Cardiol 18:241-254, 1986
27. Uemura S, Young H, Matsuoka S, Jarmakani JM: Low sodium attenuation of the Ca^{2+} paradox in the newborn rabbit myocardium. Am J Physiol 248:H345-H349, 1985
28. Zimmerman ANE, Daems W, Hülsmann WC, Snijder J, Wisse E, Durrer D: Morphological changes of heart muscle caused by successive perfusion with calcium-free and calcium-containing solutions (calcium paradox). Cardiovasc Res 1:201-209, 1967
29. Zimmerman ANE, Hülsmann WC: Paradoxical influence of calcium ions on the permeability of the cell membranes of the isolated rat heart. Nature 211:646-647, 1966

Clinical Implications

Chapter 20

Human Purine Metabolism

O. Sperling, Department of Clinical Biochemistry, Beilinson
Medical Center, Petah-Tikva, and Department of Chemical Pathology,
Sackler School of Medicine, Tel-Aviv University, Ramat Aviv, Israel

The purpose of purine metabolism in man is to maintain an optimal level of the nucleotides in the tissues. The nucleotides play an important role in nearly all biochemical processes, including energy metabolism, DNA and RNA structure, and regulation of many metabolic pathways through allosteric effects on enzymes, or through the adenylate energy charge. The nucleotides regulate their de novo and salvage synthesis and the interconversions between AMP and GMP. Uric acid is the waste, degradation endproduct of purine nucleotides in man. Uric acid is hardly soluble in physiological fluids. Therefore, hyperuricemia and hyperuricosuria are associated with precipitation of tophi in joints and other tissues and calculi in the urinary tract. Inborn errors in purine metabolism include the x-linked superactivity of 5-phosphoribosyl-1-pyrophosphate synthetase (gout and uric acid lithiasis), the complete deficiency of hypoxanthine-guanine phosphoribosyltransferase (Lesch-Nyhan syndrome) and the partial deficiency of this enzyme (gout and uric acid lithiasis), and the autosomal recessive deficiency of the following enzymes: adenine phosphoribosyltransferase (2,8-dihydroxyadenine lithiasis), adenosine deaminase (combined immunodeficiency), purine nucleoside phosphorylase (T-cell immunodeficiency), xanthine oxidase (xanthinuria) and myoadenylate deaminase (muscle disease). A transport defect for urate in the renal tubules, manifested in hypouricemia, has also been reported.

Importance

Purine nucleotides play an important role in nearly all biochemical processes (14,24,31,49). Adenine nucleotides are active in the storage and transfer of metabolically available energy, ATP being the universal currency of energy in biological systems. Another important function of the nucleotides is the carrying of a wide variety of groups, and their transfer to appropriate acceptors (glycosyl group, alcohol phosphate, sulfate, hydrogen, hydride ion, electrons, and alkyl group). In addition, the purine nucleotides form structural units. They are the activated precursors of DNA and RNA, the bulk of cellular nucleotides being in the form of these polymers. Nucleotides also participate as structural units in low molecular weight compounds, such as histidine and certain vitamins (folic acid, thiamine and riboflavin). Adenine nucleotides are components of three major coenzymes: NAD^+, FAD, and CoA, and of the important methyl donor, 5'-adenosylmethionine. Other nucleotides, such as cAMP and cGMP, have physiological functions as mediators of hormone action (regulatory signals). The nucleoside adenosine is considered to have an hormonal-like "retaliatory metabolite" role in matching energy demands to the synthesis of ATP in the heart muscle (32) and probably in some other tissues. Last but not least is the role of the nucleotides in the regulation of many catabolic pathways through allosteric effect on the enzymes, and their effect on many pathways of intermediary metabolism through the energy charge, ([ATP] + 0.5 [ADP])/([ATP] + [ADP] + [AMP]), in the cells (2).

Metabolism

The pathways of purine nucleotide metabolism operate to maintain an optimal level of the various nucleotides in the tissues. The pathways of purine metabolism include (14,24,31,38,49,58,59): synthesis of nucleotides, either from small nonpurine molecules, by the energetically costly de novo pathway, or from preformed purines by the economical salvage pathways: interconversions between the nucleotides; and degradation of excess nucleotides. The degradation of nucleotides may furnish purine bases or nucleosides for salvage nucleotide synthesis in tissues in greater need for the nucleotides than the source tissue. Not all pathways operate in all tissues, and certainly not at the same intensity (see tissue characterization, below).

De Novo Biosynthesis of Purine Nucleotides

The purine ring is assembled from glycine (C-4, C-5, and N-7), from the amino nitrogen of aspartate (N-1), from the amide nitrogen of glutamine (N-3 and N-9), from activated derivatives of tetrahydrofolate (C-2 and C-8) and from CO_2 (C-6) (Fig.1).

The purine ring structure is being assembled on a ribosyl moiety and when completed it is the nucleotide IMP. This is in contrast to the pyrimidine ring, which structure is completed before its attachment to the 5'-phosphoribosyl moiety to form OMP. The first-committed step in the de novo pathway is the formation of 5'-phosphoribosylamine from 5-phosphoribosyl-1-pyrophosphate (PRPP) and glutamine, catalyzed by glutamine-PRPP amidotransferase. The amide group of glutamine displaces the pyrophosphate group attached to the C-1 of PRPP. The other components of the purine skeleton are introduced in nine additional reactions, yielding finally IMP (Fig. 2; for full details, see refs. 14,19). IMP is the parent purine nucleotide molecule, but as a nucleotide it has no role in metabolism, except for being the precursor of AMP and GMP. AMP is synthesized from IMP in two steps. The first step is amination, accomplished through the attachment of aspartic acid to form the intermediate adenylosuccinate. This reaction is catalyzed by adenylosuccinate synthetase (EC 6.3.4.4). GTP is required in this reaction, providing a potential regulatory mechanism. Adenylosuccinate lyase (EC 4.3.2.2) cleaves fumaric acid from adenylosuccinate to form AMP. GMP is formed from IMP also in two steps. In the first step, IMP is oxidized to XMP, catalyzed by IMP dehydrogenase (EC 1.2.1.14) with NAD^+ as hydrogen acceptor. XMP is aminated by the amide group of glutamine to form GMP. ATP is consumed in this reaction, being cleaved to AMP + PPi. The conversion of GMP and AMP to their respective di- and triphosphates

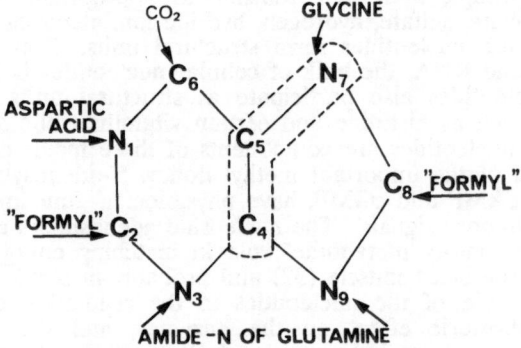

Fig. 1. The sources of the carbon and nitrogen atoms of the purine ring.

occurs in two successive steps, catalyzed by respective kinases, requiring ATP. AMP, ADP and ATP are interconvertible. Adenylate kinase (EC 2.7.4.3) catalyzes the reaction ATP + AMP $\rightleftharpoons$ 2ADP. The deoxyribonucleotides are synthesized by direct reduction at the 2'-carbon in the ribose moiety of ADP and GDP, rather than by synthesis of the molecule using the 2'-deoxy analog of PRPP. This reduction is a complex reaction, catalyzed by ribonucleotide reductase and NADPH as a cofactor (ref. 26; Fig. 3).

Interconversions between AMP, IMP and GMP

AMP and GMP can be reconverted to IMP, in one step reactions, catalyzed by other enzymes than those involved in their synthesis from IMP. GMP is converted to IMP by GMP reductase (EC 1.6.6.8), whereas AMP is converted to IMP by adenylate deaminase (EC 3.5.4.6). The last reaction plays a major role in ATP degradation, as well as in the activity of the purine nucleotide cycle.

The Purine Nucleotide Cycle

This cycle (AMP - IMP - adenylosuccinate - AMP; ref. 29) operates probably in many tissues. However, the activity and role of this cycle were investigated mainly in skeletal muscle. The roles attributed to the activity of the cycle in the muscle include (29): conservation of nucleotides in the tissues (by prevention of the formation of diffusible nucleoside and base degradation products from AMP or IMP); conservation of the adenylate energy charge (by conversion of excess AMP to IMP); activation of ATP generation by glycolysis (by the formation of ammonia) and by the tricarboxylic acid cycle (by generation of intermediates for the cycle); and neutralization of acid production due to ATP degradation (by production of ammonia). Indeed, the activity of adenylate deaminase, the rate-limiting enzyme of this cycle, is the highest in the skeletal muscle from all tissues (29). The absence

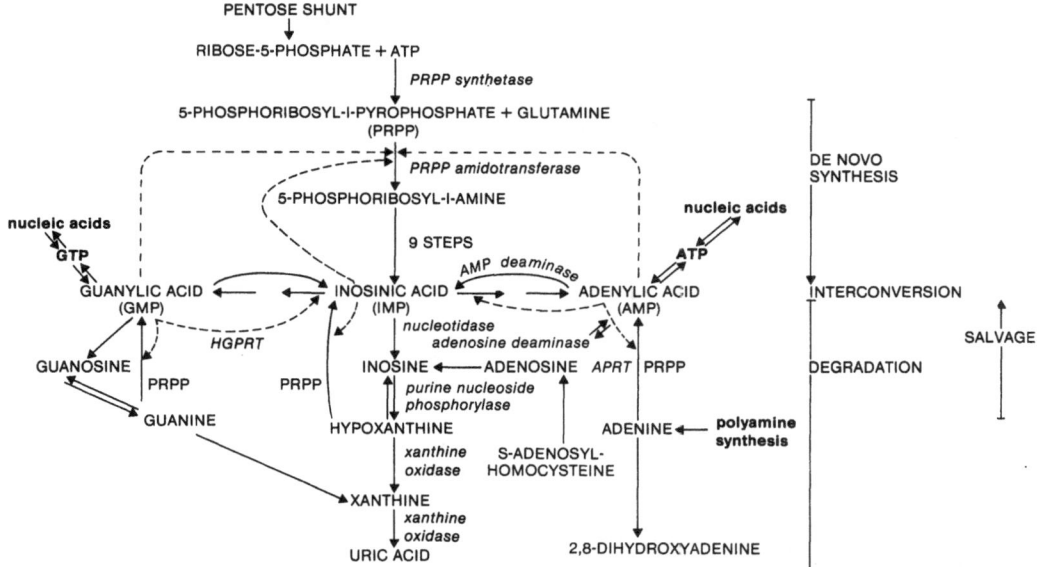

Fig. 2. Pathways of purine metabolism in the human. Not all tissues contain all pathways (see text). Interrupted lines represent regulation by feedback inhibition.

of this cycle is well manifested in the condition of hereditary deficiency of muscle adenylate deaminase (refs. 10,50; see below). The activity and role of this cycle in the heart muscle is not yet fully clarified (52,62).

The rate of activity and possible roles of the guanine nucleotide cycle (GMP - IMP - XMP - GMP) have not been evaluated.

Salvage Nucleotide Synthesis

The de novo synthesis of IMP consumes the equivalent of six high-energy phosphodiester bonds (by ATP hydrolysis). Additionally two high-energy bonds are consumed for IMP conversion to GMP, and one bond for the conversion of IMP to AMP. Thus, de novo synthesis of AMP or GMP is energetically very costly. Therefore, not surprisingly, once formed, the purine ring is recycled many times between the base (or nucleoside) and the nucleotide forms, before being further degraded to uric acid, the waste-endproduct to be excreted in the urine. The recycling of the purine bases or adenosine to the corresponding nucleotidic form is called "salvage nucleotide synthesis", since these pathways salvage the purines from the unreasonable (from the energy point of view) further degradation. Two general mechanisms perform salvage of preformed purine compounds: the phosphoribosylation of bases and the phosphorylation of nucleosides.

The phosphoribosyltransferases. Two enzymes in human tissue can phosphoribosylate purine bases. Hypoxanthine-guanine phosphoribosyltransferase (HGPRT; EC 2.4.2.8) phosphoribosylates hypoxanthine and guanine with PRPP to yield IMP and GMP, respectively. Another enzyme, adenine phosphorybosyltransferase (APRT; EC 2.4.2.7), phosphoribosylates similarly adenine to AMP. Physiologically, the salvage of hypoxanthine and guanine is much more important than that of adenine. This is indicated by the finding of PRPP accumulation in HGPRT-deficient tissues (36), but not in APRT deficiency (22,40,41,54). It is accepted that the consumption of PRPP by the salvage enzymes is greater than that by the de novo pathway, and accordingly that the proportion of nucleotides produced by salvage is greater than

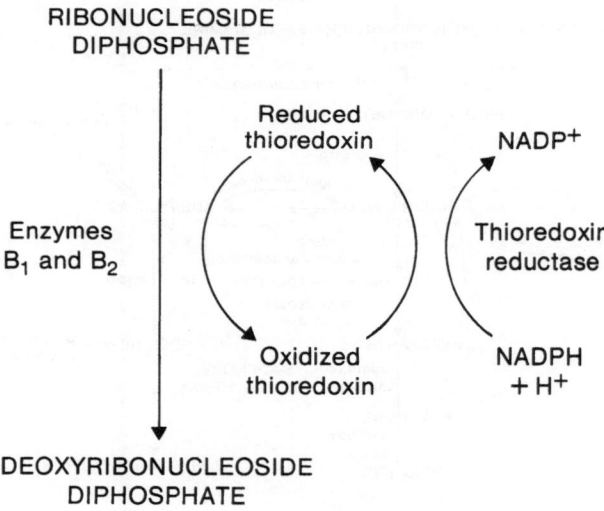

Fig. 3. Enzyme system for the reduction of ribonucleoside diphosphates to deoxyribonucleoside diphosphates.

that produced de novo. In contrast to the above, recent studies raised questions concerning the physiological efficiency of hypoxanthine salvage in various tissues, such as erythrocytes (5,34), liver (56), and skeletal (64) and heart muscle (62). The inefficient activity of HGPRT demonstrated in these tissues was attributed to low availability of substrate PRPP (34), to inhibition of HGPRT by the physiological content of cellular nucleotides (56), or to inefficient activity of adenylosuccinate synthetase (7). Evidently, the finding that a considerable quantity of hypoxanthine and guanine is degraded daily to xanthine and uric acid (see section on degradation, below), indicates a certain inefficiency of HGPRT. The physiological significance of adenine salvage is more seriously disputed, mainly in view of the questionable production of adenine in human tissues and the lack of accumulation of PRPP and absence of increase in purine production in APRT deficiency. However, the finding of 2,8-dihydroxyadenine urolithiasis in APRT deficiency (8,40) furnishes evidence that adenine is formed in the human, probably as a by-product of the synthesis of polyamines from 5'-adenosylmethionine (40), and that it is rapidly salvaged to AMP.

Adenosine kinase. The salvage of nucleoside occurs mainly for adenosine and 2'-deoxyadenosine and is catalyzed by adenosine kinase (EC 2.7.1.20). Another enzyme, deoxycytidine kinase can phosphorylate 2'-deoxyadenosine and 2'-deoxyguanosine to the respective deoxynucleotides. The salvage of adenosine is probably the main salvage operating in erythrocytes (34), heart (62), and skeletal muscle (64). The main source of adenosine is probably the transmethylation reaction (from 5'-adenosylmethionine). In some tissues, such as heart muscle, adenosine may be generated also from AMP, when this substrate accumulates during severe ATP degradation (62).

Salvage cycles. The formation of hypoxanthine, guanine and adenosine from nucleotides, and their recycling to nucleotides constitute cycles, which functions are not yet fully clarified. One possibility is that formation of hypoxanthine, guanine and adenosine in one tissue (e.g., liver and skeletal muscle) and the salvage of these purines to the respective nucleotide in another tissue (e.g., erythrocytes and brain), provide a system for transport of purines from "purine-rich" to "purine-poor" tissues (59). The adenosine futile cycle (AMP - adenosine - AMP could function in certain tissues, such as the heart during ischemia, as a mechanism for fast generation of the vasodilator adenosine. However, this role of the adenosine cycle has not been verified yet (62).

Regulation of Purine Biosynthesis

Substrate 5-phosphoribosyl-1-pyrophosphate is the most important regulator of purine synthesis. PRPP synthetase (EC 2.7.6.1) catalyzes its formation from ribose-5-phosphate and ATP. Ribose-5-phosphate is generated by the pentose phosphate pathway, by either the oxidative or non-oxidative directions, as well as from ribose-1-phosphate, produced by degradation of the purine nucleosides inosine and guanosine. There is evidence that in many tissues, the availability of ribose-5-phosphate is saturating for PRPP synthesis under physiological Pi and nucleotide concentrations (42). However, this is not the case in skeletal and heart muscle. In these tissues, probably due to low activity of the oxidative pentose phosphate pathway, ribose-5-phosphate availability is limiting for PRPP synthesis and therefore also for purine synthesis (62,63).

In accord with the main purpose of the various pathways of purine nucleotide synthesis, which is the maintenance of optimal nucleotide concentrations, the nucleotide endproducts regulate purine metabolism at several levels. The rate of the de novo pathway is controlled by feedback inhibition at several sites. The first-committed step in the pathway, catalyzed by glutamine-PRPP amidotransferase, is sensitive to inhibition by various purine nucleotides, especially by AMP and GMP, in a synergistic (cooperative) manner. These inhibitors are competitive with substrate

PRPP, which concentration is a major regulator of the amidotransferase activity (18). PRPP promotes conversion of the enzyme from the large inactive form (MW 270,000) to the small (MW 133,000) active form, whereas the purine nucleotides, alone or in combination, convert the small form of the enzyme to the large form. The activity of PRPP synthetase, which is therefore regulating the activity of the amidotransferase, is also subjected to feedback inhibition by the nucleotides, particulary AMP, ADP, GMP and GDP (16,51). At the physiological cellular content of nucleotides and Pi, the activity of PRPP synthetase is low, but there are several indications that the regulation of this enzyme, which is not committed for purine nucleotide synthesis only, is physiologically more important in the regulation of purine synthesis than that of the first-committed amidotransferase. The most conclusive indication for this is the flamboyant purine synthesis in patients with the hereditary (x-linked) superactivity of PRPP synthetase, due to resistance to feedback inhibition by nucleotides (45,47,48,60), or other molecular alterations (ref. 4; see below).

The branching point in the synthesis of AMP and GMP from IMP is also regulated by the respective nucleotides. Adenylosuccinate synthetase is subjected to feedback inhibition by AMP, whereas IMP dehydrogenase is subjected to feedback inhibition by GMP. Moreover, GMP is a substrate for the synthesis of AMP (IMP conversion to adenylosuccinate), and ATP is a substrate for the synthesis of GMP (XMP conversion to GMP). This reciprocal substrate relation balances the synthesis of adenine and guanine ribonucleotides from IMP.

Another level at which the nucleotides control their biosynthesis is the product inhibition of the salvage pathways. IMP and GMP inhibit HGPRT, whereas AMP inhibits adenosine kinase.

The activity of ribonucleotide reductase, which is essential for the generation of substrates for DNA synthesis, is subjected to a complex regulation, aimed at providing an optimal balance between the deoxyribonucleotides for DNA synthesis (refs. 26,49; Fig. 3).

Degradation of Purine Nucleotides

ATP constitutes the largest nucleotide pool. When its degradation (energy consumption) exceeds the capacity of the cell to regenerate ATP from ADP, AMP is formed. The main pathway for AMP degradation is probably through deamination to IMP. This has been demonstrated in liver (55), in erythrocytes (5), and in skeletal (64) and heart muscle (62). IMP is the main nucleotide substrate for degradation to non-nucleotide purines. Surplus IMP, formed from AMP, or from excessive de novo production (see section on inborn errors, below) is degraded by 5'-nucleotidase (EC 3.1.3.5) to inosine. The possibility that non-specific nucleotidases, such as alkaline phosphatase, have a role in nucleotide dephosphorylation has been raised, but there are several indications that the activity of the specific nucleotidases is more important in this respect than the non-specific enzymes (3). GMP can be similary dephosphorylated to guanosine. The extent of this reaction is not known, but xanthine production in xanthinuria indicates that it may have quantitative significance (see xanthinuria, below). Inosine and guanosine are degraded by purine nucleoside phosphorylase (EC 2.4.2.1) to hypoxanthine and guanine, respectively. Guanine is deaminated by guanase (EC 3.5.4.3) to xanthine, and hypoxanthine and xanthine are degraded by xanthine oxidase (EC 1.3.2.3) to uric acid, which is the waste-endproduct of purine metabolism in man (Fig. 2). The experiments of nature in humans, the hereditary deficiency of purine nucleoside phosphorylase (11,25), of adenosine deaminase (12,25), and of xanthine oxidase (17), furnish evidence supporting the above pathway of degradation. In purine nucleoside phosphorylase deficiency, the main purine metabolites in urine are inosine, deoxyinosine and guanosine (25), indicating the flow of degradation in normal subjects through these metabolites. In xanthinuria, about 79% of total oxypurines excreted are in the form of xanthine (17), deriving from hypoxanthine by the residual xanthine oxidase activity, but mainly from guanine. This last finding indicates a significant flow of degradation from GMP through guanosine to guanine. That AMP degradation does not proceed to a significant extent through adenosine to inosine is indicated by the absence of decrease in urinary uric acid in hereditary adenosine deaminase deficiency (25). This finding also indicates that adenosine, even if formed from 5'-adenosylmethionine (transmethylation reactions), does not contribute to uric acid formation, either due to insignificant formation, or more probably, due to efficient salvage.

Uric acid, the waste-endproduct of purine metabolism in the human, is not an ideal endproduct, since it is hardly soluble in body fluids. Excess uric acid in blood or urine tends to precipitate, as sodium hydrogen urate in joints or other tissues (tophi formation), or as uric acid, sodium urate or ammonium urate in calculi in the urinary tract (9,44). Interestingly, the renal handling of urate is also unlike that of classical endproducts. The renal handling of urate includes four components (43): a) urate is almost totally filtrable; b) the filtrable urate is almost totally reabsorbed in the proximal tubule; c) there is urate secretion; and d) there is subsequently postsecretory urate reabsorption. The urinary uric acid derives from the urate secretion which escapes reabsorption. In recent years, an increasing number of subjects with hereditary renal hypouricemia, due to defects in renal uric acid transport, have been reported (43). Most of these defects appear to be in the presecretory reabsorption site.

Tissue Characterization

The various tissues differ in the presence and intensity of operation of the above pathways of purine metabolism (59). The glutamine-PRPP amidotransferase, the first-committed step in the de novo pathway, is not active in erythrocytes and polymorphonuclear leukocytes, and its activity in the brain is low (1,19). In contrast, the amidotransferase is very active in liver and in skeletal muscle (6,53,63). These two tissues are probably the main source of preformed purines for salvage synthesis in other tissues. The circulating erythrocyte is devoid of many parts of purine metabolism (30). In addition to absence of almost all of the de novo pathway, it lacks adenylosuccinate synthetase. Thus, adenine nucleotides can be synthetized in this tissue only by salvage from adenosine (and adenine?). The erythrocyte lacks also xanthine oxidase. This enzyme which catalyzes the last two steps in purine degradation, is found only in liver, jejunal and rectal mucosa, and colostrum, and is absent from most other tissues (17,33). Some pathways may be markedly more active in specific tissues, due to a specific role, e.g., the purine nucleotide cycle in skeletal muscle. This cycle has an important role in energy metabolism in the skeletal muscle tissue (see below). HGPRT, the important salvage enzyme, is active in human brain markedly more (10 fold) than in other tissues (37). Indeed, the brain is the main tissue affected in HGPRT deficiency. The physiological activity of enzymes in a tissue may be deduced from the maximal activity of the enzyme in the tissue extract (V_{max}), the affinity of the enzyme to its substrates (K_m), the availability of substrates, and presence of inhibitors and modifiers. The physiological activity of pathways in the intact tissue may be gauged by fluxes of labelled precursors. Although much of the above information is known for purine metabolism in many tissues, an accurate knowledge of the actual (physiological) fluxes of the pathways of purine metabolism, especially that of de novo versus salvage synthesis in the various human tissues, is not yet available.

Inborn Errors

PRPP Synthetase Superactivity

Several families have been reported with excessive purine production due to hereditary superactivity of PRPP synthetase. In the first case reported in 1972 (45,47,48,60), the enzyme superactivity was characterized to reflect resistance to feedback inhibition by nucleotides. In the other cases, the enzyme alterations were found to be increased specific activity (increased V_{max}), increased affinity (low K_m) for substrate ribose-5'-phosphate, or combination of such abnormalities (4). The pattern of inheritance is x-linked recessive (61). The affected males exhibit uric acid lithiasis and juvenile gout. Treatment by allopurinol is very effective. The excessive purine production characteristic to all mutants of superactive PRPP synthetase, emphasizes the role of both the enzyme PRPP synthetase and product PRPP in the regulation of purine synthesis.

HGPRT Deficiency

Complete hypoxanthine-guanine phosphoribosyltransferase deficiency is causing the x-linked (recessive) Lesch-Nyhan syndrome. The clinical syndrome was first reported in 1964 (27). It is characterized by cerebral palsy with choreoathetosis and spasticity, self mutilation, mental and physical retardation, and severe purine overproduction. Purine overproduction is attributed to accumulation of PRPP, due to the absence of salvage nucleotide synthesis from hypoxanthine and guanine, and to reduced feedback inhibition, due to lower cellular content of IMP and GMP. The last mechanism has not been conclusively assessed. HGPRT deficiency, underlying the Lesch-Nyhan syndrome, was revealed in 1967 (39). Allopurinol effectively reduces the uric acid level in blood and urine, and by this prolongs the life of the patients, who otherwise are prone to develop severe renal insufficiency (gouty nephropathy). But allopurinol treatment in HGPRT deficiency is associated with increase in blood and urine oxypurines, associated in some of the patients in xanthine urolithiasis (46). No specific treatment is available for the neurological manifestations, the etiology of which is not yet clarified.

Partial HGPRT deficiency, first reported in 1967 (23), causes only excessive purine production, resulting in uric acid lithiasis and juvenile gout, but apparently does not cause significant neurological abnormalities (21).

APRT Deficiency

Complete deficiency of APRT is associated with 2,8-dihydroxyadenine urolithiasis (8,22,40,54). Adenine, which is probably produced as a by-product of the synthesis of polyamines from 5'-adenosylmethionine, is normally salvaged by APRT to AMP. In the absence of the enzyme, adenine is oxidized by xanthine oxidase to the insoluble 2,8-dihydroxyadenine. The pattern of inheritance is autosomal recessive. There is no accumulation of PRPP in tissues and purine production is normal. These findings suggest that adenine production and salvage are probably of low quantitative significance.

Muscle Myoadenylate Deaminase Deficiency

Myoadenylate deaminase deficiency is a relatively benign disorder, first described in 1978 (10), characterized clinically by muscle fatigue, following excercise. Purine production is normal. The pattern of inheritance is autosomal recessive. This disorder furnishes evidence for the importance of the nucleotide cycle in skeletal muscle. It was suggested (50) that during intensive muscle work, the AMP formed from ATP degradation, is converted to IMP, which accumulates stoichiometrically. Consequently, at rest, when ATP regeneration exceeds ATP degradation, the IMP is reconverted to AMP and to ATP. As discussed above, the activity of this cycle is of several advantages to the muscle tissue. The absence of this cycle is the cause for the manifested abnormalities.

Xanthine Oxidase Deficiency (Xanthinuria)

The enzyme deficiency is either complete or partial (17). In normal subjects the enzyme is found in liver, small intestinal and rectal mucosa, and colostrum or breast milk (17,33). Xanthine and hypoxanthine replace uric acid as the endproduct of purine metabolism. Accordingly, there is hypouricemia and hypouricosuria. Total purine production is decreased due to enhanced salvage of accumulating hypoxanthine. Xanthine urolithiasis is common. Formation of xanthine from guanine indicates the presence of GMP dephosphorylation to guanosine and subsequently to guanine (see above). Xanthine myopathy has also been reported.

Adenosine Deaminase Deficiency

The deficiency of adenosine deaminase (12,25) causes abnormalities in nucleoside metabolism which are selectively toxic to lymphocytes, resulting in combined immunodeficiency (cell-mediated T-cell- and humoral B-cell-immunodeficiency). Adenosine and deoxyadenosine levels are elevated in plasma and urine. dATP accumulates in peripheral mononuclear cells, especially in T-lymphocytes, and inhibits ribonucleotide reductase (DNA synthesis). The accumulation of adenosine results in accumulation of S-adenosylhomocysteine, inhibiting transmethylation (15,25), which is essential to lymphocyte multiplication and function. Purine production is normal, indicating absence of AMP degradation through adenosine in normal subjects. Inheritance is by an autosomal recessive pattern.

Purine Nucleoside Phosphorylase Deficiency

Children, affected by this disease, have severe cell-mediated immunodeficiency (11,25). dGTP accumulates in T-lymphocytes and inhibits ribonucleotide reductase. The disorder is inherited by an autosomal recessive trait. The degradation of purines is blocked at the nucleoside level, the main degradation products in urine being inosine, guanosine and deoxyinosine. Accordingly, there is hypouricemia and hypouricosuria. Total purine production is increased due to the absence of salvage of hypoxanthine and guanine (which formation is blocked). This disorder furnishes evidence concerning the quantitative significance of HGPRT activity and concerning the main pathway of nucleotide degradation.

Inborn Errors in other Pathways of Metabolism affecting Purine Metabolism

Glucose-6-phosphatase deficiency. Purine overproduction in Von Gierke's disease was attributed first to increased production of PRPP. The accumulation of glucose-6-phosphate results in acceleration of the oxidative pentose phosphate shunt, increasing ribose-5-phosphate generation, leading to increased production of PRPP (20). Increased nucleotide degradation, brought about by activation of adenylate deaminase, was suggested as the underlying mechanism (13,35). The AMP deaminase is activated due to the decrease in Pi (consumed by the enhanced glycogenolysis). The clarification of the mechanism(s) underlying purine overproduction in glucose-6-phosphatase deficiency is important for the understanding of the regulation of purine synthesis.

Glutathione reductase superactivity. This enzyme alteration was proposed to cause purine overproduction in some gouty black Americans by activation of the pentose shunt (28). However, the same condition was found in glucose-6-phosphate dehydrogenase deficiency without purine overproduction (57).

References

1. Allsop J, Watts RWE: Activities of amidophosphoribosyltransferase and the purine phosphoribosyltransferase and the phosphoribosylpyrophosphate content of rat central nervous system at different stages of development. J Neurol Sci 46:221-232, 1980
2. Atkinson DE: The energy charge of the adenylate pool as a regulatory parameter. Interaction with feedback modifiers. Biochemistry 7:4030-4034, 1968
3. Baer HP, Drummond GI: Catabolism of adenine nucleotides by the isolated perfused rat heart. Proc Soc Exp Biol Med 127:33-36, 1968
4. Becker MA, Losman MJ, Simmonds HA: Inherited phosphoribosylpyrophosphate synthetase superactivity due to aberrant inhibitor and activator responsiveness. Adv Exp Med Biol 195A:59-66, 1986
5. Bontemps F, Van den Berghe G, Hers HG: Pathways of adenine nucleotide catabolism in erythrocytes. J Clin Invest 77:824-830, 1986
6. Brosh S, Boer P, Zoref-Shani E, Sperling O: De novo purine synthesis in skeletal muscle. Biochim Biophys Acta 714:181-183, 1982
7. Brown AK, Raeside DL, Bowditch J, Dow JW: Metabolism and salvage of adenine and hypoxanthine by myocytes isolated from mature rat heart. Biochim Biophys Acta 845:469-476, 1985
8. Cartier MP, Hamet M: A new metabolic disease: the complete deficit of adenine phosphoribosyltransferase and lithiasis of 2,8-dihydroxyadenine. CR Acad Sci (Paris) 279:883-886, 1974
9. De Vries A, Sperling O: Uric acid stone formation: Concepts of etiology and treatment. In: Chisolm GD, Williams DI, eds: Scientific foundations of urology (2nd edn). London: Heinemann Med Books, 1982:308-314
10. Fischbein WN, Armbrustmacher VW, Griffin JL: Myoadenylate deaminase deficiency: A new disease of muscle. Science 200:545-548, 1978
11. Giblett ER, Amman AJ, Wara DW, Sandman K, Diamond LK: Nucleoside phosphorylase deficiency in a child with severely defective T-cell immunity and normal B-cell immunity. Lancet 1:1010-1013, 1977
12. Giblett ER, Anderson JE, Cohen F, Pollard B, Meuwissen HY: Adenosine deaminase deficiency in two patients with severely impaired cellular immunity. Lancet 2:1067-1069, 1972
13. Green HL, Wilson FA, Hefferon P, Terry AB, Moran JR, Slonim AE, Claus TH, Burr IM: ATP depletion. A possible role in hyperuricemia in glycogen storage disease type I. J Clin Invest 62:321-328, 1978
14. Henderson JF, Paterson ARP: Nucleotide metabolism, an introduction. New York: Academic Press, 1973
15. Hershfield MS, Kredich NM: A mechanism for adenosine cytotoxicity in adenosine deaminase deficiency. Clin Res 26:329A, 1978 (Abstr)
16. Hershko A, Razin A, Mager J: Regulation of the synthesis of 5-phosphoribosylpyrophosphate in intact red blood cells and in cell free preparations. Biochim Biophys Acta 184:64-76, 1969
17. Holmes EW, Wyngaarden JB: Hereditary xanthinuria. In: Stanbury JB, Wyngaarden JB, Fredrickson DS, Goldstein JL, Brown MS, eds: The metabolic basis of inherited disease (5th edn). New York: McGraw-Hill Book Co, 1983:1192-1201
18. Holmes EW, Wyngaarden JB, Kelley WN: Human glutamine phosphoribosylpyrophosphate amidotransferase: Two molecular forms interconvertible by purine ribonucleotides and phosphoribosylpyrophosphate. J Biol Chem 248:6035-6040, 1973
19. Howard WY, Kerson LA, Appel SH: Synthesis de novo of purines in slices of rat brain and liver. J. Neurochem 17:121-123, 1970
20. Howell RR: Hyperuricemia in childhood. Fed Proc 27:1078-1082, 1968
21. Kelley WN, Greene ML, Rosenbloom FM, Henderson JF, Seegmiller JE: Hypoxanthine-guanine phosphoribosyltransferase deficiency in gout: a review. Ann Intern Med 70:155-206, 1969

22. Kelley WN, Levy RI, Rosenbloom FM, Henderson JF, Seegmiller JE: Adenine phosphoribosyltransferase deficiency: a previously undescribed genetic defect in man. J Clin Invest 47:2281-2289, 1968
23. Kelley WN, Rosenbloom FM, Henderson JF, Seegmiller JE: A specific enzyme defect in gout associated with overproduction of uric acid. Proc Natl Acad Sci (USA) 57:1735-1739, 1967
24. Kelley WN, Weiner IM, eds: Uric acid. Berlin: Springer Verlag, 1978
25. Kredich N, Hershfield MS: Immunodeficiency diseases caused by adenosine deaminase deficiency and purine nucleoside phosphorylase deficiency. In: Stanbury JB, Wyngaarden JB, Fredrickson DS, Goldstein JL, Brown MS, eds: The metabolic basis of inherited disease (5th edn). New York: McGraw-Hill Book Co, 1983:1157-1183
26. Larsson A, Reichard P: Enzymatic reduction of ribonucleotides. Progr Nucl Acid Res Mol Biol 7:303-347, 1967
27. Lesch M, Nyhan WN: A familial disorder of uric acid metabolism and central nervous system function. Am J Med 36:561-570, 1964
28. Long WK: Glutathione reductase in red blood cells: variant associated with gout. Science 155:712-713, 1967
29. Lowenstein JM: Ammonia production in muscle and other tissues. The purine nucleotide cycle. Physiol Rev 52:382-414, 1972
30. Lowy BA, Williams MK, London IM: Enzymatic deficiencies of purine nucleotide synthesis in the human erythrocyte. J Biol Chem 237:1622-1625, 1962
31. Martin DW Jr: Nucleotides. In: Martin DW, Mayes PA, Rodwell VW, Granner DK, eds: Harper's review of biochemistry (20th edn). Los Altos, Calif: Lange Med Publ, 1987:348-375
32. Newby AC: Adenosine and the concept of 'retaliatory metabolites'. Trends Biochem Sci 9:42-44, 1984
33. Oliver I, Sperling O, Liberman UA, Frank M, De Vries A: Deficiency of xanthine oxidase activity in colostrum of a xanthinuric female. Biochem Med 5:279-284, 1971
34. Plageman PGW, Wohlhueter RM, Kraupp M: Adenine nucleotide metabolism and nucleoside transport in human erythrocytes under ATP depletion conditions. Biochim Biophys Acta 817:51-60, 1985
35. Roe TF, Kogut MD: The pathogenesis of hyperuricemia in glycogen storage disease type I. Pediat Res 11:664-669, 1977
36. Rosenbloom FM, Henderson JF, Caldwell IC, Kelley WN, Seegmiller JE: Biochemical bases of accelerated purine biosynthesis de novo in human fibroblasts lacking hypoxanthine-guanine phosphoribosyltransferase. J Biol Chem 243:1166-1173, 1968
37. Rosenbloom FM, Kelley WN, Miller J, Henderson JF, Seegmiller JE: Inherited disorder of purine metabolism: correlation between central nervous system dysfunction and biochemical defect. J Am Med Ass 202:175-177, 1967
38. Seegmiller JE: Diseases of purine and pyrimidine metabolism In: Bondy PK, Rosenberg LE, eds: Metabolic control and disease. Philadelphia: Saunders Co, 1980:777-932
39. Seegmiller JE, Rosenbloom FM, Kelley WN: Enzyme defect associated with a sex-linked human neurological disorder and excessive purine synthesis. Science 155:1682-1684, 1967
40. Simmonds HA, Van Acker KJ: Adenine phosphoribosyltransferase deficiency: 2,8-dihydroxyadenine lithiasis. In: Stanbury JB, Wyngaarden JB, Fredrickson DS, Goldstein JL, Brown MS, eds: The metabolic basis of inherited disease (5th edn). New York: McGraw-Hill Book Co, 1983:1144-1156
41. Spector EB, Hershfield MS, Seegmiller JE: Purine reutilization and synthesis de novo in long term human lymphocyte cell lines deficient in adenine phosphoribosyltransferase activity. Somatic Cell Genet 4:253-258, 1978
42. Sperling O: Ribose-5-phosphate and the oxidative pentose shunt in the regulation of purine synthesis de novo. In: Elliot K, Fitzsimons DW, eds: Purine and pyrimidine metabolism (Ciba Found Symp 48), Amsterdam: Elsevier, 1977:347-355
43. Sperling O: Hereditary hypouricemia. In: Scriver CR, Baudet AC, Sly WS, Valle D, eds: The metabolic basis of inherited disease (6th edn). New York: McGraw-Hill Book Co, in press
44. Sperling O: Uric acid nephrolithiasis. In: Wickham JEA, Buck AC, eds: Renal tract stone: Metabolic basis and clinical practice. London: Churchill Livingstone, in press
45. Sperling O, Boer P, Persky-Brosh S, Kanarek E, de Vries A: Altered kinetic property of erythrocyte phosphoribosylpyrophosphate synthetase in excessive purine production. Eur J Clin Biol Res 17:703-706, 1972

46. Sperling O, Brosh S, Boer P, Liberman UA, de Vries A: Urinary xanthine stones in an allopurinol treated gouty patient with partial deficiency of hypoxanthine-guanine phosphoribosyltransferase. Israel J Med Sci 14:288-292, 1978
47. Sperling O, Eilam G, Persky-Brosh S, de Vries A: Accelerated erythrocyte 5-phosphoribosyl-1-pyrophosphate synthesis. A familial abnormality associated with excessive uric acid production and gout. Biochem Med 6:310-316, 1972
48. Sperling O, Persky-Brosh S, Boer P, de Vries A: Human erythrocyte PRPP synthetase mutationally altered in regulatory properties. Biochem Med 7:389-395, 1973
49. Stryer L: Biochemistry. San Francisco: Freeman, 1981
50. Swain JL, Sabina RL, Holmes EW: Myoadenylate deaminase deficiency. In: Stanbury JB, Wyngaarden JB, Fredrickson DS, Goldstein JL, Brown MS, eds: The metabolic basis of inherited disease (5th edn). New York: McGraw-Hill Book Co, 1983:1184-1191
51. Switzer RL: Regulation and mechanism of phosphoribosylpyrophosphate synthetase. III. Kinetic studies of the reaction mechanisms. J Biol Chem 246:2447-2458, 1971
52. Taegtmeyer H: On the role of the purine nucleotide cycle in the isolated working rat heart. J Mol Cell Cardiol 17:1013-1018, 1985
53. Tully ER, Sheehan TG: Purine metabolism in rat skeletal muscle. Adv Exp Med Biol 122B:13-17, 1980
54. Van Acker KJ, Simmonds HA, Potter CF, Sahota A: Inheritance of adenine phosphoribosyltransferase deficiency. Adv Exp Med Biol 122A:349-353, 1980
55. Van den Berghe G, Bontemps F, Hers HG: Purine catabolism in rat hepatocytes. Influence of coformycin. Biochem J 188:913-920, 1980
56. Vincent MF, Van den Berghe G, Hers HG: Metabolism of hypoxanthine in isolated rat hepatocytes. Biochem J 222:145-155, 1984
57. Wasserzug O, Szeinberg A, Sperling O: Erythrocyte glutathione reductase in gout and in glucose-6-phosphate dehydrogenase deficiency. Monogr Human Genet 9:16-19, 1978
58. Wyngaarden JB, Kelley WN: Gout and hyperuricemia. New York: Grune and Stratton, 1976
59. Wyngaarden JB, Kelley WN: Gout. In: Stanbury JB, Wyngaarden JB, Fredrickson DS, Goldstein JL, Brown MS, eds: The metabolic basis of inherited disease (5th edn). New York: McGraw-Hill Book Co, 1983:1043-1114
60. Zoref E, de Vries A, Sperling O: Mutant feedback-resistant phosphoribosylpyrophosphate synthetase associated with purine overproduction and gout; phosphoribosylpyrophosphate and purine metabolism in cultured fibroblasts. J Clin Invest 56:1093-1099, 1975
61. Zoref E, de Vries A, Sperling O: Evidence for X-linkage of phosphoribosylpyrophosphate synthetase in man: Studies with cultured fibroblasts from a gouty family with mutant feedback resistance enzyme. Human Heredity 27:73-80, 1977
62. Zoref-Shani E, Kessler Icekeson G, Sperling O: Pathways of adenine nucleotide catabolism in primary rat cardiomyocytes. J Mol Cell Cardiol, in press
63. Zoref-Shani E, Shainberg A, Sperling O: Characterization of purine nucleotide metabolism in primary rat muscle cultures. Biochim Biophys Acta 716:324-330, 1982
64. Zoref-Shani E, Shainberg A, Sperling O: Pathways of adenine nucleotide catabolism in primary rat muscle cultures. Biochim Biophys Acta 926:287-295, 1987

Chapter 21

Diagnosis of Ischemic Heart Disease
with AMP-Catabolites

J.W. de Jong, Cardiochemical Laboratory, Thoraxcenter,
Erasmus University Rotterdam, The Netherlands

*Lack of oxygen induces rapid breakdown of myocardial energy-rich phosphates, such
as ATP. The purines adenosine, inosine, (hypo)xanthine and urate appear in the
myocardial effluent during ischemia or reperfusion. In this paper I summarize
experiments on the isolated rat heart, made temporarily ischemic; on the
anesthetized pig with a stenosis of a coronary artery; in patients undergoing an
atrial pacing stress test or coronary angioplasty. In the animal experiments, ischemic
myocardium releases mainly inosine; the human heart, on the other hand, produces
predominantly hypoxanthine during periods of ischemia. When an ischemic event
follows relatively quickly an earlier one, purine production during the second period
is substantially lower. Experimental animals or patients thus cannot serve as their
own controls in studies on drug efficacy under these conditions. In certain clinical
settings, the assay of hypoxanthine seems superior to that of lactate to demonstrate
myocardial lack of oxygen. I conclude that myocardial purine production is a
promising biochemical indicator for ischemic heart disease.*

Purines as an Indicator of Ischemia?

Among the coenzymes, nucleotides play an important role in, e.g., bioenergetic
processes (17). The nucleotide ATP is essential in heart muscle, because it is the
fuel for the pump. The product, ADP, will be rephosphorylated. Heart muscle needs
a continuous supply of large amounts of oxygen for this process - oxidative
phosphorylation. ATP turnover is very high; there is a delicate equilibrium between
ATP supply and demand (19). Figure 1 shows how energy-rich phosphates form
purines. These can leave the heart. Adenosine is a physiological vasodilator (ref. 3;
see Chapter 8).

The ischemic heart produces adenosine, inosine and hypoxanthine, a fact known for
decades. We showed that the ischemic pig heart releases inosine and hypoxanthine
(8). Relatively late the quantitative importance of xanthine and urate production
became clear (33,34). The enzyme xanthine oxidoreductase, of which the oxidase
form could play a role in atherogenesis (see ref. 34) and reperfusion damage (25), is
responsible for this conversion. In this chapter I will demonstrate - using examples
from animal and clinical studies - that AMP catabolites are promising markers of
myocardial ischemia.

Purine Production by Isolated, Ischemic Rat Heart

Figure 2 shows the time-course of coronary flow and purine release in the isolated,
rat heart (Langendorff preparation), with flow reduced temporarily by 75%. After 10
min of ischemia, the heart releases more than six times as much purine as during
control perfusion. This increases to 16 times early during reperfusion. Reactive
hyperemia occurs, possibly caused by adenosine release.

We concluded from balance studies that purine production is a good measure for ATP breakdown in ischemic heart (1,37). A caveat! During drastic reduction in coronary flow - complete cessation of flow is the extreme example - catabolite release during ischemia goes down (Fig. 3). If we reduce the cardiac output of the isolated, working rat heart by 94%, purine release does not increase, but decreases. Only during reperfusion release of adenosine plus catabolites increases in that case.

Inosine Production during Repeated Coronary Occlusion in Pig Heart

Figure 4 shows an experiment in which we reduced the left anterior descending coronary artery flow of the pig repeatedly by 60%. Arterial inosine did not show significant changes. However, during the first stenosis the heart released substantial amounts of inosine (39). During the second flow reduction, inosine release was much

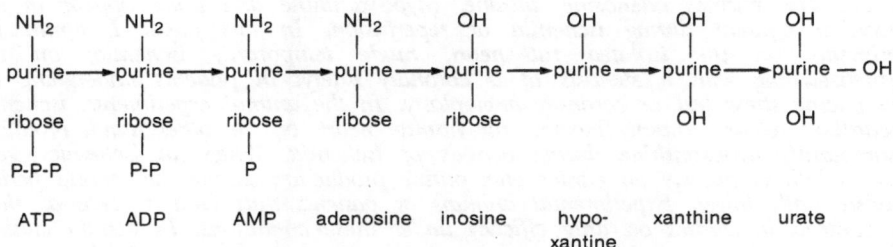

Fig. 1. Myocardial ATP breakdown. Because of their polarity, the adenine nucleotides ATP, ADP and AMP, do not pass the cell membrane. The heart releases the purine nucleosides adenosine and inosine, and the oxypurines (hypo)xanthine and urate into the blood stream. P = phosphate.

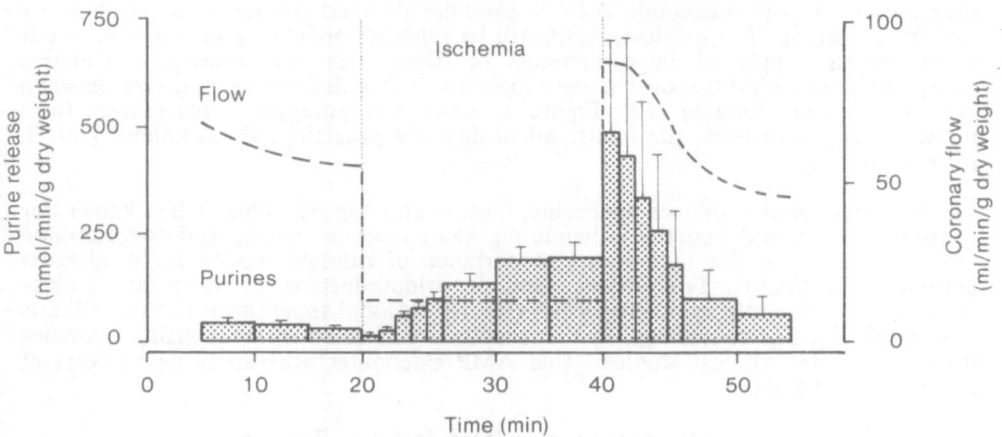

Fig. 2. Purine release from isolated non-working rat heart, paced at 300 beats/min. A roller pump restricted flow. We measured AMP-catabolites in the coronary effluent by high-pressure liquid chromatography (7,16). The release of purines (adenosine + inosine + hypoxanthine + xanthine + urate) increased during ischemia and reperfusion. Means ± SEM, n = 10 (control), n = 6 (ischemia), n = 3 (reperfusion).

less, suggesting a depleted myocardial adenine nucleotide pool. (We did a similar observation on myocardial lactate release.) We concluded that this model is not useful for interventions in which the animal serves as its own control.

Hypoxanthine Production by Human Heart during Atrial Pacing

The assay of purines in human blood is difficult, because the concentrations are so low. From Fig. 5 it is clear that the inosine concentration in human blood is lower

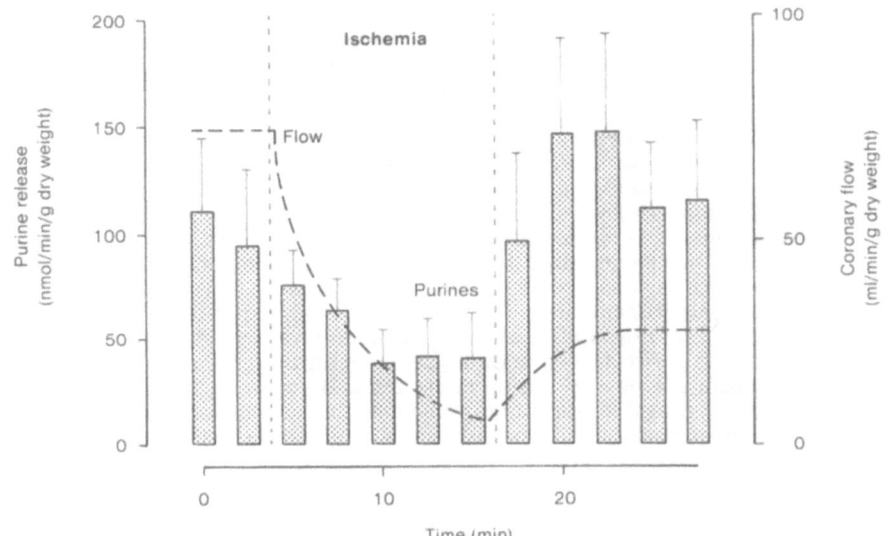

Fig. 3. Purine release from isolated working rat heart, paced at 300 beats/min. Perfusion took place with a preload of 10 cmH$_2$O and an afterload of 75 cmH$_2$O (37°C). Ischemia was induced with a ball valve in the aortic outflow tract (29). - - - Coronary flow. We measured AMP-catabolites in the coronary effluent by high-pressure liquid chromatography (7). Means ± SEM (n = 7). Note that reduction in flow in this preparation leads to increased purine efflux during reperfusion.

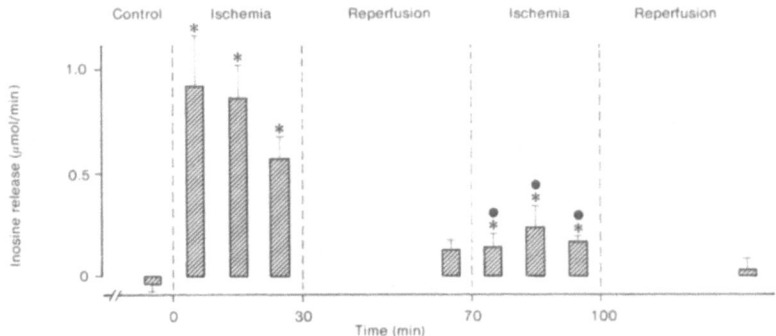

Fig. 4. Arterial concentration and myocardial balance of inosine in the anesthetized open-chested pig (39). We induced myocardial ischemia repeatedly by reduction of the flow in the left anterior descending coronary artery to 40% of control. Arterial inosine (measured enzymatically in deproteinized blood) did not change. Myocardial ATP breakdown, reflected by inosine release, during the second stenosis was much less than during the first one. Therefore, this model cannot be used for the study of interventions during myocardial ischemia in which the animal serves as its own control. Means ± SEM (n = 12). * P<0.05 vs. control. °P<0.05 vs. first flow reduction.

than 1 uM. It would be nice to prepare plasma or serum before deproteinization, because it circumvents the disturbance during chromatography by the relative high amounts of erythrocyte nucleotides (ATP etc.). However, Fig. 5 shows that during the preparation of plasma or serum the purine concentration measured increases. Therefore we deproteinize the blood immediately. We remove the disturbing nucleotides with an Al_2O_3-column, before performing high-pressure liquid chromatography. (Recently, we mixed some inhibitors of adenosine metabolism [dipyridamole and erythro-9(2-hydroxy-3-nonyl)adenine] immediately with the blood samples. This prevents the formation of AMP-catabolites during plasma preparation, and makes the tedious Al_2O_3-step unnecessary.)

During an atrial pacing stress test of patients undergoing a heart catheterization for diagnostic purposes, the arterial hypoxanthine concentration remains constant. In patients with significant narrowing of one or more coronary arteries, the coronary sinus hypoxanthine blood concentration exceeds the arterial one by a factor of two during pacing (Fig. 6). Patients with little or no angioscopically demonstrable atherosclerosis do not show such purine production. In earlier work (31) we described that patients developing angina pectoris during such a test, also show hypoxanthine production. Table 1 summarizes the effect of a repeated atrial pacing stress-test on hemodynamic and biochemical variables. Twenty-five minutes after the first test we repeated the pacing until heart frequency was maximal. The release of lactate and hypoxanthine was lower during the second pacing procedure. We concluded from that experiment, like we did from the pig experiment described in Fig. 4, that patients are not useful as their own control in this type of set-up. Kugler studied both the effect of pindolol (22) and nitroglycerine (21) on lactate and purine metabolism during two pacing stress tests. Because he failed to submit a control group (without drugs) to the same procedure, his conclusions from the lower production of lactate, inosine and hypoxanthine are dubious. It is noteworthy that Ihlen et al. (18) found reproducible lactate production, if they repeated the pacing procedure after a recovery period of 20 to 40 min. Whether a different patient population can explain the discrepancy, remains to be answered.

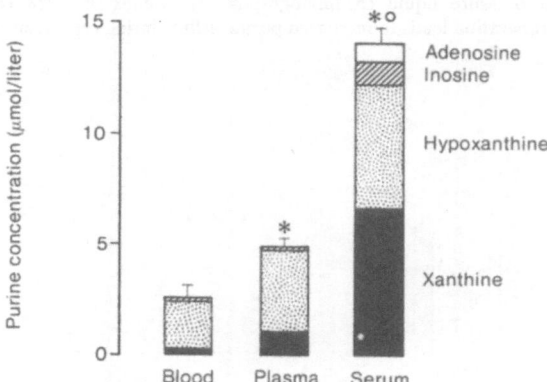

Fig. 5. Concentrations of AMP-catabolites in human blood, plasma and serum. We measured the purines by high-pressure liquid chromatography (16). In plasma and serum the total amount of these compounds was considerably higher than in blood, presumably due to nucleotide breakdown during blood processing. We omitted the urate concentration, because in human blood, plasma or serum it is much higher (200-300 uM) than that of the other purines. Data represent values of 5-6 non-resting volunteers (means ± SEM). *P<0.01 vs. blood; °P<0.05 vs. plasma.

Hypoxanthine Production during Coronary Angioplasty

Also during percutaneous transluminal coronary angioplasty marked hypoxanthine (and lactate) production is demonstrable. Immediately after deflation of the occluding balloon, the levels of these compounds increase in the coronary sinus or the great cardiac vein (Fig. 7). The amount released after subsequent balloon deflations is still large. The metabolic derangements, however, normalize rather quickly (36).

Purine Production, a Biochemical Marker for Ischemia

Table 2 summarizes studies in with AMP-catabolites used to diagnose ischemic heart disease. Various investigators did their determinations in plasma or serum, without adequate precautions against catabolism (cf. Fig. 5). Buhl et al. (6) demonstrated increased excretion of hypoxanthine and xanthine in urine of cardiac patients. However, they were unable to detect raised oxypurine levels in peripheral venous blood, as described in Sovjet studies (12,28).

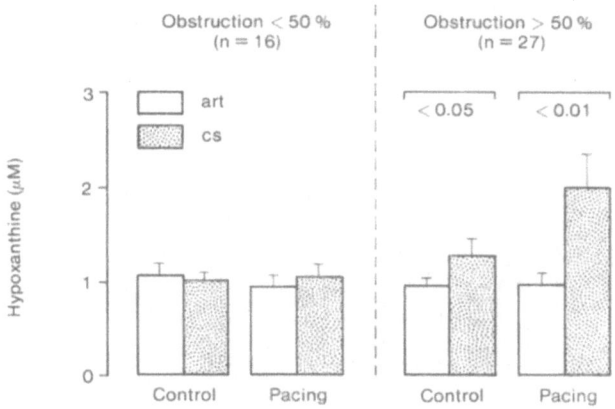

Fig. 6. Hypoxanthine production by the human heart during ischemia. An arterial (art) and coronary sinus (cs) blood sample was taken from patients undergoing heart catheterization, both at rest and during a rapid atrial pacing stress test. Pacing was stopped when the patient developed progressive chest pain or an A-V block. A stenosis of at least one coronary artery, exceeding 50%, was considered a significant obstruction. This study shows that pacing-induced myocardial hypoxanthine release is a marker for ischemic heart disease (ref. 16, see also ref. 31). Mean values ± SEM and statistically significant differences ($P < 0.05$) are presented.

Table 1. Effect of a repeated atrial pacing stress test on hemodynamic and biochemical variables

Variable	P_1	P_2
Heart frequency (beats/min)	144 ± 5	144 ± 5
Left ventricular pressure (mmHg)	137 ± 4	137 ± 4
Peak +dp/dt (mmHg/s^{-1})	2470 ± 160	2440 ± 140
Vmax (s^{-1})	70 ± 4	70 ± 4
Coronary blood flow (ml/min)	189 ± 15	186 ± 17
Lactate efflux (umol/min)	23 ± 3	14 ± 3*
Hypoxanthine efflux (nmol/min)	74 ± 10	26 ± 5

P_1 = first pacing test; P_2 = second pacing test. Means ± SEM (n=10). *$P < 0.05$ vs. P_1 (see ref. 35).

Hopefully it will be clear from this review that an increasing number of cardiological clinics show interest for these biochemical indicators. The same holds for disciplines such as obstetrics (2,4,14,15,24,27,30,38) and surgery (refs. 9,11,23,28; Chapter 22). High-pressure liquid chromatography is probably the method of choice to measure purine concentrations. Sample preparation needs improvement to make the analysis routinely applicable.

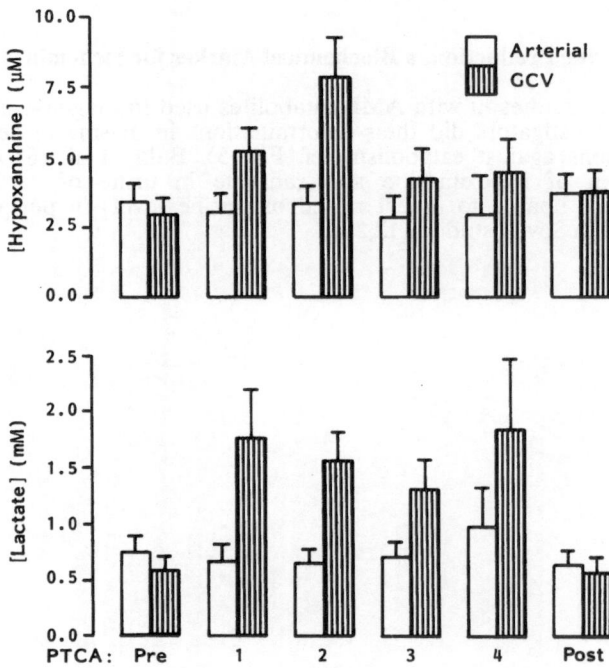

Fig. 7. Hypoxanthine and lactate release from human heart after percutaneous transluminal coronary angioplasty (36). Means ± SEM (n = 15). GCV = great cardiac vein.

Table 2. AMP-catabolites in blood and urine as marker for myocardial ischemia in patients

Purine	Body fluid	Clinical situation	Ref.
Ado	A-CS plasma	APST	10
UA	Serum	AMI, IHD	13,26
Hx	A-CS blood	APST	16,31,32
Hx	Peripheral venous plasma	AMI, ischemia, AP	12,28
Ado, Ino, Hx	A-CS plasma	ACBG	11
Ino, Hx	A-CS plasma	APST	20-22
Hx, Xan (UA)	Urine	Cardiac arrest, AP, AMI	6
Ado, Ino, Hx, Xan	A blood	ACBG	5
Hx	A-CS blood	2 x APST	35

A = arterial; ACBG = aorta-coronary bypass grafting; Ado = adenosine; AMI = acute myocardial infarction; AP = angina pectoris; APST = atrial pacing stress test; CS = coronary sinus; Hx = hypoxanthine; IHD = ischemic heart disease; Ino = inosine; UA = urate; Xan = xanthine.

In summary, I conclude that myocardial purine production is a promising biochemical indicator for ischemic conditions, encountered frequently in patients with heart disease.

Acknowledgements

I am most greatful to Ms. M.J. Kanters-Stam and Ms. M.H.A. Zweserijn for their secretarial effort.

References

1. Achterberg PW, Harmsen E, De Tombe PP, De Jong JW: Balance of purine nucleotides and catabolites in the isolated rat heart. Adv Exp Biol Med 165B:483-486, 1984
2. Bennett MJ, Carpenter KH: Experience with a simple high-performance liquid chromatography method for the analysis of purine and pyrimidine nucleosides and bases in biological fluids. Ann Clin Biochem 21:131-136, 1984
3. Berne RM: The role of adenosine in the regulation of coronary blood flow. Circ Res 47:807-813, 1980
4. Bratteby L-E, Swanstrom S: Hypoxanthine concentration in plasma during the first two hours after birth in normal and asphyxiated infants. Pediat Res 16:152-155, 1982
5. Brower RW, De Jong JW, Haalebos M, Simoons ML, Vd Bos A, De Jong DS, Bos E, Hugenholtz PG: Evaluation of cardioplegia in coronary artery bypass graft surgery. In: Just H, Tschirkov A, Schlosser V, eds: Kalziumantagonisten zur Kardioplegie und Myokardprotektion in der offenen Herzchirurgie. Stuttgart: Thieme, 1982:69-80
6. Buhl L, Vilhelmsen KN, Rokkedal Nielsen J: Oxypurine release in cardiac disease. Acta Med Scand 209:83-86, 1981
7. De Jong JW, Harmsen E, De Tombe PP, Keijzer E: Nifedipine reduces adenine nucleotide breakdown in ischemic rat heart. Eur J Pharmacol 81:89-96, 1982
8. De Jong JW, Goldstein S: Changes in coronary venous inosine concentration and myocardial wall thickening during regional ischemia in the pig. Circ Res 35:111-116, 1974
9. Flameng W, VanHaecke J, Van Belle H, Borgers M, De Beer L, Minten J: Relation between coronary artery stenosis and myocardial purine metabolism, histology and regional function in humans. J Am Coll Cardiol 9:1235-1242, 1987
10. Fox AC, Reed GE, Glassman E, Kaltman AJ, Silk BB: Release of adenosine from human hearts during angina induced by rapid atrial pacing. J Clin Invest 53:1447-1457, 1974
11. Fox AC, Reed GE, Meilman H, Silk BB: Release of nucleosides from canine and human hearts as an index of prior ischemia. Am J Cardiol 43:52-58, 1979
12. Gneushev ET, Naumova VV, Bogoslovskij VA: Content of hypoxanthin in the peripheral venous blood in infarction and ischemia of the myocardium. Terapevt Arkh 50:20-24, 1978
13. Halpern MJ, Pereira Miguel MS: Uric acid and coronary heart disease. J Am Geriat Soc 22:86-87, 1974
14. Harkness RA, Geirsson RT, McFadyen IR: Concentrations of hypoxanthine, xanthine, uridine and urate in amniotic fluid at caesarean section and the association of raised levels with prenatal risk factors and fetal distress. Br J Obstet Gynaecol 90:815-820, 1983
15. Harkness RA, Simmonds RJ, Coade SB, Lawrence CR: Ratio of the concentration of hypoxanthine to creatinine in urine form newborn infants: A possible indicator for the metabolic damage due to hypoxia. Br J Obstet Gynaecol 90:447-452, 1983
16. Harmsen E, De Jong JW, Serruys PW: Hypoxanthine production by ischemic heart demonstrated by high pressure liquid chromatography of blood purine nucleosides and oxypurines. Clin Chim Acta 115:73-84, 1981
17. Henderson JF, Paterson ARP: Nucleotide metabolism - an introduction. New York: Acad Press, 1973
18. Ihlen H, Simonsen S, Thaulow E: Myocardial lactate metabolism during pacing induced angina pectoris. Scand J Clin Invest 43:1-7, 1983
19. Kohn MC, Achs MJ, Garfinkel D: Distribution of adenine nucleotides in the perfused rat heart. Am J Physiol 232:R158-R163, 1977

20. Kugler G: Myocardial release of lactate, inosine and hypoxanthine during atrial pacing and exercise-induced angina. Circulation 59:43-49, 1079
21. Kugler G: The effect of nitroglycerin on myocardial release of inosine, hypoxanthine and lactate during pacing-induced angina. Basic Res Cardiol 73:523-533, 1978
22. Kugler G: The effect of pindolol on myocardial release of inosine, hypoxanthine and lactate during pacing-induced angina. J Pharmacol Exp Ther 209:185-189, 1979
23. Larsson J, Gidlöf A, Lewis DH, Liljedahl SO, Saugstad OD: Effect of induced ischemia on plasma hypoxanthine levels in man. In: Lewis DH, ed: Induced skeletal muscle ischemia in man. Basel: Karger, 1982:49-54
24. Manzke H, Dörner K, Grünitz J: Urinary hypoxanthine, xanthine and uric acid excretion in newborn infants with perinatal complications. Acta Pediat Scand 66:713-717, 1977
25. McCord JM: Oxygen-derived free radicals in postischemic tissue injury. N Engl J Med 312:159-163, 1985
26. McEwin R, McEwin K, Loudon B: Raised serum uric acid levels with myocardial infarction. Med J Aust 1:530-532, 1974
27. Meberg A, Saugstad OD: Hypoxanthine in cerebrospinal fluid in children. Scand J Clin Lab Invest 38:437-440, 1978
28. Naumova VV, Gneushev ET, Bogoslovskij VA: Hypoxanthine content in the peripheral venous blood of patients with ischemic heart disease with attacks of stenocardia and in patients with osteochondrosis of the cervicothoracic region of the vertebral column. Terapevt Arkh 51:22-27, 1979
29. Neely JR, Rovetto MJ, Whitmer JT, Morgan HE: Effects of ischemia on function and metabolism of the isolated working rat heart. Am J Physiol 225:651-658, 1973
30. O'Connor MC, Harkness RA, Simmonds RJ, Hytten FE: The measurement of hypoxanthine, xanthine, inosine, and uridine in umbilical cord blood and fetal scalp blood samples as a measure of fetal hypoxia. Br J Obstet Gynaecol 88:381-390, 1981
31. Remme WJ, De Jong JW, Verdouw PD: Effects of pacing-induced myocardial ischemia on hypoxanthine efflux from the human heart. Am J Cardiol 40:55-62, 1977
32. Remme WJ, Van den Berg R, Mantel M, Cox PH, Van Hoogenhuyze DCA, Krauss XH, Storm CJ, Kruyssen DACM: Temporal relation of changes in regional coronary flow and myocardial lactate and nucleoside metabolism during pacing-induced ischemia. Am J Cardiol 58:1188-1194, 1986
33. Ronca-Testoni S, Borghini F: Degradation of perfused adenine compounds up to uric acid in isolated rat heart. J Mol Cell Cardiol 14:177-180, 1982
34. Schoutsen B, De Jong JW, Harmsen E, De Tombe PP, Achterberg PW: Myocardial xanthine oxidase/dehydrogenase. Biochim Biophys Acta 762:519-524, 1983
35. Serruys PW, De Jong JW, Harmsen E, Verdouw PD, Hugenholtz PG: Effect of intracoronary nifedipine on high-energy phosphate metabolism during repeated pacing-induced angina and during experimental ischemia. In: Kaltenbach M, Neufeld HN, eds: Proc 5th Int Adalat Symp (New therapy of ischaemic heart disease and hypertension). Amsterdam: Excerpta Medica, 1983:340-353
36. Serruys PW, Piscione F, Hegge JAJ, Harmsen E, Van den Brand M, De Feyter P, Hugenholtz PG, De Jong JW: Myocardial release of hypoxanthine and lactate during percutaneous transluminal coronary angioplasty: a quickly reversible phenomenon. J Am Coll Cardiol, in press
37. Stam H, De Jong JW: Sephadex-induced reduction of coronary flow in the isolated rat heart: A model for ischemic heart disease. J Mol Cell Cardiol 9:633-650, 1977
38. Swanström S, Bratteby L-E: Hypoxanthine as a test of perinatal hypoxia as compared to lactate, base deficit, and pH. Pediat Res 16:156-160, 1982
39. Verdouw PD, Remme WJ, De Jong JW, Breeman WAP: Myocardial substrate utilization and hemodynamics following repeated coronary flow reduction in pigs. Basic Res Cardiol 74:477-493, 1979

Chapter 22

Myocardial High-Energy Phosphates during Open-Heart Surgery

G.J. van der Vusse and F.H. van der Veen,
Department of Physiology, University of Limburg,
Maastricht, The Netherlands

To obtain a quiet and bloodless operation field during open-heart surgery, periods of myocardial ischemia are deliberately introduced by aortic cross-clamping. As a consequence, an unbalance between consumption and production of cellular high-energy phosphates is created. This results in a rapid fall of creatine phosphate levels, followed by a decrease in the tissue content of ATP. Low levels of ATP are related to the loss of cellular function and the onset of cell death. The precise mechanism of action is, however, still unclear. Although tissue ATP content as a prognostic index of ischemia-induced damage recently has been challenged, most protective measures taken during open-heart surgery aim at the conservation of cardiac high-energy phosphate pools in the ischemic tissue. To this end, commonly electro-mechanical activity is abolished rapidly by intracoronary infusion of an ice-cold crystalloid or sanguineous cardioplegic solution immediately after aortic cross-clamping. All kinds of variations with respect to the application of cardioplegia such as the composition of the cardioplegic solution, addition of specific pharmaca and mode of administration, have been reported. Assessment of the tissue content of ATP and creatine phosphate in left ventricular biopsies, obtained during cardiac operations, provides a useful parameter to estimate the adequacy of the protective measures applied.

Introduction

This chapter will review selected aspects of high-energy phosphate homeostasis in the human heart with particular emphasis on the situation during open-heart surgery. To date many cardiac defects are repaired operatively. During most of the surgical procedures global cardiac ischemia is deliberately imposed to provide the surgeon a quiet and bloodless operation field. The consequent lack of oxygen will impair adequate replenishment of cardiac high-energy phosphate stores. It has been generally accepted that failure to keep intracellular ATP within safe limits will result in loss of cellular function and ultimately in cell death. Therefore, measures have to be taken to prevent cardiac high-energy phosphate depletion during the period of global ischemia during open-heart surgery. In the present overview attention will be paid to:
- production and consumption of high-energy phosphates in the normoxic heart;
- disturbed high-energy phosphate homeostasis during ischemia;
- high-energy phosphates as index for myocardial damage during open-heart surgery;
- measures to maintain cardiac high-energy phosphates at adequate intracellular levels.

J.W. de Jong (Ed.), Myocardial Energy Metabolism, Martinus Nijhoff Publishers, Dordrecht/Boston/Lancaster, 1988

Cardiac High-Energy Phosphate Homeostasis during Normoxia and Ischemia

Energy production and energy consumption are tightly coupled in the working myocardial cell. The need for ATP dictates the rate of conversion of ADP and inorganic phosphate into ATP, while substrates and molecular oxygen are utilized in amounts exactly matching that required to produce ATP. At normal workload the heart consumes approximately 5 umol O_2/min per g wet wt. If 6 mol ATP are produced per mol O_2, than about 30 umol ATP are aerobically produced per min per gram. The intracellular content of ATP is in the order of 5 umol/g wet wt., implicating that each 10 s the equivalent of the total intracellular ATP pool will turn over. This figure stresses the significance of an unimpeded supply of molecular oxygen to the myocardial cells.

As indicated in Fig. 1, exogenous carbohydrates, such as glucose and lactate, and lipids like fatty acids, serve the myocardium as readily oxidizable substrates. Under normal physiological circumstances fatty acid oxidation provides more than 50% of the energy required for cardiac function (43). Lactate appears to be the preferential substrate when the extracellular concentration of this substrate exceeds 5 mM, a situation easily encountered during heavy physical exercise. The contribution of glucose to cardiac energy production depends on both the circulating level of insulin and the workload of the heart. The supply of exogenous substrates to the heart is not rate limiting under normal circumstances. Endogenous substrates are present in the form of glycogen, highly-branched glucose polymers, and triacylglycerols, molecules in which fatty acids are complexed to glycerol. In the absence of extracellular substrates, the amount of glycogen and triacylglycerol will be sufficiently high to meet the myocardial energy demands for about 60 and 40 min, respectively (42).

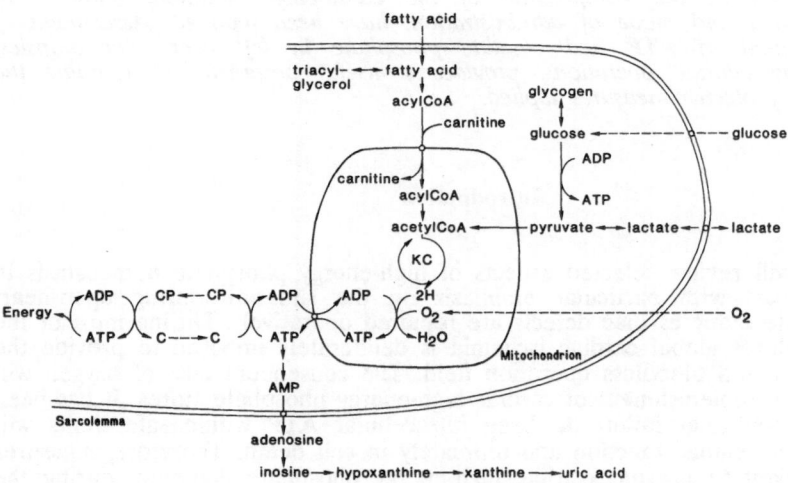

Fig. 1. Schematic representation of myocardial energy homeostasis. KC refers to Krebs-cycle activity, CP to creatine phosphate, and C to creatine.

Energy-consuming reactions take place in localized areas of the myocyte. Over 60% of the amount of ATP produced is utilized for myofibrillar contraction and relaxation; the remainder of ATP is consumed by ion pumps at the sarcolemma and sarcoplasmic reticulum and a variety of anabolic and repair processes. Under aerobic conditions approximately 98% of the cellular production of ATP takes place in the mitochondrial matrix. Quantitatively, glycolytic ATP production in the cytoplasmic compartment is of minor importance if oxygen supply to the heart is sufficient.

Traditionally the role of creatine phosphate, the intracellular concentration of which is 2-3 times that of ATP, has been considered to serve as buffer for cellular ATP. More recent findings suggest that creatine phosphate is involved in transporting energy from the site of production to the myofibrillar utilization reaction (Fig. 1; refs. 3,26).

A completely different picture emerges when supply of molecular oxygen is insufficient to maintain the production of ATP. Mechanical activity ceases rapidly. Since repetitive contraction and relaxation of the myofibrils are quantitatively the most important energy-consuming processes in the myocyte, cessation of mechanical activity can be considered as an appreciable intrinsic energy-sparing measure. The tissue content of creatine phosphate falls within one minute to very low levels. The decrease in the cardiac content of ATP is relatively slow; e.g., the overall content is halved after 20 to 30 min following the onset of ischemia (14). Mitochondrial ATP production is very low or negligible, depending on the residual supply, if any, of oxygen to the cell. Initially, the degradation of glycogen is accelerated and glycolytic activity is increased, while the majority of ATP is produced by converting glucose into lactate. Due to subsequent depletion of the endogenous glycogen pool and inhibition of the glycolytic flux by acidification of the intracellular compartment, the significance of the glycolytic production of ATP will become less during extended ischemia. In addition to reduced production of ATP in the oxygen-deprived cells, the depletion of the intracellular content of ATP is most likely further accelerated by induction of futile energy-consuming processes; e.g., accelerated turnover of the triacylglycerol pool (37) and stimulation of Ca^{2+}-dependent ATP-ases due to intracellular Ca^{2+}-overload (27). Reduction of the intracellular ATP pool in ischemic tissue is accompanied by increased levels of ADP and AMP. Concomitantly, the content of purine bases and oxypurines such as adenosine, inosine, hypoxanthine and xanthine increases as a consequence of further degradation of AMP (Fig. 1). Finally, ATP levels in the oxygen-deprived myocyte will fall beyond a critical limit. With ATP levels below this point, functional recovery of the heart is not possible despite readmission of oxygen by reperfusion. It has been suggested that the critical limit of ATP is in the order of 10 umol/g dry wt. or about 1-2 umol/g wet wt. of tissue (5,20).

It is noteworthy that the degradation products of the adenine nucleotides are released from the heart upon reperfusion. This loss hampers rapid resynthesis of the (partially) depleted adenine nucleotide pool. Commonly, creatine phosphate levels return to or even exceed pre-ischemic values upon readmission of flow. It should be kept in mind that low post-ischemic ATP levels do not necessarily imply that the turnover of ATP is impaired. Stable post-ischemic ATP values most likely reflect adequate mitochondrial activity to keep pace with the rate of ATP consumption.

Although considerable evidence has been provided that ATP is a major factor in, and possibly a determinant of, tissue injury, the precise relationship between depletion of ATP and loss of cellular function has not been elucidated. Theoretically, a variety of causal relationships can be proposed. First, ATP is required for contraction and relaxation of myofilaments. Therefore, low intracellular levels of ATP will in a direct manner hamper adequate mechanical activity. Findings of Ellis et al. (12), who observed a close relation between post-ischemic recovery of contraction and repletion of tissue ATP stores in the dog heart are, in favor of this notion. The time required for complete recovery of both indices was found to be at least 2 weeks after the start of reperfusion following 2 h of ischemia. Second, Jennings et al. (24) have emphasized the important role of ATP in maintaining fluid and ion homeostasis in the heart. Their findings suggest that the cell is incapable to control properly the cellular water and ion balance when ATP levels dropped below a certain critical level. Swelling of cells will occur, especially after reinstallation of blood flow to the previously ischemic tissue, resulting in irreversible damage of cellular structures and impediment of post-ischemic blood flow. In particular, calcium ions are thought to accumulate in ATP-depleted cells most likely resulting in hypercontraction of the myofilaments, calcium loading of mitochondria and activation of Ca^{2+}-dependent proteases and phospholipases (27). Third, ATP is required for the turnover and/or resynthesis of phospholipids, being important constituents of myocardial cellular membranes. Impaired phospholipid homeostasis in the cardiac cells might be closely related to the occurrence of irreversible damage (41). Further weight is given to the role of cellular ATP in preserving cardiac phospholipid homeostasis by the finding of Chien et al. (7) that depletion of ATP in isolated neonatal cardiac cells by metabolic blockers accelerates degradation of phospholipids.

Although the notion that the tissue content of ATP can be used as a prognostic index to estimate the extent of ischemia-induced damage and the adequacy of myocardial protection has been generally accepted, this concept was recently challenged by Buckberg and associates (6,32). They reported that the ability of cardiac tissue subjected to extended ischemia to recover contractile function, was not dependent on tissue steady-state ATP levels. In addition, they claimed that the extent of functional recovery was mainly determined by the conditions of reperfusion and the composition of the reperfusate. The duration of the preceding ischemic episode (32) and measures taken during ischemia (45) seem to be less decisive. Their experimental results, which suggest that the post-ischemic capacity to regenerate cellular ATP pools rather than the absolute level of ATP is the main determinant of cell viability, underline the complexity of high-energy phosphate homeostasis under pathophysiological conditions and the complex relationship between ATP and loss of cellular function.

The State of the Heart during Open-Heart Surgery

Routinely, the heart of the patient subdued to open-heart surgery is in an empty-beating state during the initial period of cardiopulmonary bypass. The energy demand is low. To obtain a quiet and bloodless operation field the aorta is cross-clamped just distal to the coronary ostia resulting in global cardiac ischemia. Related to the type of operation, various surgical procedures can be followed. In case of correction of congenital anomalies and valve operations, the heart is subjected to one or more prolonged periods of global ischemia. Basically two different techniques are at hand for aorto-coronary bypass surgery, i.e., intermittent aortic cross-clamping or continuous aortic cross-clamping. The former technique is characterized by repetitive short periods of ischemia with intervals for reperfusion. When the latter technique is applied, all distal anastomoses are performed while the heart is in an extended global ischemic state.

Removal of the aortic clamp allows blood, molecular oxygen and substrates to return to the previously ischemic heart. The heart fibrillates spontaneously or is empty-beating. When body temperature has returned to the physiological range and the heart has resumed normal electro-mechanical activity, the patient is weaned from cardiopulmonary bypass. To prevent loss of cardiac function and irreversible cell damage due to the preceding ischemic period a variety of protective measures are commonly taken. In general, these measures are aiming at the conservation of cardiac high-energy phosphate pools in the ischemic tissue.

Assessment of High-Energy Phosphates in Human Heart

To assess the content of high-energy phosphates in cardiac tissue of patients subjected to open-heart surgery, full thickness biopsies of the free wall of the left ventricle can routinely be obtained with the use of an air-driven drill (20) or the Tru-Cut[R] needle (Travenol Laboratoria; ref. 35). In biopsies, weighing in the range of 5-10 mg wet wt., sufficient tissue is present to assess the high-energy phosphate content. A variety of very sensitive chemical assays are available, e.g., luciferase technique, fluorometry and high-performance liquid chromatography. The advantage of the latter technique is that the whole spectrum of high-energy phosphates and related substances can be analyzed in one single run.

To avoid artificial changes in the tissue content of ATP and creatine phosphate the tissue specimen has to be immediately frozen in liquid nitrogen or comparable coolant, and stored at a temperature below -80°C until biochemical analysis.

From the two biopsies, one obtained prior the aortic cross-clamping and the second after restoration of blood flow, the effect of the intermediate period of ischemia and the efficacy of the protective measures applied can be estimated. It should be kept in mind that parts of the hearts under investigation are in a diseased state prior to the induction of global ischemia. Since the amount of tissue taken from the heart represents approximately 0.05% of total cardiac mass, there is a reasonable chance that the value measured in the biopsy is not representative for the whole left ventricle. The authors have explored this possible drawback of the biopsy technique in more detail (39,40). Figure 2 shows histograms of ATP and creatine phosphate values measured in pre-ischemic cardiac biopsies taken during cardiopulmonary bypass in 70 patients. The interindividual variation was found to be considerable, i.e., 35% for ATP and 37% for creatine phosphate. Figure 3, showing ATP and creatine phosphate levels in pairs of pre-ischemic biopsies taken from adjacent areas within 2 min prior to aortic cross-clamping in 14 patients, indicates that in some patients marked spatial differences were encountered. The intra-individual coefficient of variation for both ATP and creatine phosphate was found to

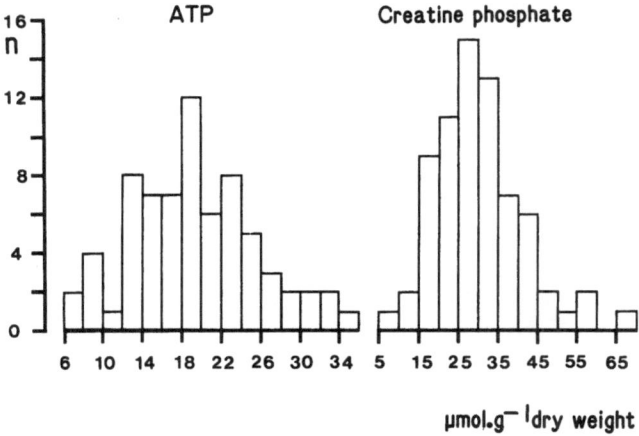

Fig. 2. Histograms representing the individual values of ATP and creatine phosphate in left ventricular biopsies obtained during the pre-aortic cross-clamp period in patients on cardiopulmonary bypass (n = 70). Tissue values are grouped in cohorts of 2 and 5 umol g dry wt. of tissue for ATP and creatine phosphate, respectively. n refers to the number of values belonging to one specific cohort (after ref. 40, with permission).

be 11%. Since the median values were not significantly different, ATP and creatine phosphate can be used as index to assess the effect of ischemia in groups of patients. Conclusions for individual casus are obviously marred by the appreciable intra-individual variation.

A disadvantage of the biopsy technique is the invasive character of this procedure. An alternative non-invasive technique to study high-energy phosphate homeostasis in the heart is the nuclear magnetic resonance (NMR) technique (23,46). Despite promising results obtained in animals (28), to our knowledge this technique has not been used to monitor cardiac high-energy phosphates during open-heart surgery.

Preservation of High-Energy Phosphates during Open-Heart Surgery

In general, reduction of the aortic cross-clamp time and, hence, duration of ischemia, and lowering the temperature of the heart, which will result in slowing the rate of energy-consuming metabolic processes, can be considered as an effective measure to preserve myocardial high-energy phosphate levels. Improved surgical skills have reduced considerably the time duration needed to perform the repair procedure. Application of cardioplegic solutions have shown to delay appreciably the intracellular depletion of high-energy phosphates (20), due to the rapid cessation of electro-mechanical activity and the low temperature of the heart.

As indicated earlier, aorto-coronary bypass surgery has been performed with the use of two different techniques, i.e., intermittent aortic cross-clamping or continuous aortic cross-clamping in combination with administration of cardioplegic solutions. An extensive study has been performed by the authors and associates (18,44) in patients operated upon with intermittent aortic cross-clamping either at 34 or 25°C body temperature, and with continuous aortic cross-clamping in combination with repetitive administration of ice-cold St. Thomas' Hospital cardioplegic solution. No significant differences could be observed regarding clinical outcome, post-operative

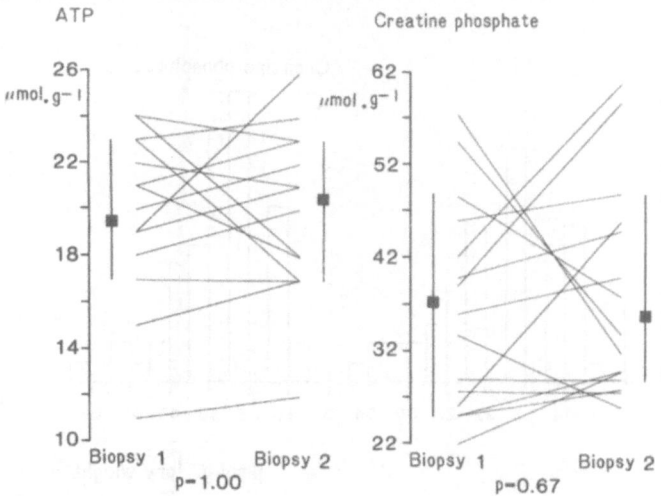

Fig. 3. The intra-individual variation of the content of ATP and creatine phosphate in the human left ventricle. Biopsy 1 and 2 were taken from adjacent areas within 2 min during pre-aortic cross-clamping in patients on cardiopulmonary bypass (n = 14). ▬ refers to median values and 95% confidence limits (after ref. 40, with permission).

hemodynamic recovery and release of cardiac enzymes between the various techniques employed. However, analysis of cardiac tissue specimen obtained during the operation revealed that the cardioplegia procedure prevented completely the reduction of ATP during the ischemic period. In contrast, the level of ATP decreased to 72 and 71% of control in the intermittent cross-clamping groups with a body temperature of 34 and 25°C during bypass, respectively. The content of creatine phosphate returned to the pre-ischemic level after a reperfusion period of about 20 min following the release of the last aortic cross-clamp. In an earlier study it was shown that the administration of the St. Thomas' Hospital cardioplegic solution during continuous aortic cross-clamping in patients who underwent isolated aortic valve replacement (15) prevented completely the reduction of creatine phosphate in the ischemic heart. In this study the second biopsy was taken within 3 min after release of the aortic clamp. A drop of 50% in cardiac creatine phosphate content occurred, when the heart was only topically cooled. The tissue level of ATP remained stable irrespective of the protective measure applied. These observations are in line with the findings of Balderman et al. (2). They observed a rapid restoration of creatine phosphate immediately upon release of the aortic clamp in previously global ischemic dog heart, when treated with a cold cardioplegic solution during the period of oxygen-deprivation. In contrast, post-ischemic creatine phosphate restoration was delayed when the temperature of the heart was not kept within safety limits during the preceding ischemic period.

Other studies have also indicated that measurements of cardiac contents of high-energy phosphates provided a sensitive metabolic indicator of ischemic damage of the human myocardium. Findings of Balderman et al. (1) have shown that ATP levels are excellently preserved in human hearts treated with a cold, crystalloid cardioplegic solution when the total ischemia did not exceed 90 min. Creatine phosphate values significantly fell by about 60%, but returned to normal levels 30 min after reperfusion. However, preservation of high-energy phosphates was incomplete when the aortic cross-clamp time exceeded 90 min. Cunningham et al. (8) explored the efficacy of a blood cardioplegic solution to preserve human cardiac ATP levels during aortic cross-clamping lasting up to 120 min. In operations without technical failures, the ATP content was found to increase by about 50% during the ischemic period and returned to pre-ischemic values after 30 min of reperfusion. Although no satisfactory explanation could be offered for the unexpected rise of ATP during the period of aortic cross-clamping, their results clearly indicate the usefulness of sanguineous cardioplegia. Kaijser et al. (25) have reported that the temperature of the heart should not decrease beyond 15°C when blood-cardioplegia is applied. Silverman et al. (36) have shown that the efficacy of crystalloid cardioplegic solutions in patients undergoing myocardial revascularization to maintain cardiac ATP at its pre-ischemic level was greatly enhanced when the cardioplegic solution was given through the completed distal left anterior descending anastomosis in addition to infusion via the aortic root.

Recently, explorations have been made whether addition of pharmaca to cardioplegic solutions stabilize cardiac adenine nucleotide homeostasis during the ischemic episode. Nifedipine used as additive to the St. Thomas' Hospital cardioplegic solution in patients undergoing valve replacements, was found to be effective to prevent accumulation of degradation products of adenine nucleotides in the oxygen-deprived heart (17). Earlier studies of the same research team (16) have shown that lidoflazine in combination with intermittent aortic cross-clamping in patients subjected to aorto-coronary bypass surgery was as effective as cardioplegia to preserve myocardial high-energy phosphate levels.

Despite the marked cardioprotective properties of most crystalloid and sanguineous cardioplegic solutions routinely used in the clinical setting, attempts have been made to interfere more directly with cardiac high-energy phosphate homeostasis. Basically, three different approaches to manipulate directly adenine nucleotide and creatine phosphate levels can be discriminated. First, addition to the cardioplegic solution of substances potentially active to promote resynthesis of high-energy phosphates. Second, the use of inhibitors of pathways involved in degradation of adenine nucleotides and related substances. Third, addition of high-energy phosphates, such as ATP and creatine phosphate, themselves to cardioplegic solutions.

Resynthesis of ATP can occur via 4 distinct pathways, one being the salvage pathway requiring inosine and/or hypoxanthine and ribose for resynthesis of AMP via adenylosuccinate, the second is phosphorylation of adenosine, the third ribophosphorylation of adenine and the fourth pathway is de novo synthesis of purines which procedes very slowly and requires substantial amounts of energy. Although promising experimentally obtained results have been described with respect to stimulation of the salvage pathway (10,13,47), data on this subject in the clinical setting are lacking.

In considering the use of inhibitors of pathways of adenine nucleotide and purine degradation experimental studies have indicated that application of these compounds may be a useful measure to preserve cardiac high-energy phosphate levels (see, for instance, ref. 9,19,22,38). There is, however, a paucity of information obtained in clinical studies to advocate the use of blockers of adenine nucleotide degradation in the operation theatre.

The effect of addition of creatine phosphate and ATP to a crystalloid cardioplegic solution has been extensively studied by Robinson et al. (29,30) in an isolated rat heart model. High-energy phosphates added to St. Thomas' Hospital cardioplegic solution significantly improved post-ischemic myocardial performance and reduced the loss of enzymes from the intracellular compartment after both 40 min of ischemia at 37°C and 270 min of oxygen deprivation at 20°C. ATP showed a relatively narrow dose-response curve. Optimal effects were obtained at 0.1 mmol ATP per liter cardioplegic solution. Higher concentrations of ATP affected negatively post-ischemic cardiac performance. Creatine phosphate exerted its maximal protective effect at relatively high concentrations, i.e., at least 10 mM (29). The beneficial effects of ATP and of creatine phosphate were found to be additive, since these substances together at their individual optimal concentrations provided significantly better myocardial protection than either substance alone. In addition to improved mechanical function and better maintenance of cellular integrity, high-energy phosphates as additives to cardioplegic solutions showed marked anti-arrhythmic effects (29).

The precise mechanisms underlying the beneficial effects of high-energy phosphates added to cardioplegic solutions remain to be established. With respect to ATP, a number of studies have indicated that extracellular ATP acts as a cardioplegic agent, depressing contractility of the heart (20). The cardioplegic action might be due to the calcium and magnesium chelating properties of ATP. Besides, it has been proposed that ATP also exerts other protective effects probably unrelated to the depression of cardiac contractility during the ischemic period. A direct intracellular site of action of ATP is not very likely. Myocardial cellular membranes are believed to be impermeable to extracellular ATP, ruling out the possibility that exogenous ATP will replenish or enhance the intracellular high-energy phosphate pool in a direct manner. Since degradation of exogenous ATP readily occurs in the coronary vascular and interstitial space (21), the beneficial effect of ATP might be mediated by its degradation products. However, Robinson et al. (30) concluded that the intact ATP molecule is required for maximal myocardial protection, since perfusion with ATP degradation products, such as adenosine and inorganic phosphate, failed to exert similar beneficial effects. The observation that the protective effects of ATP and creatine phosphate are additive (30) makes a common site of action less likely. Since creatine and inorganic phosphate, present at similar concentration in the cardioplegic solution as creatine phosphate, failed to protect the heart against loss of cellular integrity and post-ischemic depressed cardiac performance, intact creatine phosphate molecules are obviously required for the established beneficial action (29).

Extracellular creatine phosphate may exert its protective effect via an intracellular and/or extracellular site of action. It is generally believed that creatine phosphate cannot directly enter the myocyte, although findings of Breccia et al. (4) suggest that this polar compound can cross the sarcolemmal barrier under certain circumstances. In addition, Down et al. (11) have reported that intravenous injection of creatine phosphate (50 mg/kg body wt.) resulted in increased intracellular levels of creatine phosphate in the normoxic rat heart. Saks et al. (34) were unable to substantiate these findings in dogs. However, addition of creatine phosphate to a cardioplegic solution resulted in significantly enhanced post-ischemic cardiac ATP and creatine phosphate levels both in rats and human (34).

Robinson et al. (29) have pointed out that creatine phosphate may also exert a beneficial action by donating its high-energy phosphate group to plasmalemma-bound ADP with subsequent formation of extracellular ATP. The role of extracellular ATP is hypothetical, but maintenance of the content of ATP localized at the interstitial site of the sarcolemma might be qualitatively of great importance. Recently, Saks et al. (33) have hypothesized that the protective action of extracellular creatine phosphate may include 1) inhibition of sarcolemmal 5'-nucleotidase resulting in a less marked release of adenine nucleotide degradation products from the (post-)ischemic heart; 2) inhibition of accumulation of lysophospholipids. This will prevent the precipitation of arrhythmias; 3) prevention of ADP-induced aggregation of platelets. Ronca et al. (31) have further extended the list of putative beneficial modes of action. They suggest that the high-energy phosphate protects ischemic and reperfused myocardial cells by reducing free oxygen radical formation and by decreasing the peroxidative damage of cell membranes. Unfortunately human studies dealing with ATP and creatine phosphate administration during open-heart surgery are scarce. Saks et al. (34) have reported that creatine phosphate (in the range of 8-10 mM) added to a sanguineous cardioplegic solution significantly reduced the incidence of post-ischemic arrhythmias in patients who underwent valve operation. Obviously, more clinical studies are needed to rate the experimentally obtained findings at their true value.

References

1. Balderman SC, Bhayana JN, Binette P, Chan AWK, Gage AA: Perioperative preservation of myocardial ultrastructure and high energy phosphates in man. J Thorac Cardiovasc Surg 82:860-869, 1981
2. Balderman SC, Binette JP, Chan AWK, Gage AA: The optimal temperature for preservation of the myocardium during global ischemia. Ann Thorac Surg 35:605-614, 1983
3. Bessman SP, Geiger PJ: Transport of energy in muscle: the phosphorylcreatine shuttle. Science 211:448-452, 1981
4. Breccia A, Fini A, Girotti S, Gattavecchia E: Intracellular distribution of double labelled creatine phosphate in the rabbit myocardium. J Mol Cell Cardiol 16, Suppl 2:70, 1984 (Abstr)
5. Bretschneider HJ, Hubner G, Knoll D, Lohr B, Nordbeck H, Spieckermann PG: Myocardial resistance and tolerance to ischemia. Physiological biochemical basis. J Cardiovasc Surg 16:241-260, 1975
6. Buckberg GD: Studies of controlled reperfusion after ischemia. I. When is cardiac muscle damaged irreversibly? J Thorac Cardiovasc Surg 92:483-487, 1986
7. Chien KR, Sen A, Reynolds R, Chang A, Kim Y, Gunn MD, Buja LM, Willerson JT: Release of arachidonate from membrane phospholipids in cultured neonatal rat myocardial cells during adenosine triphosphate depletion. J Clin Invest 75:1770-1780, 1985
8. Cunningham JN, Adams PX, Knopp EA, Baumann FG, Snively SL, Gross RI, Nathan IM, Spencer FC: Preservation of ATP, ultrastructure, and ventricular function after aortic cross-clamping and reperfusion. J Thorac Cardiovasc Surg 78:708-720, 1979

9. Dhasmana JP, Digerness SB, Geckle JM, Ng TC, Glickson JD, Blackstone EH: Effect of adenosine deaminase inhibitors on the heart's functional and biochemical recovery from ischemia: a study utilizing the isolated rat heart adapted to ^{31}P nuclear magnetic resonance. J Cardiovasc Pharmacol 5:1040-1047, 1983

10. DeWitt DF, Jochim KE, Behrendt DM: Nucleotide degradation and functional impairment during cardioplegia: ameloriation by inosine. Circulation 67:171-178, 1983

11. Down WH, Chasseaud LF, Ballard SA: The effect of intravenously administered phosphocreatine on ATP and phosphocreatine concentrations in the cardiac musle of the rat. Arzneimittelforsch 33:552-554, 1983

12. Ellis SG, Henschke CI, Sandor T, Wynne J, Braunwald E, Kloner RA: Time course of functional and biochemical recovery of myocardium salvaged by reperfusion. J Am Coll Cardiol 1:1047-1055, 1983

13. Ely SW, Mentzer RM, Lasley RD, Lee BK, Berne RM: Functional and metabolic evidence of enhanced myocardial tolerance to ischemia and reperfusion with adenosine. J Thorac Cardiovasc Surg 90:549-556, 1985

14. Engelman RM, Rousou JH, Longo F, Auvil J, Vertrees RA: The time course of myocardial high-energy phosphate degradation during potassium cardioplegic arrest. Surgery 86:138-147, 1979

15. Flameng W, Borgers M, Daenen W, Thone F, Coumans WA, Van der Vusse GJ, Stalpaert G: St. Thomas cardioplegia versus topical cooling: ultrastructural and biochemical studies in human. Ann Thorac Surg 31:339-346, 1981

16. Flameng W, Borgers M, Van der Vusse GJ, Demeyere R, Vander Meersch E, Thone F, Suy R: Cardioprotective effects of lidoflazine in extensive aorto-coronary bypass grafting. J Thorac Cardiovasc Surg 85:758-768, 1983

17. Flameng W, Demeyere R, Daenen W, Sergeant P, Ngalikpima V, Geboers J, Suy R, Stalpaert G: Nifedipine as an adjunct to St. Thomas' Hospital cardioplegia. J Thorac Cardiovasc Surg 91:723-731, 1986

18. Flameng W, Van der Vusse GJ, de Meyere R, Borgers M, Sergeant P, Vandermeersch E, Geboers J, Suy R: Intermittent aortic cross-clamping versus St. Thomas' Hospital cardioplegia in extensive aortocoronary bypass surgery. A randomized clinical study. J Thorac Cardiovasc Surg 88:164-173, 1984

19. Foker JE, Einzig S, Wang T: Adenosine metabolism and myocardial preservation. J Thorac Cardiovasc Surg 80:506-516, 1980

20. Hearse DJ, Braimbridge MV, Jynge P, eds: Protection of the ischemic myocardium: Cardioplegia. New York: Raven Press, 1981

21. Hopkins SV: The potentiation of the action of adenosine on the guinea pig heart. Biochem Pharmacol 22:341-348, 1973

22. Humphrey SM, Seelye RN: Improved functional recovery of ischemic myocardium by suppresion of adenosine catabolism. J Thorac Cardiovasc Surg 84:16-22, 1982

23. Ingwall JS: Phosphorus nuclear magnetic resonance spectroscopy of cardiac and skeletal muscles. Am J Physiol 242:H729-H744, 1981

24. Jennings RB, Hawkins HK, Lowe JE, Hill ML, Klotman S, Reimer KA: Relation between high-energy phosphate and lethal injury in myocardial ischemia in the dog. Am J Pathol 92:187-214, 1978

25. Kaijser L, Jansson E, Schmidt W, Bomfim V: Myocardial energy depletion during profound hypothermic cardioplegia for cardiac operations. J Thorac Cardiovasc Surg 90:896-900, 1985

26. McLellan G, Weisberg A, Winegrad S: Energy transport from mitochondria to myofibril by a creatine phosphate shuttle in cardiac cells. Am J Physiol 245:C423-C427, 1983

27. Nayler WG: The role of calcium in the ischemic myocardium. Am J Pathol 102:262-270, 1981

28. Pernot AC, Ingwall JS, Menasche P, Grousset C, Bercot M, Mollet M, Piwnica A, Fossel ET: Limitation of potassium cardioplegia during cardiac ischemic arrest: a phosphorus ^{31}P Nuclear Magnetic Resonance study. Ann Thorac Surg 32:536-545, 1981

29. Robinson LA, Braimbridge MV, Hearse DJ: Creatine phosphate: an additive myocardial protective and anti-arrhythmic agent in cardioplegia. J Thorac Cardiovasc Surg 87:190-200, 1984

30. Robinson LA, Braimbridge MV, Hearse DJ: Enhanced myocardial protection with high-energy phosphates in St. Thomas' Hospital cardioplegic solution. J Thorac Cardiovasc Surg 93:415-427, 1987

31. Ronca R, Ronca-Testoni S, Zucchi R, Poddighe R: Protection of myocardial cells from oxidative

stress by exogenous creatine phosphate in isolated rat hearts. J Mol Cell Cardiol 19, Suppl III:82, 1987 (Abstr)

32. Rosenkranz ER, Okamoto F, Buckberg GD, Vinten-Johansen J, Allen BS, Leaf J, Bugyi H, Young H, Barnard RJ: Studies of controlled reperfusion after ischemia. II. Biochemical studies: failure to predict recovery of contractile function after controlled reperfusion. J Thorac Cardiovasc Surg 92:488-501, 1986

33. Saks VA, Javadov SA, Preobrazhensky AN, Krupriyanov VV, Samarenko SB, Ruda MY: Biochemical aspects of the protective action of phosphocreatine on the ischemic myocardium. J Mol Cell Cardiol 19, Suppl III:82, 1987 (Abstr)

34. Saks VA, Sharov VG, Kupriyanov VV, Kryzhanovsky SS, Semenovsky ML, Mogilevsky GM, Lakomkin VL, Ya A, Steinschneider, Preobrazhensky AN, Djavadov SA, Anyukhovsky EP, Breskrovnova NN, Rozenshtraukh LV, Kaverina HV. In: Smirnov VN, Katz AM, eds: Myocardial metabolism. New York: Gordon and Breach Publ, 1985

35. Salerno TA, Wasan SM, Charrette EJP: Prospective analysis of heart biopsies in coronary artery surgery. Ann Thorac Surg 28:436-439, 1979

36. Silverman NA, Wright R, Levitsky S, Schmitt G, Feinberg H: Efficacy of crystalloid cardioplegic solutions in patients undergoing myocardial revascularization. J Thorac Cardiovasc Surg 89:90-96, 1985

37. Trach V, Buschmanns-Denkel E, Schaper W: Relation between lipolysis and glycolysis during ischemia in the isolated rat heart. Basic Res Cardiol 81:454-464, 1986

38. Van Belle H, Wynants J, Xhonneux R, Flameng W: Changes in creatine phosphate, inorganic phosphate, and the purine pattern in dog hearts with time of coronary artery occlusion and effect thereon of mioflazine, a nucleoside transport inhibitor. Cardiovasc Res 20:658-664, 1986

39. Van der Vusse GJ, Coumans WA, Van der Veen FH, Drake AJ, Flameng W, Suy R: ATP, creatine phosphate and glycogen content in human myocardial biopsies: markers for the efficacy of cardioprotection during aorto-coronary bypass surgery. Vasc Surg 18:127-134, 1984

40. Van der Vusse GJ, Flameng W, Coumans WA, Van der Veen FH, Suy R: Myocardial ATP, creatine phosphate and glycogen as markers for the efficacy of cardioprotection during aorto coronary bypass surgery. In: Wauquier A, ed: Protection of tissues against hypoxia. Amsterdam: Elsevier Biomed Press, 1982:449-452

41. Van der Vusse GJ, Prinzen FW, Van Bilsen M, Engels W, Reneman RS: Accumulation of lipids and lipid-intermediates in the heart during ischemia. Basic Res Cardiol 82, Suppl 1:157-167, 1987

42. Van der Vusse GJ, Reneman RS: Glycogen and lipids (endogenous substrates). In: Drake-Holland AJ, Noble MIM, eds: Cardiac Metabolism. Chicester: Wiley, 1983:215-237

43. Van der Vusse GJ, Stam H: Lipid and carbohydrate metabolism in the ischaemic heart. Basic Res Cardiol 82 Suppl 1:149-153, 1987

44. Van der Vusse GJ, Van der Veen FH, Flameng W, Coumans WA, Borgers M, Willems G, Suy R, De Meyere R, Reneman RS: A biochemical and ultrastructural study on myocardial changes during aorto-coronary bypass surgery: St. Thomas Hospital cardioplegia versus intermittent aortic cross-clamping at 34 and 25°C. Eur Surg Res 18:1-11, 1986

45. Vinten-Johansen J, Rozenkranz ER, Buckberg GD, Leaf J, Gugyi H: Studies of controlled reperfusion after ischemia. VI. Metabolic and histochemical benefits of regional blood cardioplegic reperfusion without cardiopulmonary bypass. J Thorac Cardiovasc Surg 92:535-542, 1986

46. Whitman GJR, Roth RA, Kieval RS, Harken AH: Evaluation of myocardial preservation using ^{31}P NMR. J Surg Res 38:154-161, 1985

47. Zimmer HG, Gerlach E: Stimulation of myocardial adenine nucleotide biosynthesis by pentose and pentitols. Pflügers Arch 376: 223-227, 1978

Chapter 23

Apparent Inosine Incorporation and Concomitant Haemodynamic Improvement in Human Heart

W. Czarnecki, II Department of Cardiology, Medical Centre of
Postgraduate Education, Grochowski Hospital, Warsaw, Poland

Haemodynamic effects of inosine were studied in five patients suffering from prolonged, otherwise intractible cardiogenic shock, and in six patients catheterized for coronary angiography. Heart performance was improved by inosine as evidenced by enhanced cardiac index (by 63 ± 29%, p<0.002; mean ± SEM) in patients with extremely low cardiac output, and increased LV dp/dt_{max} (by 22 ± 7%) in patients with normal perfusion during angiography. Inosine exerts a positive inotropic effect in man. No data, so far, are available concerning the fate of exogenous inosine in human myocardium. Therefore, during inosine infusion in patients undergoing diagnostic cardiac catheterization, arterio-coronary sinus (a-v) differences of purine nucleosides were determined simultaneously with haemodynamic measurements. Before infusion the a-v difference of inosine, hypoxanthine and xanthine across the heart was nil, whereas urate production was evident. During infusion arterial inosine increased, exceeding the coronary sinus concentration by a maximum of 200 ± 52 uM (p<0.05) at the fourth minute. At that time the a-v difference in hypoxanthine, xanthine and urate amounted to 16 ± 11, 10 ± 3 (p<0.05), and 15 ± 14 uM, respectively. Thus there seems to be a substantial uptake of inosine by the human heart, concomitant with haemodynamic improvement.

Introduction

Severe left ventricular (LV) failure and shock following myocardial infarction (MI) lead to high mortality. This creates a continuous need for an effective drug, increasing cardiac output and improving peripheral perfusion. The efficacy of a variety of inotropic drugs currently used in the therapy of cardiogenic shock is not satisfactory. Sympathomimetics and their derivatives produce marked improvement in LV function. But they exert many known untoward effects, including increased myocardial oxygen consumption and infarct size. Digitalis, the traditional drug used in congestive heart failure, is virtually ineffective in patients with advanced pump failure. All these facts encourage an active search for other positive inotropic drugs. In view of this need, inosine, an inotropic substance naturally occurring in the body, is a promising agent. The myocardial cell produces this nucleoside by both deamination of adenosine and dephosphorylation of IMP (5). The heart releases it in large quantities after exposure to hypoxia, anoxia or ischaemia. Then ATP breakdown is stimulated (4,26,28,29). Inosine augments myocardial blood flow (3,18,19) and increases contractility of both normal (3,18,19,22) and ischaemic heart muscle (32,33) in experimental animals. Nevertheless, there are conflicting results concerning its inotropic action in different species. Its positive effect is evident in the dog (11,22,

30,32), pig (10,33), rabbit (23), and frog (6). In contrast, the effect of inosine is absent in the guinea pig (17) and even negative in the rat (17,34).

Inosine for Cardiogenic Shock in Man

Inosine limits the size of experimental MI (10,12) and counteracts experimental LV failure (30,31). This encouraged us to apply the endogenous nucleoside in the management of low cardiac output states in man. Two female and 3 male patients 25-82 years, suffering from prolonged, otherwise intractable cardiogenic shock due to MI (4 patients) and congestive cardiomyopathy (1 patient), were studied (9). All suffered from circulatory failure characterized by clammy skin, mental obtundation and anuria. Cardiac index (CI) ranged from 0.58 to 1.63 $l/min/m^2$. Before the inosine study was performed, all the patients had received conventional pharmacotherapy (dopamine, dobutamine, noradrenaline, glucagon, digoxin, hydrocortisone, etc.) with no signs of improvement, however. The haemodynamic data were obtained from a Swan-Ganz catheter. Cardiac output was measured in triplicate by the thermodilution technique and computed using a Hewlett-Packard cardiac output computer. The pressure readings were obtained by means of external transducers and recorded simultaneously with the heart rate. Inosine (Trophicardyl i.v., Laboratoire Innothéra, France) was infused in a dose of 5 mg/kg/min (4% solution). Haemodynamic variables were recorded prior to and at the 6th minute of infusion.

Inosine administered to patients who had not responded previously to conventional pharmacotherapy resulted in an increase in the CI (by 63 ± 29%, p<0.002; mean ± SEM) and the stroke volume index (by 55 ± 15%, p<0.001). Pulmonary capillary wedge pressure and heart rate were not significantly affected (Table 1). Despite a high increase in CI in each case, in 2 patients blood pressure was unobtainable throughout the whole study. In another one it was unchanged and in the remaining two slightly increased by inosine. None of the patients survived; we decided to infuse inosine at the end-stage of shock after conventional therapy had failed, to improve performance of the heart. Furthermore, the delay between the first symptoms of shock and nucleoside administration was considerable.

Table 1. Influence of inosine (5 mg/kg/min) on heart rate (HR), pulmonary capillary wedge pressure (PCWP), cardiac index (CI), and stroke volume index (SVI) in 5 patients suffering from prolonged, otherwise intractable, cardiogenic shock

Variable	Increased by		P
HR (beats/min)	5	± 6	NS
PCWP (mmHg)	0.8	± 0.8	NS
CI ($ml/min/m^2$)	680	± 14	<0.01
SVI (ml/m^2)	6.1	± 1.9	<0.001

Studies on Inosine Incorporation by Human Myocardium

Inosine exerts a positive inotropic effect in man. It improves significantly myocardial performance in patients with extremely low cardiac output. No data, so far, have been available concerning the fate of this agent in man. Therefore, we designed a study (21) in which inosine was infused to 6 patients undergoing diagnostic coronary catheterization in the course of coronary heart disease. We attempted to answer the following questions:

1. Is exogenous inosine taken up by the human heart?
2. Is the inotropic effect of inosine related to its myocardial uptake?

The protocol of inosine administration was similar to the one used in the previous study. The nucleoside (Trophicardyl) was infused at a rate of 5 mg/kg/min into the peripheral vein. The haemodynamic variables were recorded before and at the 6th minute of infusion. The first derivative of LV pressure (LV dp/dt_{max}) was obtained from a resistance-capacitance differentiating circuit (Siemens-Elema). Arterial and coronary sinus blood was sampled for a purine assay before and after 2, 4 and 6 min of the procedure. Inosine, hypoxanthine, xanthine and uric acid were determined by the high-performance liquid chromatography (HPLC) method described in detail by De Jong et al. (21) and Harmsen et al. (14).

Inosine administration had no significant effect upon heart rate, LV end-diastolic pressure (Fig. 1) as well as on LV systolic and diastolic pressures. The contractility of the heart, as measured by LV dp/dt_{max}, increased significantly by $22 \pm 7\%$ (Fig. 1), varying from 6% to 44% in individual patients. During inosine infusion, a substantial increase in the arterial nucleoside concentration as well as that of its catabolites hypoxanthine, xanthine and urate was observed (Fig. 2). Figure 3 presents the arterio-venous differences of the nucleosides under study. There was no a-v difference in inosine, hypoxanthine and xanthine in control measurements except for urate. Urate was produced by each heart, indicating the presence of xanthine oxidase in the myocardium of the patients studied. During inosine infusion part of the nucleoside was apparently converted to its catabolites. At the fourth minute the arterio-coronary sinus differences for hypoxanthine, xanthine and urate were 16 ± 11

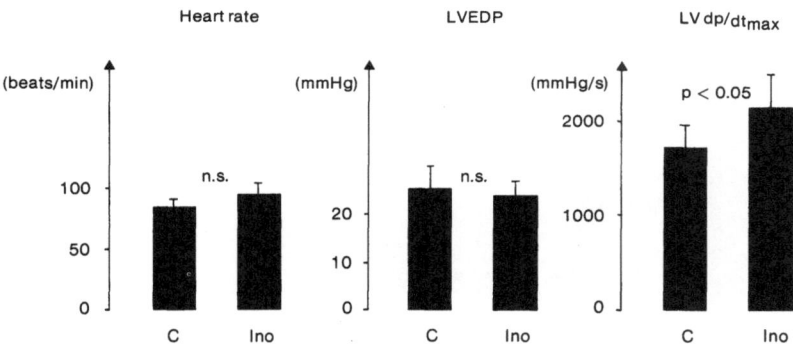

Fig. 1. Influence of inosine (5 mg/kg/min i.v.) on heart rate, left ventricular end-diastolic pressure (LVEDP) and LV dp/dt_{max}. Results obtained in five patients, catheterized for coronary angiography (means ± SEM). C = control; Ino = inosine.

uM, 10 ± 3 uM and 15 ± 14 uM, respectively (Fig. 3). The large a-v difference of inosine at the same time (200 ± 53 uM) suggests that this nucleoside is partially converted to catabolites and partially taken up by the myocardium.

Discussion

The present findings demonstrate that inosine exerts a positive inotropic effect in man. It improves significantly myocardial performance in patients with extremely low cardiac output. The same is true for patients with normal peripheral perfusion. The positive inotropic action was proved by an increase in cardiac index and stroke volume index in patients suffering from shock and by an increase in LV dp/dt_{max} in patients undergoing diagnostic cardiac catheterization. The LV end-diastolic pressure measurements (as obtained from the pulmonary capillary pressure in patients suffering from shock and from the LV cavity in patients during catheterization) suggest that inosine doesn't affect the preload. Afterload also seems unaffected by the nucleoside, since there were no significant changes in blood pressure in both studies.

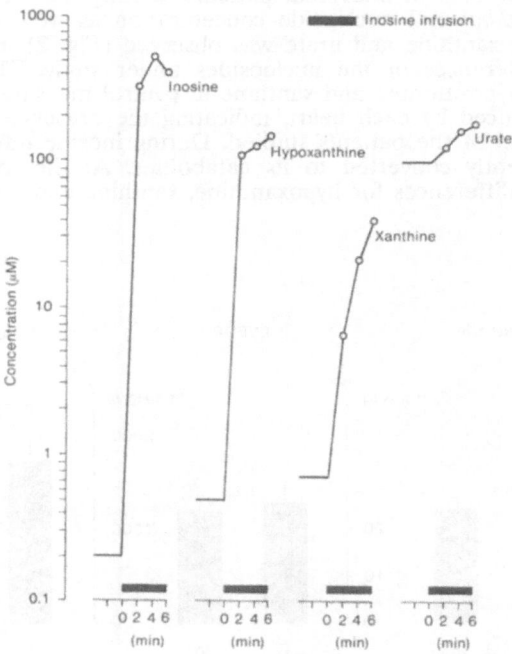

Fig. 2. Inosine-induced increase in arterial purine concentrations. The concentrations of inosine, hypoxanthine, xanthine and urate were measured prior to and at the second, fourth and sixth minute of i.v. inosine administration at a dose of 5 mg/kg/min.

The metabolic cost of inosine action is expected to be relatively low. The increase in stroke volume in patients suffering from shock, and the increase in LV dp/dt_{max} in the catheterized patients, were not accompanied by a parallel, significant increment in the heart rate. This is in agreement with our previous studies on open-chest dogs with an acutely implanted coronary flow probe (8) and on closed-chest dogs with acute LV failure (31). In these inosine enhanced contractility and myocardial blood flow without raising myocardial oxygen consumption. These results were subsequently published by Smiseth (30). Thomas and Jones (32) found in the dog model that inosine enhanced contractility of both ischaemic and nonischaemic myocardium. However, the content of high-energy phosphates remained unchanged.

The mechanism of inosine action is not completely understood. We reported (11) that the nucleoside-induced increase in contractility is mediated in part by the ß-adrenergic system and in part by an independent mechanism. To elucidate further the fate of inosine in the heart in that study, the nucleoside was administered in a dose of 5 mg/kg/min i.v. This dosage was shown to increase the inosine content in the myocardium up to the amounts observed during experimental coronary occlusion or cardiac arrest (19,20,27).

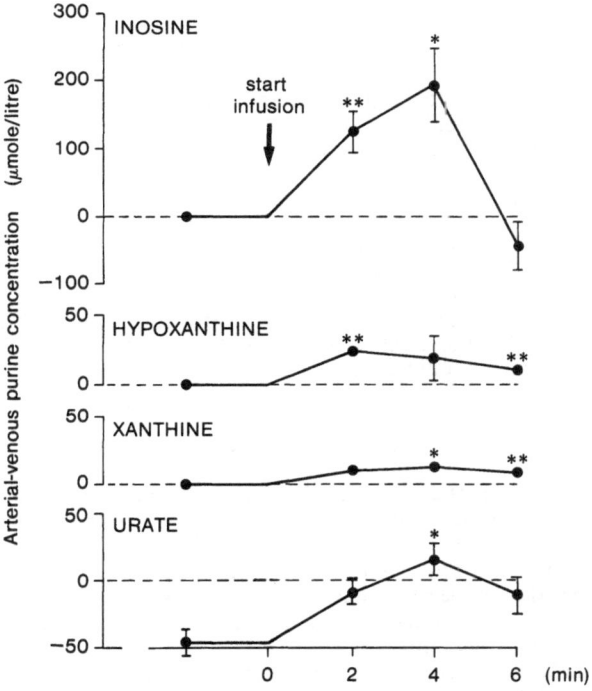

Fig. 3. Apparent uptake of inosine and its catabolites by the human heart during i.v. inosine administration. Note that before infusion the arterio-coronary sinus difference of inosine, hypoxanthine and xanthine across the heart was nil, whereas urate production was evident (lower panel). Means ± SEM (n = 5-6), *p<0.05, **p<0.01 vs. baseline value.

The present results show that under baseline conditions the heart produces substantial amounts of urate. The a-v difference across the heart is about 50 uM (Fig. 3), which is in agreement with the preliminary data of Nelson (25). The finding suggest that the human heart contains a relatively active xanthine oxidoreductase. Although this enzyme has been demonstrated in the heart of several species, there are not many data available concerning the human myocardium. The oxidase form of xanthine oxidase can generate oxygen free radicals, playing an important, causative role in the development of coronary heart disease. During inosine infusion the arterial and coronary sinus urate levels rose, apparently due to an increased supply of its precursors hypoxanthine and xanthine (Fig. 2). The a-v difference of urate-across the heart disappeared (Fig. 3), which could indicate that xanthine oxidoreductase is inhibited by inosine or its catabolites.

Simultaneously with the high a-v inosine difference at the second and fourth minute of nucleoside administration, the increase in contractility   was observed. At the sixth minute, this apparent myocardial inosine uptake was absent, but the positive inotropic effect was still the same. Whether redistribution of inosine among various intra- and extracellular compartments takes place during this phase of the infusion, remains to be studied. Inosine promotes enhanced nucleotide resynthesis in the myocardium (2,15) via breakdown to hypoxanthine, subsequent ribophosphorylation, and conversion to AMP. We therefore speculate that inosine incorporation into adenine nucleotides is responsible for the increased contractility of human heart.

It is noteworthy that no arrhythmia, ischaemia or other side effects were observed. Inosine, in contrast to catecholamines and digitalis, proved to be a cardiotonic agent without arrhythmogenic properties (1,13). Furthermore, this nucleoside could protect the myocardium against toxic effects of catecholamines (7,16,24).

In view of the need for effective therapeutic intervention to improve myocardial performance in cardiogenic shock, inosine is worth considering. It is an endogenous substance, free of side effects. Its helpful action, both in experimental myocardial infarction (10,12) and experimental acute LV failure (30,31), has been well proved. The initial clinical results obtained with inosine are promising and encourage further trials focused on its mechanism of action. Prolonged, early instituted administration of inosine seems to be the next logical step to assess the value of this therapy in the management of cardiogenic shock.

References

1. Aurosseau M, Rouge M, Huet Y, Albert O: Action d'un nucléoside, l'hypoxanthine-d-riboside ou inosine sur l'hémodynamique cardiaque de l'animal normal ou pathologique. Ann Pharm Fr 33:99-107, 1975
2. Aussedat J, Verdys M, Rossi A: Adenine nucleotide synthesis from inosine during normoxia and after ischaemia in the isolated perfused rat heart. Can J Physiol Pharmacol 63:1159-1164, 1985
3. Aviado DM: Effects of fluorocarbons, chlorinated solvents and inosine on the cardiopulmonary system. Envir Health Persp 26:207-215,1978
4. Berne RM: Cardiac nucleotides in hypoxia: possible role in regulation of coronary blood flow. Am J Physiol 204:317-322, 1963
5. Berne RM, Rubio R: Adenine nucleotide metabolism in the heart. Circ Res 34-35, Suppl 3:109-118, 1974
6. Cook MH, Greene EA, Lorber V: Effect of purine and pyrimidine ribosides on an isolated frog ventricle preparation. Circ Res: 6:735-739, 1958

7. Czarnecki A: Protective effect of inosine on adrenaline-induced myocardial necrosis. Pharmacol Res Commun 18:459-469, 1986

8. Czarnecki W: Inosine, a potent inotropic nucleoside reduces myocardial oxygen consumption in the dog. Eur Heart J 5, Suppl 1:222, 1984 (Abstr)

9. Czarnecki W, Czarnecki A: Haemodynamic effects of inosine. A new drug for failing heart? (submitted)

10. Czarnecki W, Herbaczynska-Cedro K: The influence of inosine on the size of myocardial ischaemia and myocardial metabolism in the pig. Clin Physiol 2:189-197, 1982

11. Czarnecki W, Noble MIM: Mechanism of the inotropic action of inosine on canine myocardium. Cardiovasc Res 17:735-739, 1983

12. Devous MD, Jones EE: Effect of inosine on ventricular regional perfusion and infarct size after coronary occlusion. Cardiology 64:149-161, 1979

13. Faucon G, Laverenne J, Collard M, Evreux JC: Effects d'un nucléoside, l'hypoxanthine-d-riboside, sur l'activité et l'irrigation myocardique. Thérapie 21:1239-1252, 1966

14. Harmsen E, De Jong JW, Serruys PW: Hypoxanthine production by ischemic heart demonstrated by high pressure liquid chromatography of blood purine nucleosides and oxypurines. Clin Chim Acta 115:73-84, 1981

15. Harmsen E, De Tombe PP, De Jong JW, Achterberg PW: Enhanced ATP and GTP synthesis from hypoxanthine or inosine after myocardial ischemia. Am J Physiol 246:H37-H43, 1984

16. Hiramatsu Y, Takahashi H, Izumi A: Effect of inosine on adrenaline and vasopressin induced myocardial hypoxia. Jpn J Pharmacol 21:355-360, 1971

17. Hoffmeister HM, Betz R, Fiechtner H, Seipel L: Myocardial and circulatory effects of inosine. Cardiovasc Res 21:65-71, 1987

18. Jones CE, Mayer LR: Nonmetabolically coupled coronary vasodilation during inosine infusion in dogs. Am J Physiol 238:569-574,1980

19. Jones CE, Thomas JX, Devous MD, Norris CP, Smith EE: Positive inotropic response to inosine in the in situ canine heart. Am J Physiol 233:438-443, 1977

20. Jones CE, Thomas JX, Parker JC, Parker RE: Acute changes in high energy phosphate, nucleotide derivatives and contractile force in ischaemic and nonischaemic canine myocardium following coronary occlusion. Cardiovasc Res 10:275-282, 1976

21. Jong JW de, Czarnecki W, Ruzyllo W, Huizer T, Herbaczynska-Cedro K: Apparent inosine uptake by the human heart (submitted)

22. Juhász-Nagy A, Aviado DM: Inosine as a cardiotonic agent that reverses adrenergic beta-blockade. J Pharmacol Exp Ther 202: 683-695, 1977

23. Kypson J, Hait G: Metabolic effects of inosine and uridine in rabbit hearts and skeletal muscles. Biochem Pharmacol 26:1585-1591, 1977

24. Lampa E, Imperatore A, Sasso M, Alise G, Pedone P, Ottavo R, Vacca C, Perna D: Analisi sperimentale di alcuni effetti cardiaci dell'inosina. Arch Sc Med 137:225-240, 1980

25. Nelson JA: Some clinical and pharmacological applications of high-speed liquid chromatography. Adv Chromatogr 15:273-280, 1977

26. Olsson RA: Changes in content of purine nucleosides in canine myocardium during coronary occlusion. Circ Res 26:301-306, 1970

27. Parker JC, Smith EE, Jones CE: The role of nucleoside and nucleobase metabolism in myocardial adenine nucleotide regeneration after cardiac arrest. Circ Shock 3:11-20, 1976

28. Richman HG, Wyborny L: Adenine nucleotide degradation in the rabbit heart. Am J Physiol 207:1139-1145, 1964

29. Rubio R, Berne RM: Release of adenosine by the normal myocardium in dogs and its relationship to the regulation of coronary resistance. Circ Res 25:407-415, 1969.

30. Smiseth OA: Inosine infusion in dogs with acute ischaemic left ventricular failure: favourable effects on myocardial performance and metabolism. Cardiovasc Res 17:192-199, 1983

31. Smiseth OA, Mjos OD, Czarnecki W, Herbaczynska-Cedro K: Inosine, a potent inotropic and vasodilating agent during acute ventricular failure in dogs. Eur J Clin Invest 11, Part 2:29, 1981 (Abstr)

32. Thomas JX, Jones CE: Effect of inosine on contractile force and high-energy phosphates in ischemic hearts. Proc Soc Exp Biol Med 161:468-472, 1979
33. Woollard KV, Kingaby EO, Lab MJ, Cole AWG, Palmer TN: Inosine as a selective inotropic agent on ischaemic myocardium? Cardiovasc Res 15:659-667, 1981
34. Zimmer HG, Westphal RC: Inosine, a negative inotropic agent in the closed-chest rat. Circulation 72, Suppl 4: IV-337, 1985 (Abstr)

Chapter 24

Cardioplegia and Calcium Entry Blockade

J.W. de Jong, Cardiochemical Laboratory, Thoraxcenter, Erasmus
University Rotterdam, The Netherlands

Cooling and induction of electromechanical arrest (cardioplegia) limit myocardial damage during open-heart operations. Many studies provide evidence for the use of calcium entry blockers, as adjuncts to cardioplegia to optimize the protection. This paper reviews their possible mechanism of action, application time, and efficacy during hypothermia. A major conclusion is that negative effects are virtually absent, apart from short-term negative inotropic responses. There is an increasing body of positive evidence for their efficacy. A new development is the use of these drugs for regional cardioplegia during dilation of coronary arteries (coronary angioplasty).

Necessity for Cardioprotection

In his presidential address, entitled "Intellectual creativity in thoracic surgeons", Spencer (69) analyzed 8 breakthroughs in cardiovascular surgery over the last 30 years. One of the subjects concerned myocardial preservation. Additional protection of the heart by calcium entry blockers is clear in many animal studies. The time has come for more extensive clinical work.

Buckberg et al. (9) pointed at the diminished heart function after extracorporeal circulation and the importance to protect the myocardium adequately. They asked: "Why should a heart which did not need pharmacologic or mechanical support before operation require these interventions after the cardiac lesion has been corrected?" Their methods, including avoidance of prolonged hypothermic cardiac arrest, sharply diminished the postoperative use of inotropic drugs and reduced mortality from 12 to 3% (9). Mortality within one month after aorta-coronary bypass surgery at my institution is around 1% (49). Although these figures are relatively low, one has to try to reduce mortality further. This holds a fortiori for certain other cardiovascular operations (68).

Morbidity should and can improve as well. Full restoration of myocardial function can take months after bypass grafting (66). Nowadays, induction of electro-mechanical arrest (cardioplegia) combined with hypothermia limits the initial insult. It ameliorates thereby recovery after cardiac operations. However, improvements are still possible; certain types of cardioplegia fail to protect the heart adequately from ischemic damage. In the last few years, substantial research has been carried out on interventions involving calcium metabolism. This chapter updates a review on calcium antagonists and cardioplegia (21).

Calcium

The myocardial Ca^{2+} content in mammals is about 2 mmol/kg. The largest part resides in internal stores near the endoplastic reticulum. During diastole, the cytosolic Ca^{2+} concentration is presumably lower than 0.0001 mM. However, for active tension development, a Ca^{2+} concentration of about 0.01 mM is necessary. Therefore, calcium from subcellular organelles or from outside the cell moves to the cytosol. This accomplishes the change from resting to active status. In the blood, the ionized Ca^{2+} level is about 2.5 mM.

Calcium ions can enter the cell by passive diffusion, by exchange with sodium, by voltage-activated transport, and possibly also by exchange with potassium. Voltage-activated transport takes place via the slow calcium channels. Per beat, only little calcium enters (reversibly) the cell. If there is a lack of oxygen, however, the cardiac cell membrane becomes more permeable for Ca^{2+} ions from the extracellular space. An increased influx into the ischemic tissue during reperfusion will ensue (57). Excessive accumulation leads to ischemic contracture, the "stone heart" syndrome, a life-threatening event. It is therefore imperative to prevent calcium accumulation, whatever the route. The phenomenon of the calcium paradox illustrates the importance of calcium homeostasis for the myocyte.

Calcium Paradox

The calcium paradox occurs when hearts are perfused with a solution containing less than 0.05 mM Ca^{2+}, followed by one with a normal calcium concentration (see Chapter 19). Then tissue disruption follows, with enzyme release and contracture. Abrupt massive calcium influx to intracellular compartments seems responsible for the process (29). The clinical relevance of this discovery did not become clear until later: in a number of surgical procedures, calcium supply to the heart stops. This possibly initiates the calcium paradox during reperfusion. It is unlikely that the slow channels provide a major route of Ca^{2+} entry during the calcium paradox (57). Nevertheless, calcium entry blockers can prevent some of the damage, which occurs because of the calcium influx (65).

Calcium Entry Blockers

The term *calcium antagonist* is in use since 1969 for drugs with negative inotropic properties that can be counteracted by calcium. In the definition of Fleckenstein (28), a calcium antagonist specifically blocks excitation-contraction coupling by mitigating the effects of Ca^{2+}. Drugs like verapamil, nifedipine and diltiazem are calcium antagonists that are valuable to treat angina pectoris and hypertension. They inhibit the slow, inward calcium flux, responsible for the plateau phase of the action potential. Alternative names include "slow-channel inhibitors", "calcium channel inhibitors" or "calcium entry blockers". The negative inotropic effect of these compounds can be ascribed to this decreased, slow calcium influx. Figure 1 shows in schematic form a classification of various calcium modulators. From Table 1, it is clear that cardioprotective calcium entry blockers form a heterogenous group of drugs. Nifedipine, a 1,4-dihydropyridine, is the "mother" compound of a series of drugs such as nicardipine, niludipine, nimodipine, nisoldipine, and nitrendipine. These belong to the most potent vasodilators on the market, with varying tissue specificity. Verapamil, a compound with a structural resemblance to papaverine, has as its "congeners" gallopamil, tiapamil, and anipamil.

The heart catabolizes quickly its energy-rich phosphates during lack of oxygen. Then ATP, which is essential for contraction, breaks down to the level of the purines adenosine, inosine, hypoxanthine, xanthine, and urate. These metabolites leave the cell. Catabolism of the adenine nucleotide pool by more than 75% prohibits resuscitation of an ischemic heart (42) As shown by many authors, verapamil can inhibit ATP breakdown in ischemic and reperfused heart (28,57). In addition, energy conservation is possible with diltiazem, nifedipine, nisoldipine and bepridil (22-24,41,74).

Many studies demonstrate the salutary effect of calcium antagonists on myocardial energy metabolism during transient ischemia. Nevertheless, their mechanism of action is not fully clear. Calcium overload leads to excessive ATP consumption by ATPases and reduced mitochondrial ATP production. The resulting exhaustion of energy-rich phosphates will ultimately cause myocardial necrosis. Blockade of slow Ca^{2+} channels will restore calcium homeostasis with normal ATP consumption and adequate ATP synthesis. Protection due to reduction of work, to vasodilatation, or to interference with organelles, rather than to direct action on Ca^{2+} fluxes across the cell membrane, cannot be excluded. Indeed, the degree of negative inotropy observed before ischemia, seems to determine how much ATP is catabolized during ischemia (24). And also much the heart recovers (74).

Calcium Entry Blockade during Heart Operations

With cold chemical cardioplegia during heart operations, the tolerable duration of ischemia increases from less than 1 to 3 h. In experimental studies, reversible ischemia of up to 24 h appears possible (34). Effective protection has 3 facets: 1) energy sparing by induction of instaneous electromechanical arrest with compounds such as potassium; 2) slowing down of energy-consuming and degradative reactions by hypothermia; and 3) combating harmful changes caused by ischemia and reperfusion with protective agents (34). The control of calcium movement seems to be critical to protect the heart effectively (see Chapter 22). In this setting, calcium entry blockers could be helpful. From the available literature, it is clear that many experiments with animals and some clinical trials demonstrate calcium antagonists to have a protective effect on the heart.

Animal Studies

Table 2 shows the effect of diltiazem, lidoflazine, nifedipine, and verapamil as adjuncts to potassium cardioplegia. In all studies carried out in myocardium at 37°C,

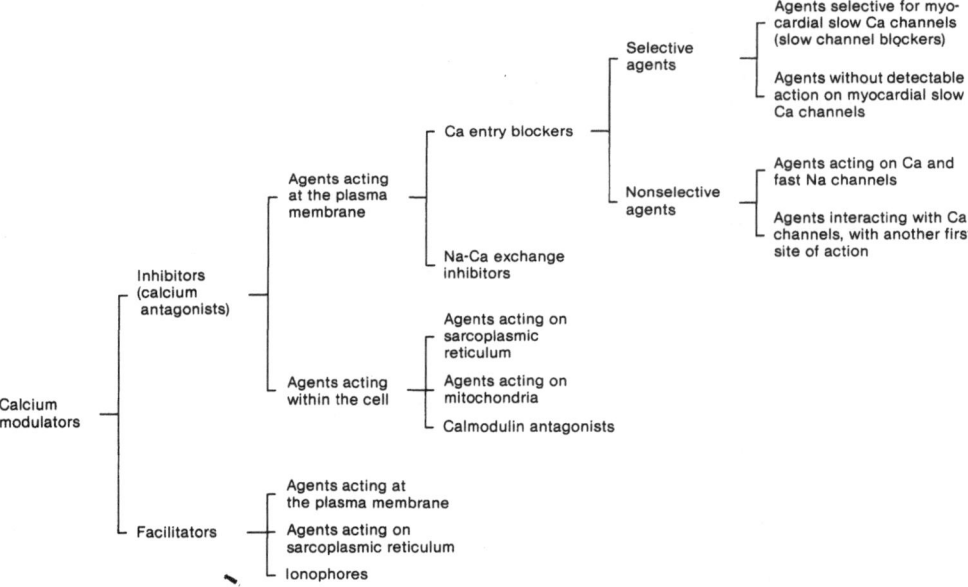

Fig. 1. Inhibition and stimulation of calcium uptake by various agents. This scheme is based on a recent classification by Godfraind (31).

calcium entry blockers protected against damage caused by ischemia. Somewhat less clear is the effect of these drugs on the hypothermic heart. Many investigators reported additional protection by calcium antagonists. In a number of studies, however, the drugs did not significantly ameliorate ischemia-induced damage during hypothermia (see Table 2). At lower temperatures, calcium antagonists also didn't prevent damage caused by the calcium paradox (3). Nayler (58), on the other hand, showed that the protective effects of hypothermia and nifedipine are additive. However, in hearts arrested with potassium, this additive effect was absent at temperatures below 28°C (57). Calcium entry blockers introduce a discontinuity around this value into the Arrhenius plot of the maximal contraction frequency vs. temperature (62).

If the ischemic damage is virtually nil or the protection by hypothermia alone already adequate, calcium antagonist administration may be redundant. However, even if the specific cardioplegic effect is diminished, the calcium antagonist may be useful by lowering the pre-ischemic energy demand (22). Also, reperfusion conditions may be improved. It is noteworthy that calcium antagonists used as adjuncts to potassium cardioplegia don't give negative effects. Moreover, hearts arrested by means other than potassium cardioplegia also showed protection by verapamil, diltiazem, nifedipine, and lidoflazine. One exception should be noted: diltiazem combined with magnesium cardioplegia was detrimental to postischemic function, despite a rapid restoration of energy stores (73). High magnesium cardioplegia could induce heart block (10). Using calcium entry blockers, one has to keep in mind that a (reversible) negative inotropic effect can persist for some time. This could cause problems in the hospital setting.

Clinical Trials

Details on studies of calcium entry blockers used during cardiac operations are scarce (Table 3). Clark et al. (19) reported the effectiveness of nifedipine in a multicenter trial. They selected 205 highest-risk patients from a large open-heart population. The investigators employed cold potassium cardioplegia. Nifedipine influenced hemodynamics in a positive sense, but the authors provided no statistical details. Nifedipine seemed to preserve myocardial levels of ATP also in this study. However, no significant reduction in myocardial creatine kinase release occurred.

Table 1. Some cardioprotective calcium antagonists, acting at the plasma membrane

	Calcium entry blocker				
	Selective			Nonselective	
Benzothiazepine	Dihydropyridine	Phenyl-alkylamine	Benzylphenylamine-pyrrolidine	Diphenylalkyl-piperazine	Phenyl-alkylamine
Diltiazem	Felodipine	Anipamil	Bepridil	Lidoflazine	Tiapamil
	Nicardipine	Gallopamil			
	Nifedipine	Ronipamil			
	Niludipine	Verapamil			
	Nisoldipine				
	Nitrendipine				

Calcium entry blockers, selective for slow calcium channels in myocardium, are also called slow channel blockers. The nonselective agents, mentioned in this table, act at similar concentrations on calcium and fast sodium channels. See also Figure 1.

Flameng et al. (27) found that the addition of nifedipine to the cardioplegic solution could prevent the ischemia-induced degradation of nucleotides, as it occurred when myocardial cooling was inadequate. However, clinical outcome didn't improve. In an earlier trial, Flameng et al. (26) showed that lidoflazine prevents in a dose-dependent way high-energy phosphates and glycogen breakdown. In addition, lidoflazine maintained ultrastructure and left ventricular stroke work index better. In this investigation, patients were pretreated with the drug. Operation conditions included intermittent aortic clamping and hypothermic conditions.

Table 2. Efficacy of calcium entry blockers given as adjuncts to potassium cardioplegia to prevent damage due to myocardial ischemia

Drug	Preparation	Temperature of heart (°C)	Variables tested	Result	Reference
Diltiazem	Isolated rat heart	20 vs. 37	Hemodynamics/mitoch.	NS vs. +	76
	Dog, CPB	7	P/hemodynamics	±	70
	Isolated rat heart	10	P/arrhythmias	+	71
	Dog, CPB	20 vs. 28	Hemodynamics	NS vs. +	48
	Dog, CPB	37	P/hemodynamics	±	1
Lidoflazine	Dog, CX ligation	30	Hemodynamics	NS	33
	Dog, CPB, CX ligation	20	Hemodynamics	+	45
	Dog, CPB	37?	Cerebral flow	+	75
Nifedipine	Dog, CPB	10	Hemodynamics/edema	+	52
	Dog, CPB	10	Hemodynamics	+	16
	Dog, CPB	10, 37	Hemodynamics	+	18
	Dog, CPB	24	Ca^{2+} homeostasis	+	7
	Dog, CPB	12, 21	P/hemodynamics	NS	43
	Isolated rabbit heart	15-37	Hemodynamics	+, $\geq$28°C	57
	Isolated rat heart	20 vs. 37	Hemodynamics/enzymes	± vs. +	77
	Isolated rat heart	15	P/hemodynamics	+	54
	Dog, CX ligation	30	Hemodynamics	+	33
	Isolated rabbit heart	14	Hemodynamics	+	46
	Isolated rat heart	20	P/hemodynamics	+/-	6
	Isolated rat heart	37	P/hemodynamics	NS/+	78
	Isolated rat heart	20-37	Hemodynamics/enzymes	+, $\geq$31°C	30
	Pig, CPB	28	Hemodynamics/lactate	±	55
	Guinea pig, CPB*	9 vs. 34	Hemodynamics	NS vs. +	14
	Dog, CPB	21	Arrhythmias	+	25
Verapamil	Dog, CPB	37?	Hemodynamics/mitoch.	+	61
	Isolated dog heart	4	Hemodynamics	+	56
	Isolated rabbit heart	14	Hemodynamics	+	46
	Dog, CPB	37?	Cerebral flow	+	75
	Isolated rat heart	20-37	Hemodynamics/enzymes	+, $\geq$34°C	37
	Dog, CPB	37	P/hemodynamics	+	51
	Dog, CX ligation	18	Mitoch.	+	79
	Isolated rat heart	37	P/hemodynamics	NS/+	78
	Dog, CPB	9	P/hemodyn./structure	+	50

+ indicates protection additive to potassium cardioplegia, and NS means protection not additive to cardioplegia. * Heart-lung model. CPB = cardiopulmonary bypass; CX = circumflex coronary artery; mitoch. = mitochondrial function; P = myocardial high-energy phosphates.

Hicks et al. (40) studied the effect of verapamil added to the cardioplegic solution and nifedipine instituted postoperatively. This regimen, for which the rationale remained unexplained, resulted in a reduction in postoperative levels of serum creatine kinase. From a more recent study, conducted without suitable controls, Hicks and DeWeese (39) concluded that verapamil potassium cardioplegia protected well against postoperative abnormalities in cardiac conduction. Kaplan et al. (44), however, did not advocate the addition of verapamil to potassium cardioplegia. In their small group of patients, it increased the need for pacing and inotropic drugs.

Vouhé et al. (72) studied the effects of diltiazem in conjunction with profound hypothermia during aortic cross-clamping. They found it safe; the drug reduced the release of myocardial enzyme and probably improved hemodynamic function immediately after bypass. Christakis et al. (15) also observed an improvement in biochemical variables and cardiac index with diltiazem cardioplegia. However, they reported potent negative inotropy and prolonged electromechanical arrest. Similarly, Barner et al. (5) noted that the addition of diltiazem to cold blood potassium cardioplegia prolonged the atrioventricular node recovery time.

From these trials and from preliminary data (35), the overall picture emerges that calcium entry blockers can contribute to protection during open-heart operations.

Calcium Entry Blockers for Regional Cardioplegia

In the last decade, percutaneous transluminal coronary angioplasty (PTCA) has established itself in the armamentarium of the cardiologist. He introduces a catheter through a systemic artery to dilate a stenotic artery by controlled inflation of a balloon. For some time only a proximal stenosis in a single vessel could be treated, provided the patients had stable angina and normal ventricular function. Today treatment is less conservative. Serruys et al. (67) reported that they use PTCA also for patients with unstable angina and diminished ventricular function. Various obstructed vessels can be dilated. The total period of occlusion can exceed 10 min.

Table 3. Prevention by calcium entry blockers of damage during/after aorto-coronary bypass grafting (clinical trials)

Drug	Variables tested	Result	Ref.
Diltiazem	Hemodynamics/enzyme release	+	72
	P/hemodynamics/enzyme release	+	15
	Hemodynamics/enzyme release	±	5
Lidoflazine	Hemodynamics/P/glycogen/EM	+	26
Nifedipine	Hemodynamics/enzyme release	+/±	19
	Hemodynamics/P	NS/+	27
Verapamil	Enzyme release	+	40
	Hemodynamics/enzyme & lactate release	NS	44
	Hemodynamics	*	32

+ indicates added protection during bypass grafting or valve replacement. * A negative effect was reported, but see criticism by Clark (17). Abbreviations: EM = electron microscopy; NS = no significant improvement; P = myocardial high-energy phosphates.

After PTCA, substantial amounts of lactate and ATP catabolites are demonstrable in the myocardial efflux (63). It is therefore obvious that protection of the heart during such an intervention is useful. Local cardioplegia with calcium antagonists is now being tested. Calcium entry blockers can prevent arterial spasm during coronary angioplasty. It seems as if the time necessary for dilation can be prolonged safely by intracoronary application of nifedipine (67). In fact, Pop et al. (63) reported that this drug reduced lactate production during the procedure.

Diltiazem, administered intracoronarily, reduced severity and time of onset of ischemic ECG changes as well as anginal pain during PTCA (47). However, this anti-ischemic effect remained unexplained. Protection of the heart with calcium entry blockers during angioplasty is only in the initial research phase.

Calcium Antagonists as Insurance Policy

Hearse (35) stressed that proper hypothermia is relatively easy under in the heart laboratory conditions. That is certainly not always possible in the clinical setting, where one observes warm spots. Also, problems exist because of a varying degree of cooling. Calcium antagonists could then act as an insurance policy against regions of poor perfusion or unexpected warming (35), when administered adjunct to hypothermic cardioplegia. Furthermore, some of these drugs protect against vasoconstriction by microscopically small particles in unfiltered cardioplegic solutions (36). This is the reason Henry (38) called them "particle antagonists". He made it clear that the action of calcium entry blockers goes beyond just another cardioplegic maneuver. These drugs seem to play an important role in protecting the heart during the periods before and after cardiopulmonary bypass. During reperfusion, the cardioplegic protection by hypothermia and potassium is no longer effective. It becomes important to interfere pharmacologically with minimal disturbance of pump activity. Calcium antagonists could exert their salutary action partly by protecting the vessel wall. Nifedipine attenuates the strong contracture of isolated coronary arteries, produced by a cardioplegic solution (14). Finally, some of these drugs could be beneficial to inhibit rhythm disturbances during reperfusion (37,71).

Administration: What, and How Much?

One has to keep in mind that certain calcium antagonists (verapamil, diltiazem) affect the electrical properties of the heart. Nifedipine and verapamil show a similar ability to protect the rat myocardium against global ischemia. However, they exhibit significant differences in mode and characteristics of recovery during early reperfusion (78).

The potency to inhibit calcium influx and to dilate coronary vessels is high for the dihydropyridines. The relative potency is nifedipine > diltiazem > verapamil. Another point of consideration is the composition of the cardioplegic solution to which a calcium antagonist is adjunctive. St. Thomas' Hospital solution contains 16 mM Mg^{2+}, a physiologic calcium blocker (2). Some cardioplegic solutions, on the other hand, are devoid of this ion.

It is sensible to give calcium antagonists in a moderate dose together with the cardioplegic solution before open-heart operation. For nifedipine, this is probably 200 to 300 ug/L, using about 1.5 L of cardioplegic solution (19,27). A high concentration of calcium antagonist strongly depresses the heart. Problems during weaning from extracorporeal circulation or in the postoperative period could occur. Difficulties could arise during the reperfusion phase if calcium antagonists wash out very slowly. The biological half-life for i.v. administered verapamil and diltiazem is 4 to 5 h, that of nifedipine 1 to 2 h (53). Verapamil and bepridil accumulate in

272   J. W. de Jong

muscle cells (60). The human heart extracts the former rapidly (32). The latter binds very strongly to membranes. It will take about 1 h to remove bepridil from myocardial tissue. In contrast, nifedipine and diltiazem permeate more slowly (60) and their displacement is rapid (see ref. 34). Bepridil and nisoldipine wash very slowly from the tubing system used in a perfusion apparatus (24). The long-lasting cardiac depression by nifedipine, observed (6) in an NMR-machine with long tubings, could be due to drug release. Therefore, it is important to check the affinity of calcium antagonists for the extracorporeal circuit to which they have access.

When, How, and to Whom?

When should calcium entry blockers be given to minimize ischemic damage effectively? Prophylactic use, i.e., administration before flow reduction, is strongly advisable (20,22,59). Calcium antagonists, given i.v. to animals before ischemia or as a component of hypothermic hyperkalemic cardioplegia, hypothermic normokalemic cardioplegia, or normothermic normokalemic cardioplegia, show salutary effects on the heart in most cases. Preusse et al. (64) claimed that pretreatment with verapamil followed by calcium-free cardioplegic solution has a "membrane-labilizing" effect. However, because of the risk of creating the calcium paradox, one shouldn't use calcium-free cardioplegia.

The calcium antagonist should be present before reperfusion starts. It is debatable whether it makes sense to treat patients per os for several weeks preoperatively. Some groups treat only high-risk surgical patients with these drugs. On the basis of their study with nifedipine, Casson et al. (12) suggested to continue the drug up to the time of operation. Verapamil pretreatment before aortic crossclamping may potentiate the myocardial preservation achieved by cardioplegia (4).

Conclusion

Many studies, mainly conducted in animals, show that calcium entry blockers are useful in open-heart surgery as adjunct of potassium cardioplegia and hypothermia. Optimal conditions have to be determined in clinical trials before their routine use can be advocated. To create local cardioplegia, calcium antagonists given intracoronary before transluminal angioplasty, are promising. Let's consider Buckberg's words: "The future does not lie in providing the value of cardioplegia, but in understanding better the deleterious effects of ischemia and reperfusion injury so that current cardioplegic techniques can be modified to avoid these effects" (8).

Acknowledgements

I am grateful to Ms. M.J. Kanters-Stam and Ms. M.H.A. Zweserijn for their secretarial assistance, and to P.W. Achterberg, Ph.D., for his criticism.

References

1. Allen BS, Okamoto F, Buckberg GD, Acar C, Partington M, Bugyi H, Leaf J: Studies of controlled reperfusion after ischemia. IX. Reperfusate composition: Benefits of marked hypocalcemia and diltiazem on regional recovery. J Thorac Cardiovasc Surg 92:564-572, 1986
2. Altura BM, Altura BT, Carella A, Gebrewold A, Murakawa T, Nishio A: Mg^{2+} - Ca^{2+} interaction in contractility of vascular smooth muscle: Mg^{2+} versus organic calcium channel blockers on myogenic tone and agonist-induced responsiveness of blood vessels. Can J Physiol Pharmacol 65:729-745, 1987
3. Baker JE, Hearse DJ: The temperature-sensitivity of slow calcium channel blockers in relation to their effect upon the calcium paradox. Eur Heart J 4, Suppl H:97-103, 1983
4. Baraka A, Usta N, Baroody M, Haroun S, Dagher I, Haddad R: Verapamil pretreatment before aortic cross-clamping in patients undergoing coronary artery bypass graft. Anesth Analg 66:560-564, 1987

5. Barner HB, Swartz MT, Devine JE, Williams GA, Janosik D: Diltiazem as an adjunct to cold blood potassium cardioplegia: A clinical assessment of dose and prospective randomization. Ann Thorac Surg 43:191-197, 1987

6. Bernard M, Menasché P, Fontanarava E, Canioni P, Grousset C, Piwnica A, Cozzone P: Effect of nifedipine in hypothermic cardioplegia: a phosphorus-31 nuclear magnetic resonance study. Clin Chim Acta 152: 43-53, 1985

7. Boe SL, Dixon CM, Sakert TA, Magovern GJ: The control of myocardial Ca^{++} sequestration with nifedipine cardioplegia. J Thorac Cardiovasc Surg 84:678-683, 1982

8. Buckberg GD: A proposed "solution" to the cardioplegic controversy. J Thorac Cardiovasc Surg 77:803-815, 1979

9. Buckberg GD, Olinger GN, Mulder DG, Maloney Jr JV: Depressed postoperative cardiac performance. Prevention by adequate myocardial protection during cardiopulmonary bypass. J Thorac Cardiovasc Surg 70:974-988, 1975

10. Buhrman C, Molter DW, German LD, Lowe E: Heart block associated with high magnesium cardioplegia. J Am Coll Cardiol 9:7a, 1987 (Abstr)

11. Bush LR, Li Y-P, Shlafer M, Jolly SR, Lucchesi BR: Protective effects of diltiazem during myocardial ischemia in isolated cat hearts. J Pharmacol Exp Ther 218:653-661, 1981

12. Casson WR, Jones RM, Parsons RS: Nifedipine and cardiopulmonary bypass. Post-bypass management after continuation or withdrawal of therapy. Anaesthesia 36:1197-1201, 1984

13. Cavero I, Boudot J-P, Feuvray D: Diltiazem protects the isolated rabbit heart from the mechanical and ultra-structural damage produced by transient hypoxia, low-flow ischemia, and exposure to Ca^{++}-free medium. J Pharmacol Exp Ther 226:258-268, 1983

14. Chiavarelli M, Chiavarelli R, Macchiarelli A, Carpi A, Marino B: Calcium entry blockers and cardioplegia: Interaction between nifedipine, potassium, and hypothermia. Ann Thorac Surg 41:535-541, 1986

15. Christakis GT, Fremes SE, Weisel RD, Tittley JG, Mickle DAG, Ivanov J, Madonik MM, Benak AM, McLaughlin PR, Baird RJ: Diltiazem cardioplegia. A balance of risk and benefit. J Thorac Cardiovasc Surg 91:647-661, 1986

16. Christlieb IY, Clark RE, Sobel BE. Three-hour preservation of the hypothermic globally ischemic heart with nifedipine. Surgery 90:947-955, 1981

17. Clark RE: Verapamil, cardioplegia and coronary artery bypass grafting. Ann Thorac Surg 41:585-586, 1986

18. Clark RE, Christlieb IY, Spratt JA, Henry PD, Fischer AE, Williamson JR, Sobel BR: Myocardial preservation with nifedipine: A comparative study at normothermia. Ann Thorac Surg 31:3-19, 1981

19. Clark RE, Magovern GJ, Christlieb IY, Boe S: Nifedipine cardioplegia experience: Results of a 3-year cooperative clinical study. Ann Thorac Surg 36:654-663, 1983

20. De Jong JW: Timely administration of nisoldipine essential for prevention of myocardial ATP catabolism. Eur J Pharmacol 118:53-59, 1985

21. De Jong JW: Cardioplegia and calcium antagonists: A review. Ann Thorac Surg 42:593-598, 1986

22. De Jong JW, Harmsen E, De Tombe PP: Diltiazem administered before or during myocardial ischemia decreases adenine nucleotide catabolism. J Mol Cell Cardiol 16:363-370, 1984

23. De Jong JW, Harmsen E, De Tombe PP, Keijzer E: Nifedipine reduces adenine nucleotide breakdown in ischemic rat heart. Eur J Pharmacol 81:89-96, 1982

24. De Jong JW, Huizer T, Tijssen JGP: Energy conservation by nisoldipine in ischaemic heart. Br J Pharmacol 118:943-949, 1984

25. Ferguson Jr TB, Damiano RJ, Smith PK, Buhrman WC, Cox JL: The electrophysiological effects of calcium channel blockade during standard hyperkalemic hypothermic cardioplegic arrest. Ann Thorac Surg 41:622-629, 1986

26. Flameng W, Borgers M, Van der Vusse GJ, De Meyere R, Van der Meersch F, Thoné F, Suy R: Cardioprotective effects of lidoflazine in extensive aorta-coronary bypass grafting. J Thorac Cardiovasc Surg 83:758-768, 1983

27. Flameng W, De Meyere R, Daenen W, Sergeant P, Ngalikpima V, Geboers J, Suy R, Stalpaert G: Nifedipine as an adjunct to St. Thomas' Hospital cardioplegia. A double-blind placebo controlled, randomized clinical trial. J Thorac Cardiovasc Surg 91:723-731, 1986

28. Fleckenstein A: Calcium antagonism in heart and smooth muscle. Experimental facts and therapeutic prospects. New York: Wiley, 1983:36-40

29. Fox KAA, Bergmann SR, Sobel BE: Pathophysiology of myocardial reperfusion. Annu Rev Med 36:125-144, 1985
30. Fukunami M, Hearse DJ: Temperature-dependency of nifedipine as a protective agent during cardioplegia in the rat. Cardiovasc Res 19:95-103, 1985
31. Godfraind T: Classification of calcium antagonists. Am J Cardiol 55:11B-23B, 1987
32. Guffin AV, Kates RA, Holbrook GW, Jones EL, Kaplan JA: Verapamil and myocardial preservation in patients undergoing coronary artery bypass surgery. Ann Thorac Surg 41:587-591, 1986
33. Guyton RA, Dorsey LM, Colgan TK, Hatcher Jr CR: Calcium-channel blockade as an adjunct to heterogeneous delivery of cardioplegia. Ann Thorac Surg 35:626-632, 1983
34. Hearse DJ: The protection of the ischemic myocardium during open heart surgery: Cold chemical cardioplegia. Eur Heart J 4, Suppl C:77, 1983 (Abstr)
35. Hearse DJ: Discussion of Symp. on Calcium Antagonists and Cardioplegia in Cardiac Surgery. Eur Heart J 4, Suppl C:83-86, 1983
36. Hearse DJ, Erol C, Robinson LA, Maxwell MP, Braimbridge MV: Particle-induced coronary vasoconstriction during cardioplegic infusion - Characterization and possible mechanisms. J Thorac Cardiovasc Surg 89:428-438, 1985
37. Hearse DJ, Yamamoto F, Shatlock MJ: Calcium antagonists and hypothermia: The temperature dependency of the negative inotropic and anti-ischemic properties of verapamil in the isolated rat heart. Circulation 70, Suppl I:I54-I64, 1984
38. Henry PD: Summary of Symp. on Calcium Antagonists and Cardioplegia in Cardiac Surgery. Eur Heart J 4, Suppl C:90-91, 1983
39. Hicks Jr GL, DeWeese JA: Verapamil potassium cardioplegia and cardiac conduction. Ann Thorac Surg 39:324-328, 1985
40. Hicks Jr GL, Salley RK, DeWeese JA: Calcium channel blockers: An intraoperative and postoperative trial in women. Ann Thorac Surg 37:319-323, 1984
41. Huizer T, De Jong JW, Achterberg PW: Protection by bepridil against myocardial ATP-catabolism is probably due to negative inotropy. J Cardiovasc Pharmacol, 10:55-61, 1987
42. Jennings RB, Steenbergen Jr C: Nucleotide metabolism and cellular damage in myocardial ischemia. Annu Rev Physiol 47:727-749, 1985
43. Johnson RG, Jacocks MA, Aretz TH, Geffin GA, O'Keefe DD, DeBoer LWV, Guyton RA, Fallon JT, Daggett WM: Comparison of myocardial preservation with hypothermic potassium and nifedipine arrest. Circulation 66, Suppl I:I73-I80, 1982
44. Kaplan JA, Guffin AV, Jones EL, Kates R, Holbrook GW: Verapamil and myocardial preservation in patients undergoing coronary artery bypass surgery. In: Althaus U, Burckhardt D, Vogt E, eds: Proc Calcium-Antagonismus/Int Symp on Calcium Antagonism, Frankfurt/Main: Universimed, 1984:228-237
45. Kates RA, Dorsey LM, Kaplan JA, Hatcher Jr CR, Guyton RA: Pretreatment with lidoflazine, a calcium-channel blocker. Useful adjunct to heterogeneous cold potassium cardioplegia. J Thorac Cardiovasc Surg 85:278-286, 1983
46. Katsumoto K, Inoue T: Experimental comparison of the efficacy of various agents on the enhancement of myocardial protection. Jpn Circ J 47:356-362, 1983
47. Kober G, Kästner R, Hopf R, Kaltenbach M: Die direkte myokardiale antiischämische Wirkung von Diltiazem beim Menschen. Z Kardiol 75:386-393, 1986
48. Krukenkamp IB, Silverman NA, Sorlie D, Pridjian A, Levitsky S: Temperature-specific effects of adjuvant diltiazem therapy on myocardial energetics following potassium cardioplegic arrest. Ann Thorac Surg 42:675-680, 1986
49. Laird-Meeter K, Van den Brand MJBM, Serruys PW, Penn OCKM, Haalebos MMP, Bos E, Hugenholtz PG: Reoperation after aortocoronary bypass procedure. Results in 53 patients in a group of 1041 with consecutive first operations. Br Heart J 50:157-162, 1983
50. Landymore RW, Marble AE, Trillo A, Faulkner G, MacAulay MA, Cameron C: Prevention of myocardial electrical activity during ischemic arrest with verapamil cardioplegia. Ann Thorac Surg 43:534-538, 1987
51. Lupinetti FM, Hammon Jr JW, Huddleston CB, Boucek Jr RJ, Bender Jr HW: Global ischemia in the immature canine ventricle. Enhanced protective effect of verapamil and potassium. J Thorac Cardiovasc Surg 87:213-219, 1984
52. Magovern GJ, Dixon CM, Burkholder JA: Improved myocardial protection with nifedipine and potassium-based cardioplegia. J Thorac Cardiovasc Surg 82:239-244, 1981

53. McAllister Jr RG, Hamann SR, Blouin RA: Pharmacokinetics of calcium-entry blockers. Am J Cardiol 55:30B-40B, 1985
54. Menasché P, Groussett C, Piwnica A: Bases ioniques des solutions cardioplégiques. II. Influence de la formule ionique d'une solution cardioplégique sur la préservation métabolique et fonctionelle du myocarde ischémique. Evalution expérimentale par résonance magnétique nucléaire du phosphore 31 et applications à la chirurgie cardiaque. Arch Mal Coeur 76:1465-1474, 1983
55. Moores WY, Mack JW, Dembitsky WP, Heydorn WH, Daily PO: Quantitative evaluation of the myocardial preservative characteristics of nifedipine during hypothermic myocardial ischemia. J Thorac Cardiovasc Surg 90:912-920, 1985
56. Morishita Y, Saigenji H, Umebayashi Y, Higashi T, Taira A, Goto M: Potassium-verapamil cardioplegia for myocardial protection: An experimental evaluation through preservation and transplantation. Jpn J Surg 13:524-529, 1983
57. Nayler WG: Cardioprotective effects of calcium ion antagonists in myocardial ischemia. Clin Invest Med 3:91-99, 1980
58. Nayler WG: Protection of the myocardium against postischemic reperfusion damage. The combined effect of hypothermia and nifedipine. J Thorac Cardiovasc Surg 84:897-905, 1982
59. Nayler WG, Panagiotopoulos S, Elz JS, Sturrock WJ: Fundamental mechanisms of action of calcium antagonists in myocardial ischemia. Am J Cardiol 59:75B-83B, 1987
60. Pang DC, Sperelakis N: Nifedipine, diltiazem, bepridil and verapamil uptakes into cardiac and smooth muscles. Eur J Pharmacol 87:199-207, 1983
61. Pinsky WW, Lewis RM, McMillin-Wood JB, Hara H, Hartley CJ, Gillette PC, Entman ML: Myocardial protection from ischemic arrest: Potassium and verapamil cardioplegia. Am J Physiol 240:H326-H335, 1981
62. Piper HM, Huetter JF, Spieckermann PG: Temperature dependence of calcium antagonist action. Arzneimittelforsch 35:1495-1498, 1985
63. Pop G, Serruys PW, Piscione F, De Feyter PJ, Van den Brand M, Huizer T, De Jong JW, Hugenholtz PG: Regional cardioprotection by subselective intracoronary nifedipine is not due to enhanced collateral flow during coronary angioplasty. Int J Cardiol 15:27-41, 1987
64. Preusse CJ, Gebhard MM, Schnabel A, Ulbricht LJ, Bretschneider HJ: Post-ischemic myocardial function after pre-ischemic application of propranolol or verapamil. J Cardiovasc Surg 25:158-164, 1984
65. Ruigrok TJC, Boink ABTJ, Slade A, Zimmerman ANE, Meijler FL, Nayler WG: The effect of verapamil on the calcium paradox. Am J Pathol 98:769-782, 1980
66. Serruys PW, Brower RW, Ten Katen HJ, Meester GT: Recovery from circulatory depression after coronary artery bypass surgery. Eur Surg Res 12:369-382, 1980
67. Serruys PW, Van den Brand M, Brower RW, Hugenholtz PG: Regional cardioplegia and cardioprotection during transluminal angioplasty, which role for nifedipine? Eur Heart J 4, Suppl C:115-121, 1983
68. Shahian DM: Concepts and techniques of myocardial protection for adult open heart surgery. Surg Clin North Am 65:323-346, 1985
69. Spencer FC: Intellectual creativity in thoracic surgeons. J Thorac Cardiovasc Surg 86:163-179, 1983
70. Standeven JW, Jellinek M, Menz LJ, Kolata RJ, Barner HB: Cold blood potassium diltiazem cardioplegia. J Thorac Cardiovasc Surg 87:201-212, 1984
71. Van Gilst WH, Boonstra PW, Terpstra JA, Wildevuur CRH, De Langen CDJ: Improved recovery of cardiac function after 24 h of hypothermic arrest in the isolated rat heart: Comparison of a prostacyclin analogue (ZK 36 374) and a calcium entry blocker (diltiazem). J Cardiovasc Pharmacol 7:520-524, 1985
72. Vouhé PR, Loisance DY, Aubry P, Piszker G, Cazor JL, Cachera J-P: Double-blind evaluation of the intraoperative use of diltiazem during coronary artery surgery. In: Just H, Schroeder JS, eds: Advances in Clinical Applications of Calcium Antagonist Drugs. Int Diltiazem Workshop. Düsseldorf: Excerpta Med, 1985:91-104
73. Wallis DE, Gierke LW, Scanlon PJ, Wolfson PM, Kopp SJ: Sustained postischemic cardiodepression following magnesium-diltiazem cardioplegia. Proc Soc Exp Biol Med 182:375-385, 1986
74. Watts JA, Maiorano LJ, Maiorano PC: Comparison of the protective effects of verapamil, diltiazem, nifedipine, and buffer containing low calcium upon global myocardial ischemic injury. J Mol Cell Cardiol 18:255-263, 1986

75. White BC, Winegar CD, Wilson RF, Krause GS: Calcium blockers in cerebral resuscitation. J Trauma 23:788-793, 1983
76. Yamamoto F, Manning AS, Braimbridge MV, Hearse DJ: Calcium antagonists and myocardial protection: Diltiazem during cardioplegic arrest. Thorac Cardiovasc Surg 31:369-373, 1983
77. Yamamoto F, Manning AS, Braimbridge MV, Hearse DJ: Nifedipine and cardioplegia: Rat heart studies with the St. Thomas' cardioplegic solution. Cardiovasc Res 17:719-727, 1983
78. Yamamoto F, Manning AS, Crome R, Braimbridge MV, Hearse DJ: Calcium antagonists and myocardial protection: A comparative study of the functional, metabolic and electrical consequences of verapamil and nifedipine as additives to the St. Thomas' cardioplegic solution. Thorac Cardiovasc Surg 33:354-359, 1985
79. Yoon SB, McMillin-Wood JB, Michael LH, Lewis RM, Entman ML: Protection of canine cardiac mitochondrial function by verapamil-cardioplegia during ischemic arrest. Circ Res 56:704-708, 1985

Chapter 25

The Effect of Anesthetics on
Myocardial Adenine Nucleotides

R.G. Merin, Department of Anesthesiology, University of
Texas Medical School at Houston, Houston, Texas, U.S.A.

The major interest in the effect of anesthetics on high-energy phosphates (adenine nucleotides) in the heart stems from the search for the mechanism of the negative inotropic effect of anesthetics and the interest in preserving myocardial energy during cardiac ischemia. Although the number of studies published is rather small, the results when the preparation has been stable are consistent. Anesthetics do not appear to change the concentration of high-energy phosphates in normal hearts. There is some evidence that barbiturates in high concentrations may interfere with myocardial energy synthesis and decrease high-energy phosphates in hypoxic or ischemic hearts and additional evidence that perhaps cardiac depressant anesthetics such as halothane may preserve high-energy phosphates in ischemic hearts. However, much more investigation is necessary in the ischemic heart in order to come to definite conclusions.

Introduction

Why should there be interest in the effect of anesthetics on high-energy phosphates (adenine nucleotides) in the heart? In my opinion, there are basically two reasons. First, all potent inhalation anesthetics are dose-related myocardial depressants. However, the mechanism of this depression remains elusive (12,18). It seems likely that one or more sites lie in the supply of energy for myocardial muscle contraction or in the use of that energy as depicted in Fig. 1. Note that the myocardial high-energy phosphates, adenosine triphosphate (ATP) and creatine phosphate (CP), lie in the middle of these two opposing processes. Consequently, investigators interested in the mechanism of the negative inotropic effect of anesthetics have sought to measure the concentration of the storage form of energy for the heart, these high-energy phosphates, in an attempt to determine whether interference in the energy supply for the work of the heart by anesthetics might result in a decrement in the major energy source for heart muscle contraction, the ATP and CP.

The prevalence of patients needing anesthesia and surgery with ischemic heart disease in the western world dictates the second reason for the interest in the effect of anesthetics on myocardial high-energy phosphates. Inasmuch as the continuing high energy demand of the heart dictates that oxidative metabolism be maintained for adequate synthesis of the energy supply of the heart (ATP and CP), the ultimate pathophysiologic defect produced by myocardial ischemia must be a deficit of these energy sources (6). In order to decrease the mortality and morbidity of surgery and anesthesia in these high-risk patients, anesthetics which preserve high-energy phosphates, particularly in the face of ischemia, would be desirable. Consequently, all published reports on the effects of anesthetics on myocardial high-energy phosphates have been related to either the negative inotropic effect of anesthetics or the relationship between anesthetics, ischemia and myocardial high-energy phosphates.

Anesthetic Effects

The paucity of published investigations of the effect of anesthetics on myocardial high-energy phosphates is certainly due to the difficulty in assay. Until recently (see Chapters 12-14), accurate measurement necessitated tissue excision and rapid freezing to avoid the breakdown of ATP and CP. In addition, the biochemical analysis was tedious and demanding. More recently, the availability of nuclear magnetic resonance (NMR) and positron emission tomography has enabled studies to be done noninvasively. However, again, the techniques are expensive and cumbersome.

The earliest published study of the effect of anesthetics on myocardial high-energy phosphates investigated the effect of halothane in cultured neonatal rat heart cells. Stong et al. (19) found no change in ATP pool size in their cultured heart cells under the influence of as high as a 5% concentration of halothane. However, they did document a decrease in the incorporation of inorganic phosphate (P_i) into the ATP and suggested that halothane was decreasing the rate of synthesis of ATP as a possible mechanism for the negative inotropic effect. However, they admitted that the tight feedback control between energy supply and energy demand could also account for the decreased incorporation of P_i into ATP under the influence of halothane (13,20,21). The author, working together with the editor of this book, studied the effect of halothane on myocardial function and metabolism in intact open-chest swine (14). A marked depression in left ventricular function as indicated by more than 50% decrease in peak contractile element velocity and maximum velocity of shortening, and a 40% decrease in cardiac output and stroke volume, was accompanied by absolutely no change in tissue concentrations of ATP or CP. In a study using one of the new techniques, magnetic resonance spectroscopy, Murray et

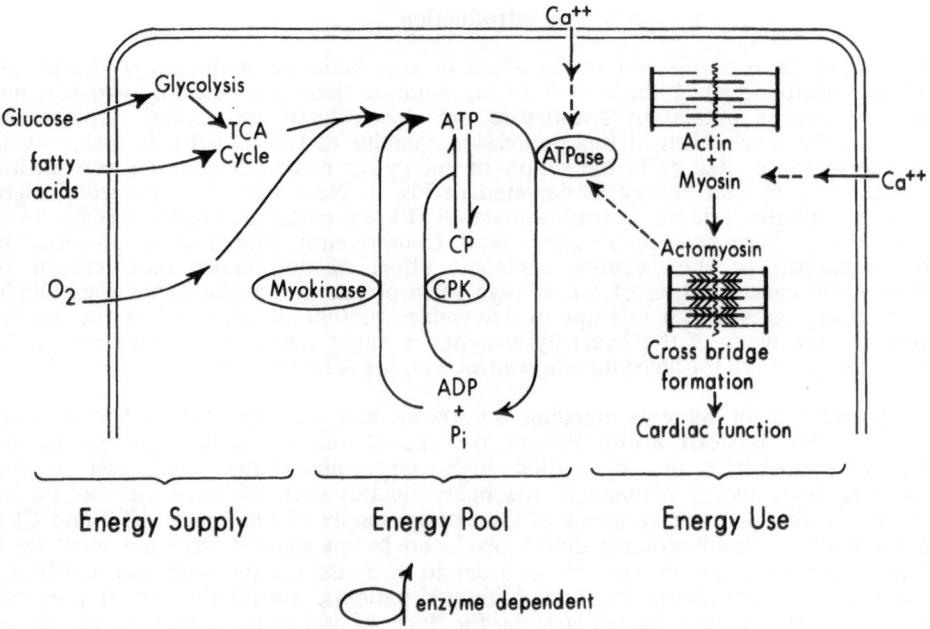

Fig. 1. Cardiac energetics: TCA = tricarboxylic acid; ATP = adenosine triphosphate; CP = creatine phosphate; ADP = adenosine diphosphate; P_i = inorganic phosphate; CPK = creatine kinase. Used with permission from ref. 13.

al. examined the effect of halothane on high-energy phosphates in a rabbit Langendorff preparation (15). In addition, they also examined intracellular pH and the creatine kinase reaction. Clinically relevant concentrations of halothane (0.016 mM) produced a significant decrease in left ventricular developed pressure, beta ATP and CP. However, 60 min of perfusion in a separate group of rat hearts without halothane also produced a significant decrease in high-energy phosphates. Although the authors reported a decrease in the forward rate constant for the transfer of phosphate from CP to ATP by halothane, no mention was made of whether that effect was also time dependent.

In a far superior study, McAuliffe and Hickey also examined the effect of halothane on high-energy phosphates in the heart (11). Inasmuch as they are pediatric anesthetists, they used neonatal rabbit hearts, prepared as a Langendorff preparation with an intraventricular balloon for measurement of performance and varying of preload. 1.5% halothane produced a 50% decrease in left ventricular pressure and dP/dt, but there was no change in ATP, CP, P_i or intracellular pH. In addition, in a separate series of hearts, they correlated measured oxygen consumption with varying oxygen demand induced by changing the volume of the left ventricular balloon. Both with and without halothane, there was a good correlation between oxygen demand and oxygen consumption. Consequently, they concluded that the decreased oxygen demand produced by halothane was matched by decreasing high-energy phosphate supplies and that there was no evidence of a primary effect of halothane on energy supply in the neonatal rabbit heart.

Preliminary studies from our laboratory at the University of Texas Medical School in Houston have also suggested that anesthetics do not produce an effect on myocardial high-energy phosphates. Using an isolated perfused working rat heart, we have previously shown that halothane produced a dose-related decrease in glucose utilization along with the negative inotropic effect (2). In further ongoing investigations, we have also measured the concentrations of ATP and CP under the influence of negative inotropic doses of both halothane (1.6 volumes %) and isoflurane (2.4 volumes %). These equipotent anesthetic concentrations depressed aortic pressure, cardiac output, coronary flow, and myocardial oxygen consumption (Table 1). Note that the effect of halothane was greater than that of isoflurane. However, even with the marked depression of cardiac function, there was no change in tissue ceoncentrations of ATP, CP or the calculated energy charge ([ATP] + 0.5 [ADP])/([ATP] + [ADP] + [AMP]) (Table 2).

It would appear, therefore, that the potent inhalation anesthetics, halothane and isoflurane, do not interfere in the maintenance of normal myocardial high-energy phosphate levels as a mechanism for their negative inotropic effect.

Table 1. Effects of halothane and isoflurane on working rat heart

Variable	Control	1.6% H	2.4% I
MAP (mmHg)	82 ± 17	49 ± 10[*][+]	61 ± 13[*][+]
Cardiac output (ml/min)	54 ± 8	39 ± 10[*][+]	50 ± 10
Heart rate (beats/min)	259 ± 40	246 ± 48	238 ± 33
Coronary flow (ml/min)	16 ± 4	11 ± 5[*][+]	14 ± 5
O_2 consumption (min/g)	8.2 ± 0.6	3.6 ± 0.3[*]	6 ± 10

Means ± S.D. MAP = mean arterial pressure; [*]$p < 0.05$ vs. Control; [+]$p < 0.05$ vs. 0.8% halothane (H) or 1.2% isoflurane (I).

Anesthetic Effects on the Ischemic Heart

The cardiac anesthesia/surgery laboratories from Duke University Medical Center investigated the interaction of potent inhalation anesthetics and myocardial ischemia. Halothane produced no change in high-energy phosphate levels either before or after ischemia in a working rat-heart preparation (16). Time to ischemic contracture was also unchanged in this preparation by halothane. On the other hand, in a Langendorff preparation, isoflurane increased tissue content of ATP before ischemia and at the time of ischemic contracture, although the time to ischemic contracture was not different (4). Following reperfusion with the Krebs-Henseleit buffer, functional recovery and metabolic restitution of ATP levels were greater in the hearts treated with isoflurance. In their last publication (and only full paper), the same group investigated the effects of enflurane in the Langendorff perfused rat heart (5). The effect was intermediate between isoflurane and halothane. There was no effect of enflurane on pre-ischemic CP or ATP concentrations nor on high-energy phosphates at the time of ischemic contracture. In addition, there was no prolongation of the time to ischemic contracture. However, following reperfusion, both recovery of peak pressure and tissue ATP and CP were improved in the animals treated with enflurane.

Another author in this volume investigated the effect of thiopental on myocardial tissue levels of ATP and CP during total ischemia (17). Ruigrok and co-workers showed that 100 mg·L^{-1} of thiopental significantly decreased myocardial concentrations of CP before ischemia and at various times during ischemia, and concentrations of ATP only during ischemia. They also examined creatine kinase release and the myocardial ultrastructure of hearts with and without thiopental treatment during hypoxia and low flow ischemia. Thiopental decreased creatine kinase release during hypoxia and during low-flow ischemia at a normal pH, but had no effect in total ischemia or low-flow ischemia at a low pH. The authors hypothesized that increased hydrogen ions produced by a low pH perfusate or total ischemia favored the entry of thiopental into the cell where the well-known barbiturate inhibition of mitochondrial function reduced ATP production. At normal pH, the cardiodepressant effect of the anesthetic decreased myocardial oxygen demand so that there was no net effect on high-energy phosphates.

From the laboratories of Kumazawa in Yamanashi Medical College Japan, came a series of investigations looking at the interaction between various anesthetics and ischemia in a rat-heart/lung preparation. Neither morphine (2 mg·kg^{-1}) nor fentanyl (100 ug·kg^{-1}) had an effect on myocardial ATP concentration (8). In a companion study, the same group examined the effect of fentanyl (1.2 ug·ml^{-1}) and the same dose of fentanyl plus diazepam (3 ug·ml^{-1}) in both an ischemic and non-ischemic heart/lung preparation (10). ATP concentrations were measured after 10 min of global ischemia and 20 min of reperfusion. There was no difference in the ATP concentrations between the control and the two anesthetic groups. Likewise, there were minimal differences in cardiac function although in the fentanyl group, recov-

Table 2. No effect of halothane and isoflurane on high-energy phosphates in working rat heart

Condition	Adenine triphosphate (umol/g protein)	Creatine phosphate (umol/g protein)	Energy charge
Control	23.8 ± 0.2	21.1 ± 1.2	0.817 ± 0.005
1.6% Halothane	19.3 ± 2.0	20 ± 2	0.76 ± 0.03
2.4% Isoflurane	25.6 ± 1.2	21.9 ± 1.2	0.818 ± 0.009

The energy charge is ([ATP] + 0.5[ADP])/([ATP] + [ADP] + [AMP]).

ery from ischemia to a cardiac output of 30 ml·min⁻¹ and an aortic pressure of 80 mmHg was faster. In a similar preparation, the effects of two concentrations of thiopental (10 and 100 ug·ml⁻¹) were investigated (7). Although there were no significant changes in ATP and CP concentrations produced by either dose of thiopental, the 100 ug·ml⁻¹ dose did decrease the concentrations of both high-energy phosphates and significantly increased concentrations of ADP and AMP with a resultant decrease in the myocardial energy charge (see above). In addition, recovery was delayed by this concentration of thiopental. Thus, their results agreed with those reported by Ruigrok et al. (17). Finally, the effects of halothane and enflurane in the same model were investigated (20). The animals were given 1% halothane and 2% enflurane from calibrated vaporizers to the excised heart/lung preparation. Neither anesthetic produced a change in the concentration of ATP, ADP or AMP compared to the control.

Another Japanese worker investigated the effect of thoracic epidural anesthesia in a basally anesthetized open-chest dog preparation before and after complete occlusion of the left anterior descending coronary artery (22). Although thoracic epidural anesthesia decreased the ischemia-produced elevation of the ST segment and increased myocardial pH, there were no changes in ATP or CP either before or after the coronary artery occlusion. Occlusion did markedly decrease the concentration of both ATP and CP in the control basally anesthetized animal and after thoracic epidural anesthesia.

Summary and Conclusions

Although the number of investigations has been rather limited, and predominantly has examined the effect of halothane, the results are consistent concerning the effect of the potent inhalation anesthetics on myocardial high-energy phosphates. In spite of marked depression of cardiac function produced by these compounds, no change in tissue level of ATP or CP has been ascribed to the pharmacologic effects of the anesthetics themselves. However, it is becoming apparent that measuring tissue homogenate concentrations of the high-energy phosphates may not accurately reflect pharmacologic effects. Several groups of investigators have recently demonstrated that glycolytically produced ATP and the ATP synthesized as a result of oxidative phosphorylation may subserve different functions (1,3,23,24). Glycolytic ATP appears to be necessary for the maintenance of electrical integrity of the cardiac sarcolemma and particularly governs the state of the potassium channel (24), while oxidatively based ATP may be more important for the major demands of the contractile proteins (23). Thus, analysis of total ATP and CP may not reveal a specific effect mediated through glycolysis. This may be relevant for anesthetic effects inasmuch as our laboratory has demonstrated a specific interference in glycolysis produced by halothane (2). Obviously, further investigation along these lines with anesthetics would appear to be fruitful.

The development of ischemic contracture has been used to quantitate the degree of ischemia. Bricknell et al. demonstrated that maintenance of glycolytically produced ATP was more important in delaying or preventing ischemic contracture than oxidative ATP (1). They speculated that the energy necessary for sarcoplasmic reticular Ca⁺⁺ uptake might be preferentially derived from glycolysis, especially during ischemia. There is considerable ongoing controversy about the differential effects of the volatile anesthetics halothane, enflurane and isoflurane on the cardiac sarcoplasmic reticulum (18). These differences might explain the different effects of the three anesthetics on ischemia and ischemic contracture (4,5,16). However, considerably more studies are necessary to document and explain these effects.

References

1. Bricknell OL, Daries PS, Opie LH: A relationship between adenosine triphosphate, glycolysis and ischemic contracture in the isolated rat heart. J Mol Cell Cardiol 13:941-945, 1981
2. Cronau LH, Merin RG, De Jong JW: Effect of halothane on glucose utilization in the perfused working rat heart. J Cardiovasc Pharmacol, in press.
3. Doorey AJ, Barry WH: The efects of inhibition of oxidative phosphorylation and glycolysis on contractility and high energy phosphate content in cultured chick heart cells. Circ Res 53:192-201, 1983
4. Freedman B, Christian C, Ham D, Everson C, Wechsler A: Isoflurane and myocardial protection. Anesthesiology 59:A25, 1983 (Abstr)
5. Freedman BM, Ham D, Everson CT, Wechsler AS, Christian CM: Enflurance enhances post-ischemic functional recovery in the isolated rat heart. Anesthesiology 62:29-33, 1985
6. Hearse DJ, Crome R, Yellon DM, Wyse R: Metabolic and flow correlates of myocardial ischemia. Cardiavasc Res 17:452-458, 1983
7. Kashimoto S, Hinohara S, Tanaka Y, Kumazawa T: Effects of thiopental on cardiac energy metabolism in post-ischemic reperfusion in rat. J Anesth 1:77-81, 1987
8. Kashimoto S, Tanaka Y, Manabe M, Kumazawa T: Comparison of hemodynamic and metabolic effects of fentanyl and morphine in isolated rat heart lung preparations. Hiroshima J Anesth 21:19-25, 1985
9. Kashimoto S, Tsuji Y, Kumazawa T: Effects of halothane and enflurane on myocardial metabolism during post-ischemic reperfusion in the rat. Acta Anaesth Scand 31:44-47, 1987
10. Kashimoto S, Tsuji Y, Miyaji T, Kumazawa T: Effects of fentanyl and fentanyl-diazepam on myocardial metabolism in isolated rat heart lung preparation. Hiroshima J Anesth 22:297-304, 1986
11. McAuliffe JJ, Hickey PR: Effect of halothane on the steady state levels of high energy phosphates in the neonatal heart. Anesthesiology 67:231-235, 1987
12. Merin RG: Inhalation anesthetics and myocardial metabolism: Possible mechanisms of functional effects. Anesthesiology 39:216-255, 1973
13. Merin RG: Myocardial metabolism. In: Kaplan JA, ed: Cardiac anesthesia. Vol 2, Cardiovascular pharmacology. New York: Grune and Stratton, 1983:243-266
14. Merin RG, Verdouw, PD, De Jong JW: Dose-dependent depression of cardiac function and metabolism by halothane in swine. Anesthesiology 46:417-423, 1977
15. Murray PA, Blanck TJJ, Rogers MC, Jacobus WE: Effects of halothane on myocardial high-energy phosphate metabolism and intracellular pH utilizing ^{31}PNMR spectroscopy. Anesthesiology 67:649-653, 1987
16. Peyton R, Christian C, Fargraeus L, van Trigt P, Spray T, Pellom G, Pasque M, Wechsler A: Halothane and myocardial protection. Anesthesiology 57:A9, 1982 (Abstr)
17. Ruigrok TJC, Slade AM, Van der Meer P, De Moes D, Sinclair DM, Poole-Wilson PA, Meijler FL: Different effects of thiopental in severe hypoxia, total ischemia and low flow ischemia in rat heart muscle. Anesthesiology 63:172-178, 1985
18. Rusy BF, Komai H: Anesthetic depression of myocardial contractility: A review of possible mechanisms. Anesthesiology 67:745-766, 1987
19. Stong LJ, Hartzell CR, McCarl RL: Halothane and the beating response in ATP turnover rate of heart cells in tissue culture. Anesthesiology 42:123-132, 1975
20. Taegtmeyer H: Six blind men explore an elephant: Aspects of fuel metabolism in the control of tricarboxylic acid cycle activity in heart muscle. Basic Res Cardiol 79:322-366, 1984
21. Taegtmeyer H: Carbohydrate interconversions and energy production. Circulation 72, Suppl IV:1-8, 1985
22. Tsuchida H: Experimental study on the effects of thoracic epidural anesthesia on myocardial pH decrease and metabolic change induced by acute coronary artery occlusion. Sapporo Med J 56:143-155, 1987
23. Weiss J, Hiltbrand B: Functional compartmentation of glycolytic versus oxidative metabolism in isolated rabbit heart. J Clin Invest 75:436-447, 1985
24. Weiss JN, Lamp ST: Glycolysis preferentially inhibits ATP-sensitive K^+ channels in isolated guinea pig cardiac myocytes. Science 238:67-70, 1987

Chapter 26

Regeneration of Adenine Nucleotides in the Heart

P. van der Meer and J.W. de Jong, Cardiochemical Laboratory,
Thoraxcenter, Erasmus University Rotterdam, The Netherlands

The fast adenine nucleotide degradation, which takes place in the ischemic heart, contrasts with the slow regeneration after ischemia. There are four pathways for adenine nucleotide regeneration:
1. De novo synthesis. This is a slow process. IMP is built from small molecules in a number of steps. The process can be accelerated to some extent by administering ribose. Ribose is incorporated in 5-phosphoribosyl-1-pyrophosphate (PRPP) and the amount of PRPP is rate-limiting. Administration of 5-amino-4-imidazolecarboxamide riboside (AICAriboside), which can be converted to an intermediate of this pathway, has also been tried for a faster regeneration, but this compound produced a deterioration of heart function. The conversion of IMP, end product of both de novo synthesis and salvage pathway, to AMP is very slow.
2. Salvage of hypoxanthine or inosine. Inosine has to be catabolized to hypoxanthine before it is incorporated. IMP is formed from hypoxanthine in one step, with subsequent conversion to AMP. PRPP is rate-limiting, so adding ribose enhances the incorporation rate. Inosine has hemodynamic effects.
3. Adenine ribophosphorylation. Adenine is incorporated in AMP in one step. PRPP is again rate-limiting. No nucleotide conversion is necessary. Toxic effects of long-term adenine administration have been described.
4. Adenosine phosphorylation. This is the fastest pathway for adenine nucleotide regeneration. Adenosine causes renal vasoconstriction together with systemic vasodilation (hypotension). These adverse reactions limit the practical value of adenosine.
Even though there is limited clinical experience with some of these compounds that enhance postischemic ATP pool repletion, none has been examined for this purpose in humans.

Introduction

Ischemia leads to a rapid fall in the adenine nucleotide content of the heart. The rate of ATP decrease depends on factors like tissue perfusion (collaterals), heart function, temperature, species. The dephosphorylated breakdown products of ATP, the purine nucleosides and oxypurines, leak out of the myocardial cell. A simple scheme of ATP breakdown is shown in Fig. 1. The amount of adenine nucleotides will be roughly halved within the first hour of ischemia. A return of flow will cause a restoration of the adenine nucleotide levels (if the cells are not irreversibly damaged), but this takes days to accomplish.

The effects of ischemia on a cardiac myocyte are complex and their mechanisms have not fully been unravelled. There is a good correlation between the effect of myocardial protection during ischemia and the postischemic content of adenine nucleotides. Thus it is plausible that a fast restoration of ATP levels will promote the recovery of the postischemic heart. To regenerate ATP in the heart, four pathways are available, as was pointed out by Goldthwait (14) 30 years ago:

1. De novo synthesis;
2. Salvage of hypoxanthine or inosine;
3. Adenine ribophosphorylation;
4. Adenosine phosphorylation.

A scheme of these pathways is shown in Fig. 2.

1. De Novo Synthesis

This pathway involves the production of the purine skeleton from small molecules in 10 steps (see Fig. 3). The first step is the formation of 5-phosphoribosyl-1-amine (PRA) from 5-phosphoribosyl-1-pyrophosphate (PRPP) and an amino group donor like glutamine. The last step is the formation of IMP. There are no branching points between PRA and IMP. The conversion of IMP, end product of both de novo synthesis and salvage pathway, to AMP or GMP is slow.

Little is known about the value of enhanced guanine nucleotide regeneration. Adenylosuccinate synthetase catalyzes the reaction: IMP + aspartate + GTP → adenylosuccinate + GDP + P_i. The activity of this enzyme is very low (5). Adenylosuccinate

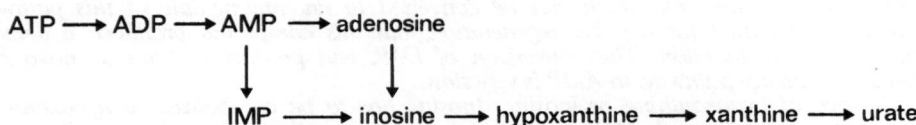

Fig. 1. Adenine nucleotide catabolism: ATP loses its phosphate groups. AMP can be converted to either adenosine or IMP. They are both broken down to inosine. This is catabolized to hypoxanthine, xanthine and urate.

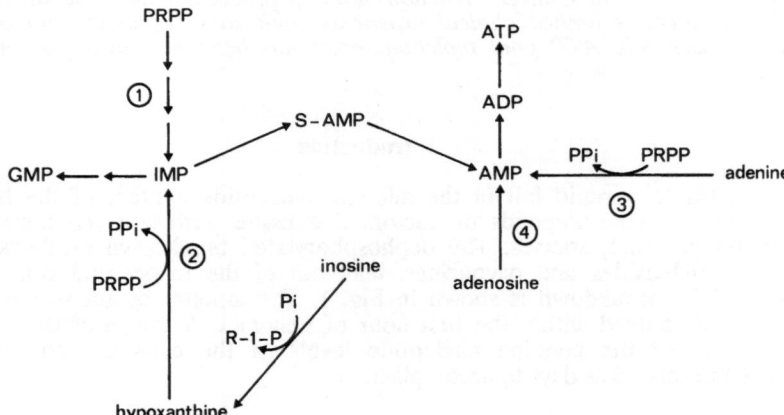

Fig. 2. Pathways for adenine nucleotide regeneration: 1. De novo synthesis; 2. Salvage of hypoxanthine and inosine; 3. Adenine ribophosphorylation; 4. Adenosine phosphorylation. Pathways 1 to 3 use 5-phosphoribosyl-1-pyrophosphate (PRPP). Pi = inorganic phosphate, PPi = inorganic pyrophosphate, R-1-P = ribose-1-phosphate, S-AMP = adenylosuccinate.

lyase catalyzes the reaction: Adenylosuccinate → AMP + fumarate; the same enzyme is also involved in one of the last steps of the de novo synthesis. Zimmer et al. (39) estimated that 0.04% of the nucleotide pool can be regenerated per hour this way. The rate of ATP synthesis is enhanced after ischemia (26), and limited by the concentration of PRPP (11). Ribose induces a considerable enhancement of the PRPP-concentration and of the rate of adenine nucleotide synthesis in heart (38), although the ribokinase activity in most tissues is low (2). There is some clinical experience with ribose. Zöllner et al. (40) described the successful symptomatic treatment of a patient suffering from an AMP-deaminase deficiency with high oral doses of ribose (maximum 60 g per day). Segal et al. (29) described a hypoglycemic effect of ribose. Human lymphocytes, exposed to extremely high ribose concentrations (25-50 mM), showed inhibition of their DNA repair synthesis (41).

Another possibility is to give 5-amino-4-imidazolecarboxamide riboside (AICAriboside), which can be phosphorylated to its ribotide (AICAR). This intermediate in the pathway becomes IMP in two steps. Swain et al. (31) showed that a 24-hour infusion of AICAriboside caused an accelerated repletion of ATP and GTP pools in postischemic canine myocardium. In an isolated cat heart, Mitsos et al. (27) described an

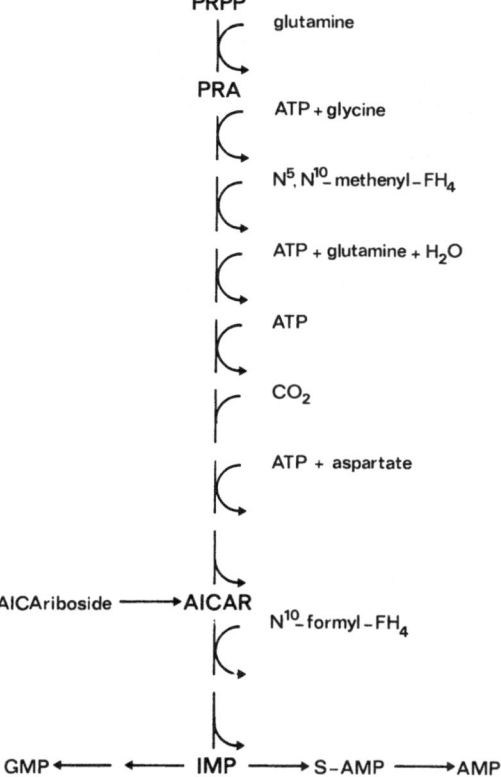

Fig. 3. De novo synthesis: Formation of IMP from small molecules in 10 steps. IMP is converted to AMP. AICAR = 5-amino-4-imidazolecarboxamide riboside, FH4 = tetrahydrofolate, PRA = 5-phosphoribosyl-1-amine, PRPP = 5-phosphoribosyl-1-pyrophosphate, S-AMP = adenylosuccinate.

increase in postischemic left ventricular developed pressure, but this was accompanied by a considerable increase in left ventricular compliance. In an open-chest dog model, Hoffmeister et al. (18) observed a marked impairment of regional function by postischemic AICAriboside. This deterioration progressed in time. We conclude that the value of accelerating de novo synthesis is limited.

2. Salvage of Hypoxanthine or Inosine

Hypoxanthine can be utilized to restore the ATP content of the heart. The enzyme hypoxanthine guanine phosphoribosyltransferase catalyzes the reaction: Hypoxanthine + PRPP → IMP + PP$_i$. PRPP is rate-limiting (38). Harmsen et al. (16) found that adding ribose stimulates the hypoxanthine incorporation rate. (For a discussion about the conversion of IMP to the other nucleotides, we refer to the part on de novo synthesis.) Although the rate of adenine nucleotide production from this pathway is faster than the de novo synthesis, it is limited by the availability of PRPP and the activity of adenylosuccinate synthetase. Hypoxanthine enters the cell by simple diffusion. As a substrate for xanthine oxidase (XO), hypoxanthine may generate free oxygen radicals in species showing XO activity in the heart. Its presence in the human heart has to be clearly demonstrated. Addition of a xanthine oxidase inhibitor could be important.

Inosine has been described as a cardioprotective agent (1,13,32). Mammalian cells lack inosine kinase (35). Therefore inosine has to be broken down by nucleoside phosphorylase: Inosine + P$_i$ → hypoxanthine + ribose-1-phosphate, before it is incorporated in the adenine nucleotides. As nucleoside phosphorylase is not located in the cardiomyocyte (28), the ribose-1-phosphate can be used for PRPP production, but presumably not in the myocyte (See Fig. 4). It is possible that inosine incorporation in the myocyte can be enhanced by ribose addition, but as far as we know this has never been tried.

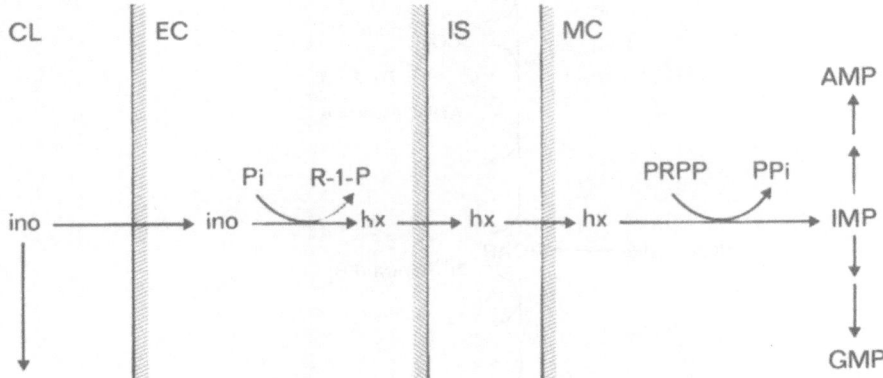

Fig. 4. Inosine has to be broken down to hypoxanthine within the endothelium before incorporation in adenine nucleotides of the myocytes. cl = capillary lumen, ec = endothelial cell, hx = hypoxanthine, ino = inosine, is = interstitial space, mc = myocyte, Pi = inorganic phosphate, PRPP = 5-phosphoribosyl-1-pyrophosphate.

In contrast to hypoxanthine, inosine does effect contractility, heart rate and blood flow. Juhász-Nagy and Aviado (22) found an increase in myocardial contractility in dog heart, while Hoffmeister et al. (17) observed a decrease in rat heart. There is conflicting evidence about the effect on heart rate (21,22). Coronary vasodilation has been reported by many authors (21,32). Inosine inhibits the uptake of adenosine in the heart (25), both nucleosides use the same carrier to enter the cell (36). Although inosine is used in Mediterranean countries as a cardiotonic, the effects of this nucleoside on humans have not extensively been investigated. Recently, De Jong et al. (7) described the apparent inosine uptake by the human heart together with an increase in dP/dt. We conclude that inosine and hypoxanthine are possibly valuable to the postischemic heart, but more research on this subject is necessary.

3. Adenine Ribophosphorylation

Little research has been devoted to this pathway. The enzyme adenine phosphoribosyltransferase catalyzes the reaction: Adenine + PRPP $\rightarrow$ AMP + PP$_i$. The adenine concentration in blood is very low (24). If adenine is administered, the amount of PRPP becomes rate-limiting. Addition of ribose can enhance the incorporation rate. Ward et al. (33) gave adenine + ribose to a dog heart, in which the ATP levels had been halved. They repleted ATP in 24 hours, while it took 5 days without treatment. In isolated ventricular cells, Dow et al. (8) compared the incorporation rate of adenosine, of adenine + ribose, and of hypoxanthine + ribose. They showed that adenosine incorporation is twice as fast as adenine + ribose, and twenty times as fast as hypoxanthine + ribose is. Adenine lacks hemodynamic effects; it has no effect on adenosine transport (25). Clifford and Story (6) described growth disturbances in young animals fed a high adenine diet. Recently Yokozawa et al. (37) showed that metabolic abnormalities resembling chronic renal failure could be produced by long-term feeding of adenine to rats. Human experience with adenine is available from its use as an erythrocyte preservative in blood banks. We conclude that adenine could be a valuable compound, provided its safety for postischemic ATP regeneration can be proved.

4. Adenosine Phosphorylation

This pathway provides the fastest possibility to regenerate adenine nucleotides. The enzyme adenosine kinase catalyzes the reaction: Adenosine + ATP $\rightarrow$ AMP + ADP. Adenosine has been used for cardioprotection with varying success (9,15,18,19,20,23, 30). Gerlach et al. showed that adenosine incorporation is very fast in endothelial cells (12). Breakdown of adenosine in blood is very fast: a half life of 10 seconds has been estimated. It is possible that most of the infused adenosine will be either catabolized or incorporated in the endothelial cells, with little adenosine incorporated in the myocardial cells. Infused adenosine is also taken up very rapidly by other cells (especially in the lungs). Adenosine transport is mediated by a carrier. Dow et al. (8) reported that in isolated cardiac myocytes adenosine transport is rate-limiting for adenosine incorporation.

Adenosine exerts other effects mediated by purinergic receptors. It is a potent vasodilator and plays presumably an important role in coronary vasoregulation (4). Some groups described the slowing of atrioventricular impulse conductance and a decrease in heart rate (3,34). Foker et al. (10) reported that adenosine caused a vasoconstriction of the renal vessels. In humans, adenosine is used against arrhythmias (superventricular tachycardias). We conclude that adenosine is potentially useful although it has many side effects.

Conclusion

There is no clinical experience with enhanced ATP regeneration in the postischemic heart. For this purpose inosine, or adenine in combination with ribose, seem the most promising compounds.

Acknowledgment

The authors wish to thank Ms. M.J. Kanters-Stam and Ms. C.E. Zandbergen-Visser for their secretarial assistance.

References

1. Aussedat J, Verdys M, Rossi A: Adenine nucleotide synthesis from inosine during normoxia and after ischaemia in the isolated perfused rat heart. Can J Physiol Pharmacol 63:1159-1164, 1985
2. Becker MA, Raivio KO, Seegmiller JE: Synthesis of phosphoribosylpyrophosphate in mammalian cells. Adv Enzymol 49:281-306, 1979
3. Belardinelli L, Shryock J, West A, Clemo HF, DiMarco JP, Berne RM: Effects of adenosine and adenine nucleotides on the atrioventricular node of isolated guinea pig hearts. Circulation 70:-1083-1091, 1984
4. Berne RM: Cardiac nucleotides in hypoxia: Possible role in regulation of coronary blood flow. Am J Physiol 204:317-322, 1963
5. Brown AK, Raeside DL, Bowditch J, Dow JW: Metabolism and salvage of adenine and hypoxanthine by myocytes isolated from mature rat heart. Biochim Biophys Acta 845:469-476, 1985
6. Clifford AJ, Story DL: Levels of purines in foods and their metabolic effects in rats. J Nutr 106:435-442, 1976
7. De Jong JW, Czarnecki W, Ruzyllo W, Huizer T, Herbaczynska-Cedro K: Apparent inosine uptake by the human heart. J Mol Cell Cardiol 19, Suppl III: S 16, 1987 (Abstr)
8. Dow JW, Bowditch J, Nigdikar SV, Brown AK: Salvage mechanisms for generation of adenosine triphosphate in rat cardiac myocytes. Cardiovasc Res 21:188-196, 1987
9. Ely SW, Mentzer RM Jr, Lasley RD, Lee BK, Berne RM: Functional and metabolic evidence of enhanced myocardial tolerance to ischemia and reperfusion with adenosine. J Thorac Cardiovasc Surg 90:549-556, 1985
10. Foker JE, Einzig S, Wang T: Adenosine metabolism and myocardial preservation. Consequences of adenosine catabolism on myocardial high-energy compounds and tissue blood flow. J Thorac-Cardiovasc Surg 80:506-516, 1980
11. Fox IH, Kelly WN: Phosphoribosylpyrophosphate in man: Biochemical and clinical significance. Ann Intern Med 74:424-433, 1971
12. Gerlach E, Nees S, Becker BF: The vascular endothelium: A survey of some newly evolving biochemical and physiological features. Basic Res Cardiol 80:459-474, 1985
13. Goldhaber SZ, Pohost GM, Kloner RA, Andrews E, Newell JB, Ingwall JS: Inosine: A protective agent in an organ culture model of myocardial ischemia. Circ Res 51:181-188, 1982
14. Goldthwait DA: Mechanisms of synthesis of purine nucleotides in heart muscle extracts. J Clin Invest 36:1572-1578, 1957
15. Haas GS, DeBoer LWV, O'Keefe DD, Bodenhamer RM, Geffin GA, Drop LJ, Teplick RS, Daggett WM: Reduction of postischemic myocardial dysfunction by substrate repletion during reperfusion. Circulation 70, Suppl I:I65-I74, 1984
16. Harmsen E, De Tombe PP, De Jong JW, Achterberg PW: Enhanced ATP and GTP synthesis from hypoxanthine or inosine after myocardial ischemia. Am J Physiol 246:H37-H43, 1984
17. Hoffmeister HM, Betz R, Fiechtner H, Seipel L: Myocardial and circulatory effects of inosine. Cardiovasc Res 21:65-71, 1987
18. Hoffmeister HM, Mauser M, Schaper W: Effect of adenosine and AICAR on ATP content and regional contractile function in reperfused canine myocardium. Basic Res Cardiol 80:455-458, 1985
19. Humphrey SM, Seelye RN: Improved functional recovery of ischemic myocardium by suppression of adenosine catabolism. J Thorac Cardiovasc Surg 84:16-22, 1982

20. Humphrey SM, Seelye RN, Gravin JB: The influence of adenosine on the no-flow phenomenon in anoxic and ischemic hearts. Pathology 14:129-133, 1982
21. Jones CE, Mayer LR: Nonmetabolically coupled coronary vasodilatation during inosine infusion in dogs. Am J Physiol 238:H569-H574, 1980
22. Juhász-Nagy A, Aviado DM: Inosine as a cardiotonic agent that reverses adrenergic beta blockade. J Pharmacol Exp Ther 202:683-695, 1977
23. Kao RL, Magovern GJ: Prevention of reperfusional damage from ischemic myocardium. J Thorac Cardiovasc Surg 91:106-114, 1986
24. Kelley WN, Levy RI, Rosenbloom FM, Henderson JF, Seegmiller JE: Adenine phosphoribosyltransferase deficiency: A previously undescribed genetic defect in man. J Clin Invest 47:2281-2289, 1968
25. Kolassa N, Pfleger K, Rummel W: Specificity of adenosine uptake into the heart and inhibition by dipyridamole. Eur J Pharmacol 9:265-268, 1970
26. Mauser M, Hoffmeister HM, Nienhaber C, Schaper W: Influence of ribose, adenosine and "AICAR" on the rate of myocardial adenosine triphosphate synthesis during reperfusion after coronary artery occlusion in the dog. Circ Res 56:220-230, 1985
27. Mitsos SE, Jolly SR, Lucchesi BR: Protective effects of AICAriboside in the globally ischemic isolated cat heart. Pharmacology 31:121-131, 1985
28. Rubio R, Wiedmeier T, Berne RM: Nucleoside phosphorylase: Localization and role in the myocardial distribution of purines. Am J Physiol 222:550-555, 1972
29. Segal S, Foley J, Wyngaarden JB: Hypoglycemic effect of D-ribose in man. Proc Soc Exp Biol Med 95:551-555, 1957
30. Silverman NA, Kohler J, Feinberg H, Levitsky S: Beneficial metabolic effect of nucleoside augmentation on reperfusion injury following cardioplegic arrest. Chest 83:787-792, 1983
31. Swain JL, Hines JJ, Sabina RL, Holmes EW: Accelerated repletion of ATP and GTP pools in postischemic canine myocardium using a precursor of purine de novo synthesis. Circ Res 51:102-105, 1982
32. Thomas JX Jr, Jones CE: Effect of inosine on contractile force and high-energy phosphates in ischemic hearts. Proc Soc Exp Biol Med 161:468-472, 1979
33. Ward HB, St Cyr JA, Cogordan JA, Alyono D, Bianco RW, Kriett JM, Foker JE: Recovery of adenine nucleotide levels after global myocardial ischemia in dogs. Surgery 96:248-253, 1984
34. Watt AH, Routledge PA: Transient bradycardia and subsequent sinus tachycardia produced by intravenous adenosine in healthy subjects. Br J Clin Pharmacol 21:533-536, 1986
35. Wiedmeier T, Rubio R, Berne RM: Inosine incorporation into myocardial nucleotides. J Mol Cell Cardiol 4:445-452, 1972
36. Woo YT, Manery JF, Riordan JR, Dryden EE: Uptake and metabolism of purine nucleosides and nucleotides in isolated frog skeletal muscle. Life Sci 21:861-876, 1977
37. Yokozawa T, Zheng PD, Oura H, Koizumi F: Animal model of adenine-induced chronic renal failure in rats. Nephron 44:230-234, 1986
38. Zimmer H-G, Gerlach E: Stimulation of myocardial adenine nucleotide biosynthesis by pentoses and pentitols. Pflügers Arch 376:223-227, 1978
39. Zimmer H-G, Trendelenburg C, Kammermeier H, Gerlach E: De novo synthesis of myocardial adenine nucleotides in rat. Acceleration during recovery from oxygen deficiency. Circ Res 32:635-642, 1973
40. Zöllner N, Reiter S, Gross M, Pongratz D, Reimers CD, Gerbitz K, Paetzke I, Deufel T, Hübner G: Myoadenylate deaminase deficiency: Successful symptomatic therapy by high dose oral administration of ribose. Klin Wochenschr 64:1281-1290, 1986
41. Zunica G, Marini M, Brunelli MA, Chiricolo M, Franceschi C: D-ribose inhibits DNA repair synthesis in human lymphocytes. Biochem Biophys Res Commun 138:673-678, 1986

Index

Acetate 130
Acetoacetate 17, 25, 158
Acetoacetyl-CoA 17, 25, 174
Acetyl-CoA, Acetyl-coenzyme A 17, 19, 37, 174
Acetylcarnitine 130
Acetylcholine, ACh 99
Acidosis 17, 47, 133, 181
Actin 55, 189
Action potential 93, 187
Activation, Fatty acid, *see* F.a.a.
Actomyosin 84
Acyl-CoA, Acyl-coenezyme A, Palmitoyl-c.,
  Stearoyl-c 17, 37, 58, 147, 156
Acylcarnitine, Palmitoylcarnitine 17, 25, 157
Adenine 5, 45, 53, 105, 225, 247, 283
Adenine nucleotide breakdown, A.n. catabolism,
  A.n. degradation 105, 204, 216, 229, 266, 283
Adenine nucleotide interconversion 227
Adenine nucleotide pool 105, 225, 239, 266
Adenine nucleotide regeneration 283
Adenine nucleotide synthesis, A.n. biosynthesis
  105, 229
Adenine nucleotide translocator, A.n. translocase
  3, 41
Adenine phosphoribosyltransferase, APRT 3, 60,
  225, 287
Adenosine 3, 9, 15, 45, 53, 67, 83, 93, 119, 163,
  185, 206, 225, 237, 247, 257, 266, 283
Adenosine 5'-diphosphate, *see* ADP
Adenosine 5'-monophosphate, *see* AMP
Adenosine 5'-triphosphate, *see* ATP
Adenosine antagonist 93, 94
Adenosine deaminase 54, 61, 87, 93, 132, 206,
  225
Adenosine kinase 3, 50, 60, 67, 206, 229, 287
Adenosine receptor 87
Adenosylhomocysteine, S- 4, 62, 85, 229
Adenosylhomocysteinase, S-, Adenosylhomocys-
  teine hydrolase, S-, SAH 4, 85, 86
Adenosylmethionine, S- 86, 225
Adenylate cyclase 87, 93
Adenylate deaminase, *see* AMP d.
Adenylate kinase, *see* Myokinase 3, 53, 55, 67,
  119, 139, 205, 227
Adenylosuccinase, Adenylosuccinate hydrolase,
  Adenylosuccinate lyase 55, 61, 229, 285
Adenylosuccinate 3, 55, 227, 284
Adenylosuccinate synthetase 55, 229, 284
ADP formation 227

ADP, Free 133
ADP, Regulation by 181
ADP-ATP translocation 182
ADP/O ratio, *see* P/O ratio
Adrenal cortex 108
Adrenergic stimulation, beta- 87
Adrenergic system, beta- 261
Adult cell 171
Age 9, 157
Alanine 17, 23 130
Albumin, *see* Fatty acid, albumin complex 173
Alkaline phosphatase 230
Alkylxanthine 87, 93
Allopurinol 232
Amino acid 17, 26
Amino-4-imidazolecarboxamide riboside, 5- 50,
  205, 283
Amino-4-imidazolecarboxamide ribotide, 5-,
  AICAR 5, 205, 283
Amino-5'-deoxyadenosine, 5'- 61
Aminophylline 94
Ammonia, *see* Ammonium ion 227
Ammonium ion, NH4+ 17, 53
AMP breakdown, A. catabolism, A. degradation
  230, 237, 247
AMP, cyclic, cAMP 17, 22, 95, 226
AMP deaminase, Adenylate d. 53, 54, 55, 67, 112,
  225, 227, 285
AMP, dibutyryl-c 94
AMP synthesis, A. synthesis 247, 283
Anesthetic 195, 277
Angina pectoris 151, 240, 266
Angina, unstable 208
Angiography 257
Angioplasty, *see* Percutaneous transluminal
  coronary a. 237, 265
Anipamil 134, 266
Anti-arrhythmic effect 253
Aortic cross-clamping 245
Aortic pressure, *see* Pressure, Aortic
APRT, *see* Adenine phosphorbosyltransferase 225
Arachidonate 164
Arrhythmia 191, 287
Arterial pressure, *see* Pressure, Arterial
Aspartate, Aspartic acid 130, 226, 284
Asphyxia 105
Atherosclerosis, Altherogenesis 195, 240
ATP activation 85
ATP catabolism 196, 237, 267

DEVELOPMENTS IN CARDIOVASCULAR MEDICINE

Recent volumes

Perry, H.M., ed.: Lifelong management of hypertension. 1983. ISBN 0-89838-582-2.

Jaffe, E.A., ed.: Biology of endothelial cells. 1984. ISBN 0-89838-587-3.

Surawicz, B., Reddy, C.P., Prystowsky, E.N., eds.: Tachycardias. 1984.
ISBN 0-89838-588-1.

Spencer, M.P., ed.: Cardiac Doppler diagnosis. 1983. ISBN 0-89838-591-1.

Villarreal, H., Sambhi, M.P., eds.: Topics in pathophysiology of hypertension. 1984.
ISBN 0-89838-595-4.

Messerli, F.H., ed.: Cardiovascular disease in the elderly. 1984. ISBN 0-89838-596-2.

Simoons, M.L., Reiber, J.H.C., eds.: Nuclear imaging in clinical cardiology. 1984.
ISBN 0-89838-599-7.

Ter Keurs, H.E.D.J., Schipperheyn, J.J., eds.: Cardiac left ventricular hypertrophy. 1983.
ISBN 0-89838-612-8.

Sperelakis, N., ed.: Physiology and pathophysiology of the heart. 1984.
ISBN 0-89838-615-2.

Messerli, F.H., ed.: Kidney in essential hypertension. 1984. ISBN 0-89838-616-0.

Sambhi, M.P., ed.: Fundamental fault in hypertension. 1984. ISBN 0-89838-638-1.

Marchesi, C., ed.: Ambulatory monitoring: Cardiovascular system and allied applications.
1984. ISBN 0-89838-642-X.

Kupper, W., MacAlpin, R.N., Bleifeld, W., eds.: Coronary tone in ischemic heart disease.
1984. ISBN 0-89838-646-2.

Sperelakis, N., Caulfield, J.B., eds.: Calcium antagonists: Mechanisms of action on cardiac muscle and vascular smooth muscle. 1984. ISBN 0-89838-655-1.

Godfraind, T., Herman, A.S., Wellens, D., eds.: Calcium entry blockers in cardiovascular
and cerebral dysfunctions. 1984. ISBN 0-89838-658-6.

Morganroth, J., Moore, E.N., eds.: Interventions in the acute phase of myocardial infarction. 1984. ISBN 0-89838-659-4.

Abel, F.L., Newman, W.H., eds.: Functional aspects of the normal, hypertrophied, and
failing heart. 1984. ISBN 0-89838-665-9.

Sideman, S., Beyar, R., eds.: Simulation and imaging of the cardiac system. 1985.
ISBN 0-89838-687-X.

Van der Wall, E., Lie, K.I., eds.: Recent views on hypertrophic cardiomyopathy. 1985.
ISBN 0-89838-694-2.

Beamish, R.E., Singal, P.K., Dhalla, N.S., eds.: Stress and heart disease. 1985.
ISBN 0-89838-709-4.

Beamish, R.E., Panagio, V., Dhalla, N.S., eds.: Pathogenesis of stress-induced heart disease. 1985. ISBN 0-89838-710-8.

Morganroth, J., Moore, E.N., eds.: Cardiac arrhythmias. 1985. ISBN 0-89838-716-7.

Mathes, E., ed.: Secondary prevention in coronary artery disease and myocardial infarction. 1985. ISBN 0-89838-736-1.

Lowell Stone, H., Weglicki, W.B. eds.: Pathology of cardiovascular injury. 1985.
ISBN 0-89838-743-4.

Meyer, J., Erbel, R., Rupprecht, H.J., eds.: Improvement of myocardial perfusion. 1985.
ISBN 0-89838-748-5.

Reiber, J.H.C., Serruys, P.W., Slager, C.J.: Quantitative coronary and left ventricular
cineangiography. 1986. ISBN 0-89838-760-4.

Fagard, R.H., Bekaert, I.E., eds.: Sports cardiology. 1986. ISBN 0-89838-782-5.

Reiber, J.H.C., Serruys, P.W., eds.: State of the art in quantitative coronary arteriography. 1986. ISBN 0-89838-804-X.

Roelandt, J., ed.: Color Doppler Flow Imaging. 1986. ISBN 0-89838-806-6.

Van der Wall, E.E., ed.: Noninvasive imaging of cardiac metabolism. 1986.
ISBN 0-89838-812-0.

Liebman, J., Plonsey, R., Rudy, Y., eds.: Pediatric and fundamental electrocardiography.
1986. ISBN 0-89838-815-5.

Hilger, H.H., Hombach, V., Rashkind, W.J., eds.: Invasive cardiovascular therapy. 1987.
ISBN 0-89838-818-X

Serruys, P.W., Meester, G.T., eds.: Coronary angioplasty: a controlled model for ischemia. 1986. ISBN 0-89838-819-8.

Tooke, J.E., Smaje, L.H.: Clinical investigation of the microcirculation. 1986.
ISBN 0-89838-819-8.

Van Dam, R.Th., Van Oosterom, A., eds.: Electrocardiographic body surface mapping.
1986. ISBN 0-89838-834-1.

Spencer, M.P., ed.: Ultrasonic diagnosis of cerebrovascular disease. 1987.
ISBN 0-89838-836-8.

Legato, M.J., ed.: The stressed heart. 1987. ISBN 0-89838-849-X.

Safar, M.E., ed.: Arterial and venow systems in essential hypertension. 1987.
ISBN 0-89838-857-0.

Roelandt, J., ed.: Digital techniques in echocardiography. 1987. ISBN 0-89838-861-9.

Dhalla, N.S. et al., eds.: Pathophysiology of heart disease. 1987. ISBN 0-89838-864-3.

Dhalla, N.S. et al., eds.: Heart function and metabolism. 1987. ISBN 0-89838-865-1.

Dhalla, N.S. et al., eds.: Myocardial Ischemia. 1987. ISBN 0-89838-866-X.

Beamish, R.E. et al., eds.: Pharmacological aspects of heart disease. 1987.
ISBN 0-89838-867-8.

Ter Keurs, H.E.D.J., Tyberg, J.V., eds.: Mechanics of the circulation. 1987.
ISBN 0-89838-870-8

Sideman, S., Beyar, R., eds.: Activation, metabolism and perfusion of the heart. 1987.
ISBN 0-89838-871-6.

Aliot, E., Lazzara, R., eds.: Ventricular tachycardias. 1987. ISBN 0-89838-881-3.

Schneeweiss, A. et al., eds.: Cardiovascular drug therapy in the elderly. 1987.
ISBN 0-89838-883-X.

Chapman, J.V., Sgalambro, A., eds.: Basic concepts in Doppler echocardiography. 1987.
ISBN 0-89838-888-0

Chien, S. et al., eds.: Clinical hemocheology. 1987. ISBN 0-89838-807-4.

Morganroth, J., ed.: Congestive heart failure. 1987. ISBN 0-89838-955-0.

Messerli, F.H., ed.: Cardiovascular disease in the elderly. 2nd ed. 1988.
ISBN 0-89838-962-3.

Heintzen, P.H., Bürsch, J.H., eds.: Progress in digital angiocardiography. 1988.
ISBN 0-89838-965-8.

Scheinman, M.A., ed.: Catheter ablation of cardiac arrhythmias. 1988.
ISBN 0-89838-967-4.

Spaan, J.A.E., Bruschke, A.V.G., Gittenberger-de Groot, A.C., eds.: Coronary circulation. 1987. ISBN 0-89838-978-X.

Visser, C., Kan, G., Meltzer, R., eds.: Echocardiography in coronary artery disease. 1988.
ISBN 0-89838-979-8.

Bayés de Luna, A., Betriu, A., Permanyer, G., eds.: Therapeutics in cardiology. 1988.
ISBN 0-89838-981-X.

Mirvis, D.M., ed.: Body surface electrocardiographic mapping. 1988.
ISBN 0-89838-983-6.

Konstam, M.A., Isner, J.M., eds.: The right ventricle. 1988. ISBN 0-89838-987-9.

Kappagoda, C.T., Greenwood, P.V., eds.: Long-term management of patients after myocardial infarction. 1988. ISBN 0-89838-352-8.

Gaasch, W.H., Levine, H.J., eds.: Chronic aortic regurgitation. 1988.
ISBN 0-89838-364-1.

Singal, P.K., ed.: Oxygen radicals in the pathophysiology of heart disease. 1988.
ISBN 0-89838-375-7.

Reiber, J.H.C., Serruys, P.W., eds.: New developments in quantitative coronary arteriography. 1988. ISBN 0-89838-377-3.

Morganroth, J., Moore, E.N., eds.: Silent myocardial ischemia. 1988.
ISBN 0-89838-380-3.

Ter Keurs, H.E.D.J., Noble, M.I.M., eds.: Starling's law of the heart revisited. 1988.
ISBN 0-89838-382-X.

Sperelakis, N., ed.: Physiology and pathophysiology of the heart. 1988.
ISBN 0-89838-388-9